AF770454

National Center for Construction Education and Research

Mobile Crane Operations Level Three

Annotated Instructor's Guide

PEARSON
Prentice Hall

Upper Saddle River, New Jersey
Columbus, Ohio

contren®
Learning Series

National Center for Construction Education and Research

President: Don Whyte
Director of Product Development and Revision: Daniele Dixon
Mobile Crane Operations Project Manager: Tara Cohen
Production Manager: Jessica Martin
Product Maintenance Supervisor: Debie Ness
Editor: Natalie Smith
Desktop Publisher: Laura Parker

The NCCER would like to acknowledge the contract service provider for this curriculum:
Topaz Publications, Liverpool, New York.

This information is general in nature and intended for training purposes only. Actual performance of activities described in this manual requires compliance with all applicable operating, service, maintenance, and safety procedures under the direction of qualified personnel. References in this manual to patented or proprietary devices do not constitute a recommendation of their use.

Copyright © 2005 by the National Center for Construction Education and Research (NCCER), Gainesville, FL 32614-1104, and published by Pearson Education, Inc., Upper Saddle River, NJ 07458. All rights reserved. Printed in the United States of America. This publication is protected by Copyright and permission should be obtained from the NCCER prior to any prohibited reproduction, storage in a retrieval system, or transmission in any form or by any means, electronic, mechanical, photocopying, recording, or likewise. For information regarding permission(s), write to: NCCER Product Development, P.O. Box 141104, Gainesville, FL 32614-1104.

PEARSON Prentice Hall

nccer

contren Learning Series

10 9 8 7 6 5 4
ISBN 0-13-109870-5

Preface

This volume was developed by the National Center for Construction Education and Research (NCCER) in response to the training needs of the construction, maintenance, and pipeline industries. It is one of many in NCCER's *Contren® Learning Series*. The program, covering training for close to 40 construction and maintenance areas, and including skills assessments, safety training, and management education, was developed over a period of years by industry and education specialists.

NCCER also maintains a National Registry that provides transcripts, certificates, and wallet cards to individuals who have successfully completed modules of NCCER's *Contren® Learning Series*, when the training program is delivered by an NCCER Accredited training Sponsor.

The NCCER is a not-for-profit 501(c)(3) education foundation established in 1995 by the world's largest and most progressive construction companies and national construction associations. It was founded to address the severe workforce shortage facing the industry and to develop a standardized training process and curricula. Today, NCCER is supported by hundreds of leading construction and maintenance companies, manufacturers, and national associations, including the following partnering organizations:

PARTNERING ASSOCIATIONS

- American Fire Sprinkler Association
- American Petroleum Institute
- American Society for Training & Development
- American Welding Society
- Associated Builders & Contractors, Inc.
- Association for Career and Technical Education
- Associated General Contractors of America
- Carolinas AGC, Inc.
- Citizens Development Corps
- Construction Industry Institute
- Construction Users Roundtable
- Design-Build Institute of America
- Electronic Systems Industry Consortium
- Merit Contractors Association of Canada
- Metal Building Manufacturers Association
- National Association of Minority Contractors
- National Association of State Supervisors for Trade and Industrial Education
- National Association of Women in Construction
- National Insulation Association
- National Ready Mixed Concrete Association
- National Systems Contractors Association
- National Utility Contractors Association
- National Technical Honor Society
- North American Crane Bureau
- North American Technician Excellence
- Painting & Decorating Contractors of America
- Portland Cement Association
- SkillsUSA
- Steel Erectors Association of America
- Texas Gulf Coast Chapter ABC
- U.S. Army Corps of Engineers
- University of Florida
- Women Construction Owners & Executives, USA
- Youth Training and Development Consortium

Some features of NCCER's *Contren® Learning Series* are:

- An industry-proven record of success
- Curricula developed by the industry for the industry
- National standardization providing portability of learned job skills and educational credits
- Credentials for individuals through NCCER's National Registry
- Compliance with Apprenticeship, Training, Employer, and Labor Services (ATELS) requirements for related classroom training (CFR 29:29)
- Well-illustrated, up-to-date, and practical information

Acknowledgments

This curriculum was revised as a result of the farsightedness and leadership of the following sponsors:

Crane Inspection and Certification Bureau
Southwest Contractors
Turner Industries Holding Co.
Austin Bridge & Road, Inc.
Zachry Construction Corporation
Bay Ltd., A Berry Company

This curriculum would not exist were it not for the dedication and unselfish energy of those volunteers who served on the Authoring Team. A sincere thanks is extended to:

John Cole
Norm Fox
Tony Guidry
Jim Harbor
Samuel Rogers
Michael J. Eggenberger
Gene Couch

Contents

Module 21301-05

Load Charts

NCCER STANDARDIZED CRAFT TRAINING PROGRAM

The National Center for Construction Education and Research (NCCER) provides a standardized national program of accredited craft training. Key features of the program include instructor certification, competency-based training, and performance testing. The program provides trainees, instructors, and companies with a standard form of recognition through a National Craft Training Registry. The program is described in full in the *Guidelines for Accreditation*, published by the NCCER. For more information on standardized craft training, contact the NCCER by writing us at P.O. Box 141104, Gainesville, FL 32614-1104; calling 352-334-0911; or e-mailing info@nccer.org. More information may be found at our Web site, www.nccer.org.

HOW TO USE THIS ANNOTATED INSTRUCTOR'S GUIDE

Each page presents two sections of information. The larger section displays each page exactly as it appears in the Trainee Module. The narrow column ties suggested trainee and instructor actions to each page and provides icons (detailed below) to call your attention to material, safety, audiovisual, or testing requirements. The bottom of each page includes space for your notes.

The **Audiovisual** icon indicates an appropriate time to show a transparency or other audiovisual aid.

The **Classroom** icon prompts you to define a term, stress a point, ask trainees to explain a concept, or give examples.

The **Demonstration** icon directs you to show trainees how to perform tasks.

The **Examination** icon tells you to administer the written module examination.

The **Homework** icon is placed where you may wish to assign reading for the next class, to assign a project, or to advise trainees to prepare for an examination.

The **Laboratory** icon is used when trainees are to practice performing tasks.

The **Materials** icon is a reminder for you to gather materials needed for classes, labs, and testing.

The **Performance Testing** icon tells you to administer a performance test or a portion thereof.

The **Safety** icon is used to emphasize safety issues. It is often keyed to *Caution* and *Warning* statements in the Trainee Module.

The **Teaching Tip** icon indicates additional guidance is available, such as how to conduct an exercise, get the most educational value from a field trip, or encourage class participation. Teaching Tips may expand on a feature (*Think About It*, *Did You Know?*) or provide Quick Quizzes or similar exercises. You will be referred to the Teaching Tips section at the back of the module if there is additional material.

The **Combination** icon indicates that the laboratory listed corresponds with a performance task. If desired, you can note the proficiency of the trainees during the laboratory and use it to satisfy performance testing requirements.

PREPARATION

Before teaching this module, you should review the Objectives, Performance Tasks, Materials and Equipment List, and the Module Outline. Be sure to allow ample time to prepare your own training or lesson plan and gather all required materials and equipment.

MODULE OVERVIEW

This module discusses the importance of load charts and charts that apply to different configurations. It includes on-rubber, on-outrigger, jib, and deduction charts, as well as range diagrams and operational notes. In addition, parts of line and capacity calculations are covered.

PREREQUISITES

Prior to training with this module, it is recommended that the trainee shall have successfully completed *Core Curriculum; Mobile Crane Operations Level One;* and *Mobile Crane Operations Level Two.*

OBJECTIVES

Upon completion of this module, the trainee will be able to do the following:

1. Define the concepts of center of gravity and leverage.
2. Define the importance of using a load/capacity chart for lifting operations.
3. Define the terms on a load/capacity chart to indicate boom angle, load radius, and boom length.
4. Calculate crane capacity using a load/capacity chart.
5. Calculate parts of line.
6. Identify the differences between on-rubber and on-outrigger charts.
7. Identify the difference between lattice boom, hydraulic boom, and boom truck, and boom attachment charts.
8. Describe different crane counterweight configurations.
9. Explain the importance of reviewing setup and operational procedures.

PERFORMANCE TASKS

Under the supervision of the instructor, the trainee should be able to do the following:

1. Identify boom angle, boom length, and load radius on a load/capacity chart.
2. Identify the requirements of the on-rubber load/capacity chart.
3. Identify the requirements of the on-outrigger load/capacity chart.
4. Properly identify load/capacity charts that are used in different configurations.
5. Identify parts of line and counterweight considerations in load/capacity chart information.
6. Calculate parts of a line.
7. Calculate crane capacities using load/capacity charts.

MATERIALS AND EQUIPMENT LIST

Overhead projector and screen
Transparencies
Blank acetate sheets
Transparency pens
Whiteboard/chalkboard
Markers/chalk
Pencils and scratch paper
Manufacturer's literature on different types of cranes
ASME B30.5
ANSI/SAE J765
ANSI/SAE J987
Load/capacity charts for different machines
Copies of site safety manual or procedures
Videotape: *Lift Calculations*
TV and VCR
Load moment indicator
Copies of the Quick Quiz*
Module Examinations**
Performance Profile Sheets**
Performance Profile Examination Worksheets**

*Located in the back of this module.

**Located in the Test Booklet.

SAFETY CONSIDERATIONS

Ensure that the trainees are equipped with appropriate personal protective equipment and know how to use it properly. This module may require trainees to visit construction sites. Brief trainees on site safety procedures.

TEACHING TIME FOR THIS MODULE

An outline for use in developing your lesson plan is presented below. Note that each Roman numeral in the outline equates to one session of instruction. Each session has a suggested time period of 2½ hours. This includes 10 minutes at the beginning of each session for administrative tasks and one 10-minute break during the session. Approximately 25 hours are suggested to cover *Load Charts*. You will need to adjust the time required for hands-on activity and testing based on your class size and resources. Because laboratories often correspond to Performance Tasks, the proficiency of the trainees may be noted during these exercises for Performance Testing purposes.

Topic	Planned Time
Sessions I and II. Introduction, Leverage, Load Charts, and Types of Cranes	
A. Introduction	____________
B. Center of Gravity and Leverage	____________
C. Load/Capacity Charts and Manufacturers	____________
D. Types of Cranes	____________
E. Importance of Load/Capacity Charts for Lifting Operations	____________
Sessions III and IV. Crane Configuration I	
A. Boom Length, Boom Angle, and Load Radius	____________
B. Laboratory – Trainees practice identifying boom angle, boom length, and load radius on a load/capacity chart. This laboratory corresponds to Performance Task 1.	____________
C. Quadrants of Operation	____________
D. Configuration of the Crane Base	____________
E. Laboratory – Trainees practice identifying requirements of the on-rubber load/capacity chart and the on-outrigger load/capacity chart. This laboratory corresponds to Performance Tasks 2 and 3.	____________
F. Tower and Ring Attachments	____________
Sessions V and VI. Crane Configuration II	
A. Counterweight Configurations	____________
B. Laboratory – Trainees practice identifying load/capacity charts that are used in different configurations. This laboratory corresponds to Performance Task 4.	____________
C. Deduction Charts	____________
D. Laboratory – Trainees practice identifying parts of line and counterweight considerations in load/capacity chart information. This laboratory corresponds to Performance Task 5.	____________
E. Laboratory – Trainees practice calculating parts of line. This laboratory corresponds to Performance Task 6.	____________
Sessions VII–IX. Calculating Crane Capacity	
A. Calculating Crane Capacity	____________
B. Range Diagram	____________
C. Laboratory – Trainees practice calculating crane capacities using load/capacity charts. This laboratory corresponds to Performance Task 7.	____________

Session X. Review, Module Examination, and Performance Testing

A. Review

B. Module Examination

1. Trainees must score 70 percent or higher to receive recognition from NCCER.
2. Record the testing results on Craft Training Report Form 200, and submit the results to the Training Program Sponsor.

C. Performance Testing

1. Trainees must perform each task to the satisfaction of the instructor to receive recognition from NCCER. If applicable, proficiency noted during laboratory exercises can be used to satisfy the Performance Testing requirements.
2. Record the testing results on Craft Training Report Form 200, and submit the results to the Training Program Sponsor.

Module 21301-05

Load Charts

Instructor's Notes:

COURSE MAP

This course map shows all of the modules in the third level of the *Mobile Crane Operations* curriculum. The suggested training order begins at the bottom and proceeds up. Skill levels increase as you advance on the course map. The local Training Program Sponsor may adjust the training order.

Assign reading of Module 21301-05.

MOBILE CRANE OPERATIONS LEVEL THREE

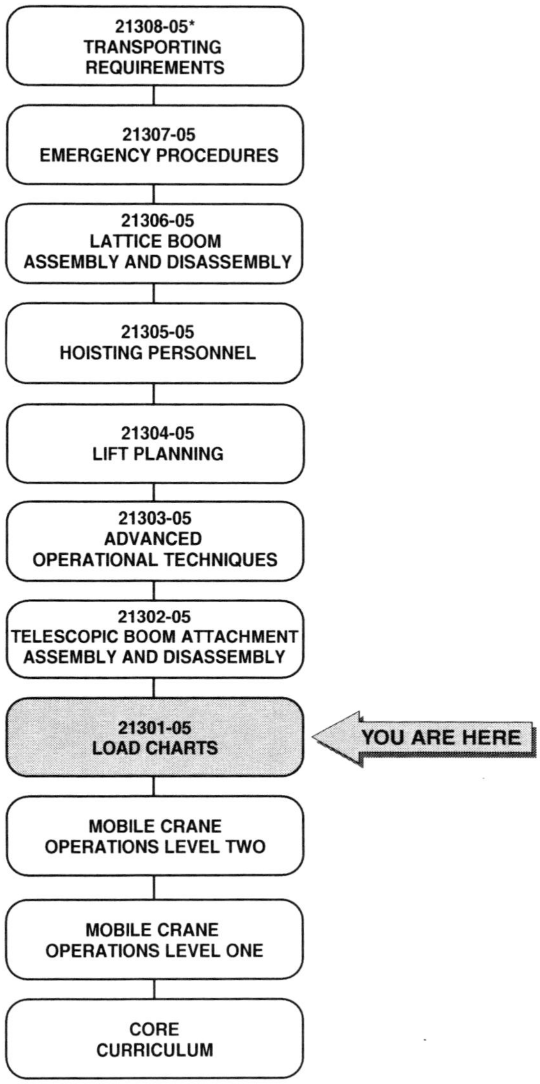

*Please note that 21308-05 is an elective module. It is not necessary to complete this module in order to advance in your training.

301CMAP.EPS

Copyright © 2005 National Center for Construction Education and Research, Gainesville, FL 32614-1104. All rights reserved. No part of this work may be reproduced in any form or by any means, including photocopying, without written permission of the publisher.

LOAD CHARTS 1.iii

Instructor's Notes:

MODULE 21301-05 CONTENTS

LOAD CHARTS 1.v

Figures

Instructor's Notes:

Tables

LOAD CHARTS 1.vii

MODULE 21301-05

Load Charts

Objectives

When you have completed this module, you will be able to do the following:

1. Define the concepts of center of gravity and leverage.
2. Describe the importance of using a load/capacity chart for lifting operations.
3. Define the terms on a load/capacity chart to indicate boom angle, load radius, and boom length.
4. Calculate crane capacity using a load/capacity chart.
5. Calculate parts of line.
6. Identify the differences between on-rubber and on-outrigger charts.
7. Identify the difference between lattice boom, hydraulic boom and boom truck, and boom attachment charts.
8. Describe different crane counterweight configurations.
9. Explain the importance of reviewing setup and operational procedures.

Prerequisites

Before you begin this module, it is recommended that you successfully complete *Core Curriculum; Mobile Crane Operations Level One*; and *Mobile Crane Operations Level Two.*

Required Trainee Materials

1. Pencil and paper
2. Calculator

1.0.0 ◆ INTRODUCTION

This module begins the last level in the *Mobile Crane Operations* program. In the past, operators used tipping as a capacity indicator, but there are now safer methods due to technological developments in crane construction. The importance of understanding load/capacity charts cannot be emphasized enough. It is one of the least understood areas of crane operations. Thousands of man hours go into developing load/capacity charts for cranes to ensure safety and productivity. It is imperative that operators are able to distinguish between the different categories and countless numbers associated with a load/capacity chart.

This module presents information on various aspects relative to the use of load/capacity charts. A description of methods used to determine gross and net capacity will also be discussed.

2.0.0 ◆ CENTER OF GRAVITY AND LEVERAGE

The principles of center of gravity and leverage play a critical role in crane operations. These principles are integral to any lifting operation.

A key concept associated with lifting ability is the center of gravity of the crane. The center of gravity of any object is the point where the object's weight can be assumed to be concentrated. The location of a crane's center of gravity depends on the weight and location of its components, as shown in *Figure 1*. The heavier components include the boom, carrier, upperworks, and counterweight.

The center of gravity can change depending on the crane's configuration. When the crane is in

Ensure that you have everything required to teach the course. Check the Materials and Equipment list at the front of this module.

Explain that terms shown in bold are defined in the Glossary at the back of this module.

See the general Teaching Tip at the end of this module.

Show Transparency 1, Objectives, and Transparency 2, Performance Tasks. Review the goals of the module, and explain what will be expected of trainees.

Discuss the importance of understanding load/capacity charts. Ask a trainee to explain center of gravity.

Show Transparency 3 (Figure 1). Explain that the location of a crane's center of gravity depends on the weight and location of its components. Discuss how the center of gravity changes when the load's position changes.

Show Transparency 4 (Figure 2). Discuss crane leverage, load leverage, and relative tipping force. Explain how load leverage increases as the distance from the tipping axis increases.

Explain the importance of using the specific load/capacity chart information given by the manufacturer for the crane being used. Emphasize that load charts are never interchangeable between cranes.

Provide and compare two load charts for different machines. Have trainees note their differences.

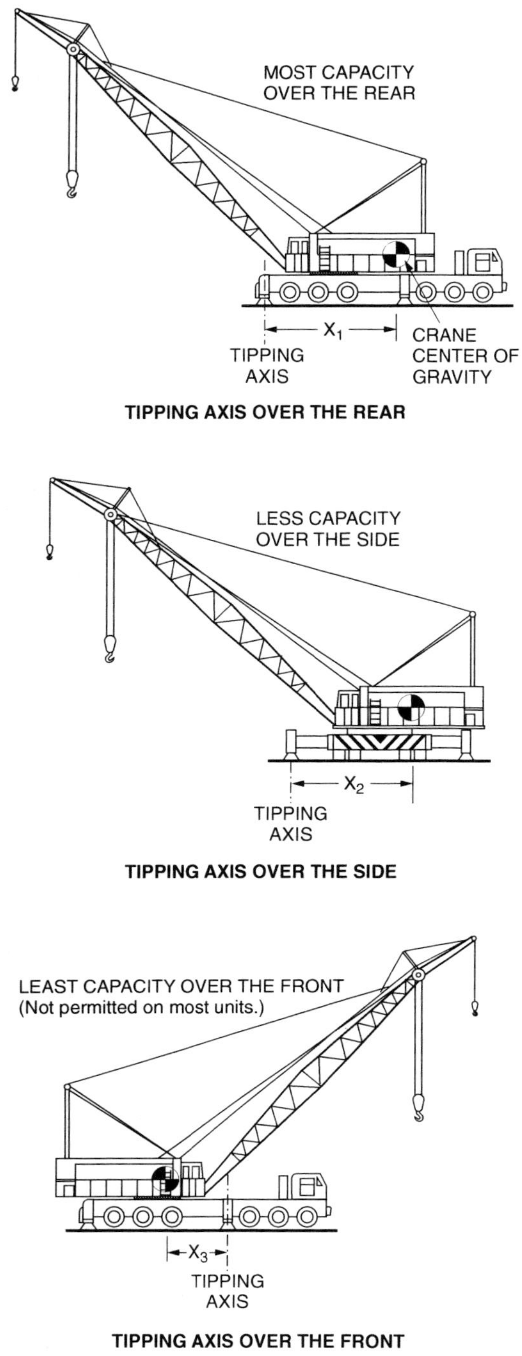

Figure 1 ◆ Typical center of gravity and leverage distances for each position of a truck-mounted crane.

over-the-side, over-the-front, or over-the rear configurations, the center of gravity is shifted either closer or farther away from the **tipping axis (point)**. This, in turn, may increase or decrease the crane's leverage (*Figure 1*).

Leverage is another important concept associated with cranes. A crane's leverage (crane weight × horizontal distance from the center of gravity of the crane to the tipping axis) has to be greater than the load's leverage (load weight × horizontal distance from the center of gravity of the load to the tipping axis) to make a lift, as shown in *Figure 2*. If it isn't, a tipping situation results. In addition to the configuration of the crane, booming down or telescoping out increases the lifting radius and moves the load's center of gravity further from the tipping axis. This, in turn, increases the load's leverage.

For example, place a twenty-pound cement brick directly in front of you. Bend over and lift it up. It does not require much effort. Now, put the brick on the ground about a foot in front of you. Lifting it requires more effort because it is farther away from you. An understanding of these principles will enable you to see why load/capacity charts are important for lifting operations.

> **WARNING!**
> For the protection of personnel and equipment, operators must follow load/capacity chart information specified by the manufacturer for the crane being used.

3.0.0 ◆ LOAD/CAPACITY CHARTS AND MANUFACTURERS

The crane manufacturer designs a load/capacity chart for each model that is sold. These charts may even be serial-number specific, meaning a different load/capacity chart is required for each crane, even if a manufacturer only makes one model. This is done because there could be some slight differences between cranes due to design changes, manufacturing techniques, materials used, permissible boom configurations, or many other reasons. The load/capacity chart will bear the model number and serial number or other identifier (the part number, for example) of the crane to which it applies. These charts cannot be interchanged, because they are identified with one crane only.

If the load/capacity chart becomes lost or illegible, it can only be replaced by the manufacturer. No crane may be operated without a load/capacity chart.

Instructor's Notes:

There are many different models of cranes in use today. They differ on structural base, lift capacity, boom configuration, and maneuverability. The next section will give a brief overview of these models.

4.0.0 ◆ TYPES OF CRANES

Mobile cranes are categorized into hydraulic (telescopic) boom cranes and lattice boom cranes. Hydraulic cranes include industrial hydraulic, rough-terrain, all-terrain, and truck-mounted types. Lattice boom cranes include lattice boom truck cranes and crawler-mounted lattice boom cranes.

4.1.0 Industrial Hydraulic Cranes

Industrial hydraulic cranes include fixed and rotating boom (*Figure 3*) crane types. They have lower lifting capacities compared to larger cranes, and they are commonly found in industrial yard work and factory operations. These cranes generally have lifting capabilities of 2 to 35 tons. Some models are operated from a standing position behind the controls (the operator station does not contain a seat). They can have fixed or rotating cabs.

4.2.0 Rough-Terrain Cranes

Rough-terrain cranes (*Figure 4*) are categorized as such to describe their unique job-site maneuverability. These cranes are designed to be operated within one percent of level on a firm, supporting surface. This type of crane normally has both four-wheel crab and tram steering, as well as the normal two-wheel front or rear capabilities.

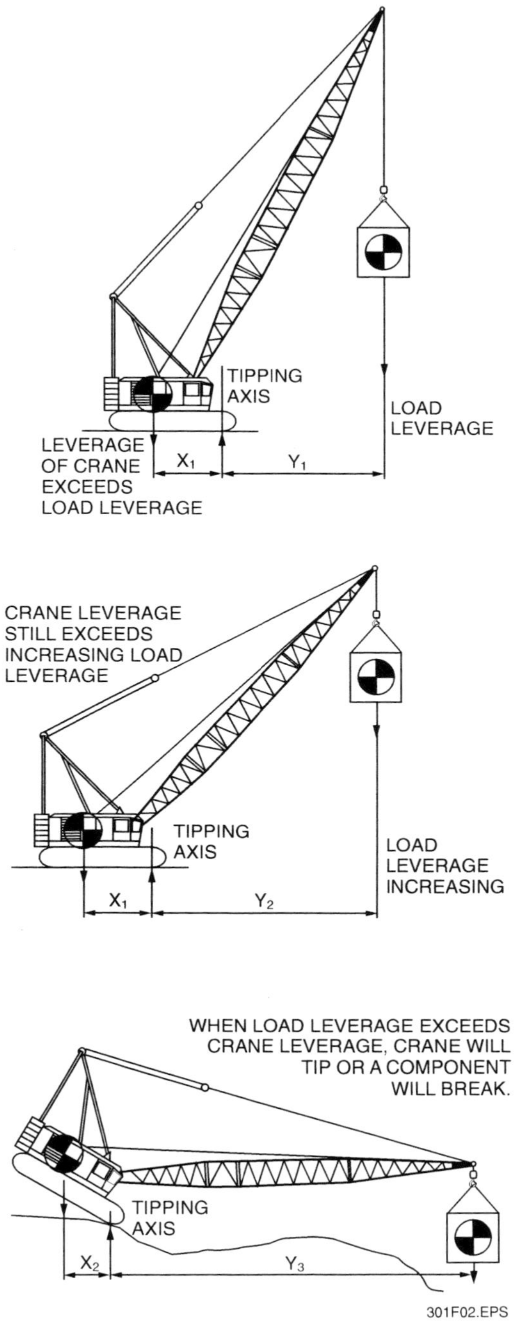

Figure 2 ◆ Relative tipping force.

Figure 3 ◆ Industrial hydraulic crane with rotating boom.

Explain that cranes are categorized into hydraulic boom cranes and lattice boom cranes. Describe the different cranes in each category.

See that trainees can describe the features and capacities of an industrial hydraulic crane and a rough-terrain crane.

See the Teaching Tip for Section 4.0.0 at the end of this module.

Provide manufacturer's literature on different types of cranes for the trainees to examine.

LOAD CHARTS 1.3

Explain how all-terrain cranes have the capabilities of both rough-terrain and truck-mounted cranes.

Describe the features of truck-mounted telescopic cranes and lattice boom cranes.

Explain that manufacturers are required to post load rating charts in a location accessible to the operator at the controls.

See the Teaching Tip for Section 5.0.0 at the end of this module.

Provide a copy of *ASME B30.5* for the trainees to examine.

301F04.EPS

Figure 4 ◆ Rough-terrain crane.

There are countless models of rough-terrain cranes by numerous manufacturers in use today. Lifting capacities range from 7 to over 160 tons.

4.3.0 All-Terrain Cranes

These cranes, with highway tires, low center of gravity, improved suspension, and all-wheel-drive capabilities, are a blend of both the rough-terrain and truck-mounted cranes (*Figure 5*). The combination of these capabilities gives the all-terrain crane the ability to transport itself from job site to job site without any major component removal.

Depending on the model of the crane, capacities in different configurations can range from several tons to over 175 tons.

301F05.EPS

Figure 5 ◆ All-terrain crane.

4.4.0 Truck- Mounted Telescopic Crane

Truck-mounted telescopic cranes consist of a hydraulic upper works placed on a carrier specifically designed and engineered for the extreme duty of transporting the unit on the highway. Truck cranes (*Figure 6*) range in capacity from 18 to over 450 tons. Disassembly of major components may be required for highway transport.

301F06.EPS

Figure 6 ◆ Truck-mounted telescopic crane.

4.5.0 Lattice Boom Cranes

Lattice boom cranes can be truck mounted or crawler mounted (*Figure 7*). In comparison to their hydraulic counterparts, these cranes are capable of much longer **boom lengths**. Site preparation and erection of lattice boom cranes may take time—from a day for smaller rigs to weeks for larger, heavy lift units with their various lift attachments. Lifting capacities, depending on the model and configuration, can range from a few tons to several hundred tons.

5.0.0 ◆ IMPORTANCE OF LOAD/CAPACITY CHARTS FOR LIFTING OPERATIONS

According to the American Society of Mechanical Engineers (ASME*) B30.5, Mobile and Locomotive Cranes*, manufacturers are required to provide specific information in load rating charts for a crane. The load rating charts must be posted in a location accessible to the operator at the controls.

The information displayed in the rating chart must include the following:

Instructor's Notes:

301F07.EPS

Figure 7 ◆ Lattice boom cranes – truck-mounted and crawler-mounted.

- The complete load rating for all operating configurations of the crane, including any optional equipment that may be installed that may affect the ratings
- A working area chart, which is often referred to as a quadrant-of-operation diagram, based on those specified in ***ASME/SAE J1028***
- Load limitation based on structural and/or mechanical strength of materials, stability, hydraulics, and/or other factors
- Work areas and load ratings where no loads are permitted to be handled
- Recommended reeving of hoist line

Load/capacity charts indicate the gross lifting capacity of the crane. This is not the weight of the load that can be suspended from the crane hook. **Gross capacity** is the value listed in the load/capacity charts. **Net capacity** is the gross capacity minus deductions (*Figures 8* and *9*).

301F08.EPS

Figure 8 ◆ Typical deductions for a hydraulic crane.

Describe the information on the load rating chart.

Have a trainee explain the difference between gross capacity and net capacity.

Show Transparency 5 (Figure 9). Discuss various deductions.

Write an example of gross pay, common deductions, and net pay on the board. Use the example to explain gross capacity and net capacity.

LOAD CHARTS 1.5

Show Transparency 6 (Figure 10). Explain that the load chart is developed based on specific boom length, boom angle, and load operating radius.

Explain that load charts are based on a crane operating at ideal conditions. Discuss allowances that should be made for less than ideal conditions.

Reiterate the importance of following a specific load chart for a particular crane.

Provide a copy of *ANSI/SAE J765* for the trainees to examine.

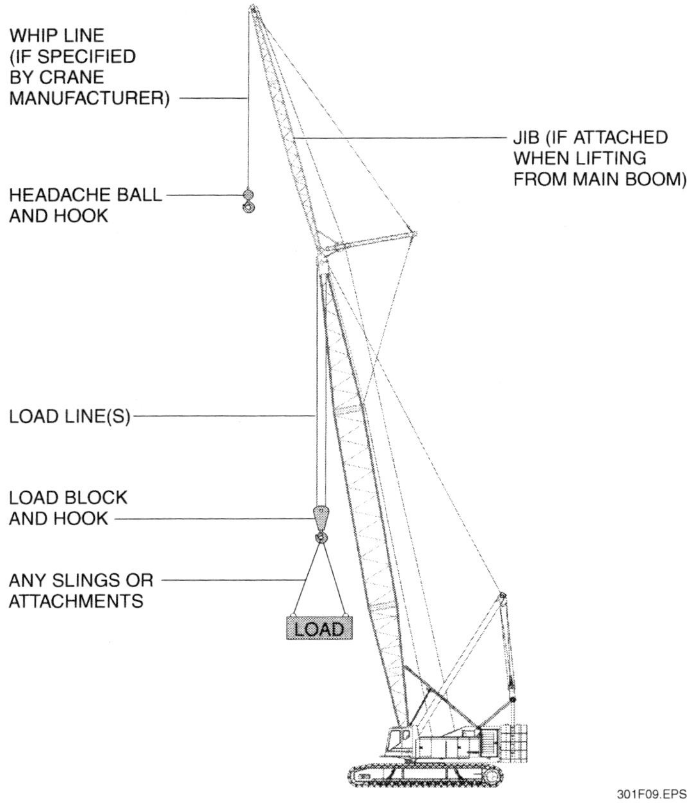

Figure 9 ◈ Typical deductions for a lattice boom crane.

It is important to remember that the manufacturer bases the load/capacity chart capacities on the maximum allowable freely suspended loads at a given boom length, **boom angle**, and **load operating radius**, as shown in *Figure 10*. Remember that boom length can vary according to different manufacturers. Boom length is the length of the main boom and, in some cases, boom extensions.

Boom length and load operating radius are not necessarily in a direct relationship. They have been determined for a machine standing level on a firm, supporting surface under the following ideal operating conditions:

- Equipment maintained in like-new condition
- Equipment operated by an experienced, qualified operator
- No inclement weather during operation or with a suspended load
- Known load weight
- Correct rigging and counterweights used
- Boom length, operating radius, and boom angle known
- Load block or hook over the center of gravity of the load

Refer to the manufacturer's load charts for notes concerning allowances for any conditions other than ideal. The operator must limit or derate loads to allow for adverse conditions.

Also note that load chart capacities do not exceed the percentage of tipping loads as determined in accordance with ***ANSI/SAE J765***.

> **WARNING!**
> For the protection of personnel and equipment, operators must follow load/capacity chart information according to manufacturer's specifications.

Instructor's Notes:

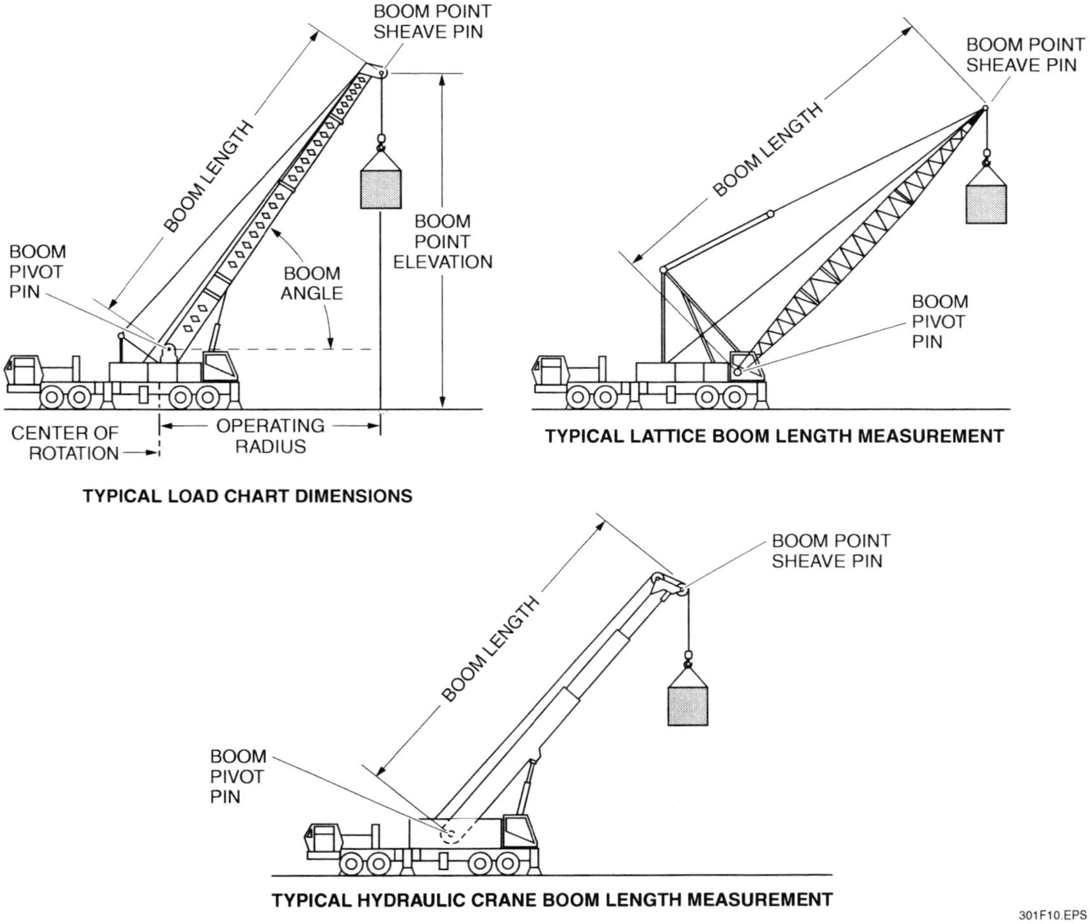

Figure 10 ◆ Boom length, boom angle, load operating radius, and boom point elevation.

The ability to understand and use the crane's load/capacity chart correctly is critical to the safe operation of a mobile crane. When operators are unable to do so, they dangerously rely on guesswork or the sensation of a crane tipping to gauge the load limit. This practice can result in serious accidents and fatalities.

It is critical to remember that tipping cannot be relied upon as an indication that the crane has reached capacity. A crane may be overloaded before any signs of tipping are evident. A crane can be overstressed or can fail structurally before tipping occurs.

If there is no capacity listing for a particular boom length or radius, the operator is prohibited from lifting in that configuration. In some cases, a boom cannot be lowered below a minimum angle in some quadrants because, even without a load, the weight of the boom and attachments may cause tipping.

The capacities on a load/capacity chart are limited by either the structural strength of the crane, as determined by *ANSI/SAE J987* guidelines, or stability (the crane's ability to resist tipping). The load/capacity chart is generally divided with shading, bold lines, or asterisks to indicate these two areas. *Figure 11* shows examples of charts that use bold lines, shading, or asterisks to separate the areas.

Discuss the importance of using a load chart to operate a crane safely. Discuss the dangers of relying on feel to gauge the load limit. Explain that a crane may be overloaded and fail structurally before any signs of tipping are evident. If applicable, discuss some real-life accidents and how they could have been prevented by using load charts.

Explain that a crane's capacities are limited by either its structural strength or its stability.

Discuss how these are differentiated on a load chart.

Provide a copy of *ANSI/SAE J987* for the trainees to examine.

Have trainees review Sections 6.0.0–6.4.0.

LOAD CHARTS 1.7

Ensure that you have everything required for teaching this session.

Explain the three basic crane configurations. Discuss the principal factors that influence the crane's capacity and the reading of the load chart.

Main Boom Capacities* – 110 ft — Open Throat Boom

Load Radius (ft)	Boom Angle (deg)	360° Rotation				Over End Blocked	
		135K + LWR (lb)	135K + 0 (lb)	83K + 0 (lb)	31K + 0 (lb)	135K + LWR (lb)	135K + 0 (lb)
18.5	82.0	346,500	346,500	329,000	283,900	346,500	346,500
19	81.7	344,400	344,400	320,000	276,000	344,400	344,400
20	81.2	341,100	341,100	303,900	246,900	341,100	341,100
25	78.6	276,100	276,100	233,200	158,900	276,100	276,100
30	75.9	229,100	226,500	171,200	116,000	229,100	229,100
35	73.2	192,600	178,500	134,500	90,500	195,200	195,200
40	70.4	158,500	146,800	110,300	73,700	169,700	169,700
50	64.8	116,200	107,400	80,100	52,900	133,700	133,000
60	58.9	90,900	83,900	62,200	40,400	109,700	102,800
70	52.6	74,100	68,300	50,200	32,100	90,700	83,100
80	45.7	62,100	57,200	41,600	26,100	75,800	69,300
90	37.9	53,100	48,800	35,200	21,700	64,700	59,100
100	28.3	46,100	42,200	30,200	18,100	56,100	51,100
110	13.8	40,400	36,900	26,100	15,200	49,200	44,700

*Ratings are not more than 75% of tipping loads

SHADED AREA RATINGS BASED ON STRUCTURAL STRENGTH

UNSHADED AREA RATINGS BASED ON STABILITY

Boom and Jib Length	Jib Radius (ft)	5.0° Offset		15.0° Offset		25.0° Offset	
		Boom Angle	Ratings (lbs)	Boom Angle	Ratings (lbs)	Boom Angle	Ratings (lbs)
40' (12.2M) JIB & 200' (61.0M) BOOM	45	81.0	19,380*				
	50	79.8	18,840*				
	60	77.3	17,740*	78.9	16,740*	80.4	15,650*
	70	74.9	16,670*	76.5	15,890*	77.9	15,050*
	80	72.4	15,640*	74.0	15,050*	75.4	14,440*
	90	69.9	13,330	71.5	13,330	72.9	13,330
	100	67.4	11,190	68.9	11,200	70.3	11,200
	110	64.7	9,480	66.3	9,480	67.6	9,490
	120	62.1	8,070	63.6	8,070	64.9	8,070
	130	59.3	6,900	60.8	6,900	62.1	6,900
	140	56.5	5,900	58.0	5,900	59.3	5,900
	150	53.6	5,040	55.1	5,040	56.3	5,040
	160	50.6	4,290	52.0	4,290	53.2	4,290
	170	47.4	3,640	48.8	3,640	49.9	3,650
	180	44.0	3,070	45.4	3,080	46.5	3,080
	190	40.4	2,560	41.8	2,560	42.8	2,570
	200	36.6	2,110	37.9	2,110	38.8	2,120

RATINGS WITH AN ASTERISK BASED ON STRUCTURAL STRENGTH

RATINGS WITHOUT AN ASTERISK BASED ON STABILITY

301F11A.EPS

Figure 11 ◆ Structural strength and stability charts. (1 of 2)

6.0.0 ◆ CRANE CONFIGURATION

The load/capacity charts of mobile cranes are based on the configuration of the crane at the time of the lift. These configurations start with one of three basic situations:

- Jibs and boom extensions not installed
- Jib and/or boom extensions installed but load lifted from main boom
- Jibs and/or boom extensions installed and load lifted from jib or boom extension

All other aspects of the crane's configuration are considered in relation to these three basic cases. The principal factors influencing the machine's capacity and how the chart is read include these considerations:

- Boom length, boom angle, and load radius
- Type of boom
  - Hydraulic
  - Lattice
- Operational quadrant in which the superstructure is working
- Configuration of crane base
  - On rubber
  - On outriggers (stabilizers)
  - On crawlers
- Hydraulic boom modes (A or B)
- Effective/stowed weight of jib or boom extension (or other attachments)
- Ring or tower attachment
- Type, size, and weight of load block and headache ball
- Weight of all required rigging
- Weight of rope per manufacturer's requirement
- Counterweight configuration

Instructor's Notes:

Rated Lifting Capacities in Pounds (35–110 ft Boom) on Outriggers Fully Extended – 360°

Radius in Feet	#0001								
	Main Boom Length in Feet								
	35	40	50	*60	70	80	90	100	110
10	100,000 (63.5)	80,400 (66.5)	74,400 (71.5)	44,600 (75.5)					
12	88,050 (60)	79,050 (63.5)	70,900 (69)	44,600 (74)	@35,600 (75.5)				
15	74,500 (54)	67,450 (59)	63,350 (65.5)	44,600 (71)	35,600 (74)	@33,000 (75.5)			
20	54,700 (43)	53,850 (50.5)	50,900 (59)	44,600 (66)	35,600 (70)	33,000 (72.5)	25,500 (75)	@23,300 (75.5)	
25	41,450 (29)	41,150 (40.5)	40,700 (52.5)	40,350 (60.5)	35,550 (65.5)	33,000 (69)	25,500 (71.5)	23,300 (74)	@18,500 (75.5)
30		32,450 (28)	32,050 (45)	31,750 (55)	30,550 (61)	28,950 (65)	25,500 (68)	23,300 (71)	18,500 (73)
35			25,950 (36.5)	25,650 (48.5)	26,500 (56.5)	24,900 (61)	23,000 (64.5)	21,200 (68)	18,500 (70.5)
40	See Note 16		21,400 (25)	21,150 (41.5)	22,000 (51.5)	21,750 (57)	20,000 (61)	18,450 (65)	18,000 (67.5)
45				17,600 (33.5)	18,500 (46)	19,100 (53)	17,600 (57.5)	16,300 (61.5)	15,750 (65)
50				14,600 (23)	15,250 (39.5)	15,700 (48)	15,650 (53.5)	14,400 (58)	13,950 (62)
55					12,650 (32.5)	13,100 (43)	13,550 (49.5)	12,850 (54.5)	12,450 (59)
60					10,500 (23)	11,000 (37.5)	11,450 (45)	11,550 (51)	11,150 (55.5)
65						9,350 (31)	9,780 (40.5)	10,200 (47)	10,050 (52.5)
70						8,870 (22)	8,370 (35)	8,780 (43)	9,090 (49)
75							7,180 (28.5)	7,590 (38.5)	7,980 (45)
80							6,120 (20)	6,560 (33)	6,950 (41)
85								5,680 (27)	6,060 (37)
90								4,910 (19)	5,280 (32)
95									4,600 (26)
100									3,990 (18.5)
Minimum boom angle (deg.) for indicated length									0
Maximum boom length (ft.) at 0 deg. boom angle (no load)									110

Note: ( ) Boom angles are in degrees.

RATINGS ABOVE BOLD LINE IN EACH COLUMN BASED ON STRUCTURAL STRENGTH

RATINGS BELOW BOLD LINE IN EACH COLUMN BASED ON STABILITY

301F11B.EPS

Figure 11 ◆ Structural strength and stability charts. (2 of 2)

6.1.0 Boom Length, Boom Angle, and Load Radius

For hydraulic (telescopic) cranes, the load radius is located on the left-hand side, and sometimes on the far right-hand side, of load/capacity charts. For lattice boom cranes, the load radius is usually located only on the left-hand side of the chart. The location of the boom angle information can vary. It may be in a column to the right of the load radius, or it may be set off in parentheses under the gross capacities.

Understanding where these terms are on a load/capacity chart is key to reading and interpreting a load/capacity chart properly. Operators must demonstrate their ability to read, comprehend, and write in the language of the crane manufacturer's operation and maintenance instruction materials. They must also exhibit arithmetic skills and the ability to use the manufacturer's load/capacity charts.

Most hydraulic load/capacity charts (*Figure 12*) start with the shortest boom length at the upper left side and read to the right and down for progressively longer booms. First locate the number that corresponds to the length in feet of the boom in the boom length heading. Then, locate the operating radius of the crane in the left- or right-hand column. The capacity where the operating radius line and a particular operating quadrant under the boom length column intersect is the gross capacity that can be lifted safely with a defined counterweight configuration.

Lattice boom crawler crane load/capacity charts (*Figure 13*) can be presented differently, depending on the manufacturer. In this example, the boom length is located in the chart heading, and the load radius is in the left-hand column.

Explain how to find information on the load radius and boom angle on the load charts. Discuss the importance of being able to understand these terms and their implications.

Refer to Figure 12. Explain how to read the boom length on a hydraulic load chart. Explain how to find the operating radius to determine capacity.

Refer to Figure 13. Explain that the load charts for lattice boom cranes may be presented differently. Explain how to read them.

Give some example problems and have trainees use Figures 12 and 13 to determine capacity for hydraulic boom and lattice boom cranes.

LOAD CHARTS 1.9

REQUIRED COUNTERWEIGHT CONFIGURATION

BOOM LENGTH HEADING (1 OF 4)

BOOM MODE

LOADED BOOM ANGLE COLUMN

LOAD RADIUS COLUMN

CAPACITY IN ANY QUADRANT

CAPACITY IN REAR QUADRANT ONLY

Rated Lifting Capacities In Pounds
Fully Extended Outriggers
See Set Up Note 2

4,700 lbs.

FULL EXTENSION

MAIN BOOM "A"

Load Radius (Ft.)	33 Ft. ∡°	33 Ft. 360	33 Ft. Over Rear	40 Ft. ∡°	40 Ft. 360	40 Ft. Over Rear	Load Radius (Ft.)
9	68.0	80,000	80,000				9
10	66.0	72,300	72,300	70.5	72,300	72,300	10
12	62.0	65,500	65,500	67.5	65,200	65,200	12
15	55.5	55,600	55,600	62.5	55,100	55,100	15
20	43.5	38,400	38,500	54.0	37,900	37,900	20
25	26.5	25,600	25,600	43.5	25,300	25,300	25
30				31.0	18,100	18,100	30
Min.Bm Ang/Cap	0 (27.5)	18,400	18,400	0 (34.5)	13,700	13,700	Min.Bm Ang/Cap

Load Radius (Ft.)	50 Ft. ∡°	50 Ft. 360	50 Ft. Over Rear	57 Ft. ∡°	57 Ft. 360	57 Ft. Over Rear	Load Radius (Ft.)
10	75.0	67,500	67,500	77.0	43,800	43,800	10
12	73.0	61,200	61,200	75.0	43,800	43,800	12
15	69.0	53,400	53,400	72.0	42,100	42,100	15
20	62.5	37,300	37,300	66.5	34,300	34,300	20
25	55.5	24,900	24,900	60.5	24,600	24,600	25
30	48.0	17,900	17,900	54.5	17,600	17,600	30
35	39.0	13,200	13,200	47.5	13,100	13,100	35
40	27.5	9,800	10,000	40.0	9,600	9,900	40
45				30.5	7,200	7,600	45
50				16.0	5,200	5,700	50
Min.Bm Ang/Cap	0 (44.5)	7,400	7,800	0 (51.5)	4,700	5,200	Min.Bm Ang/Cap

Note: Refer to page 5 for (Capacity Deductions for Auxiliary Load Handling Equipment).

∡° Loaded Boom Angle in Degrees

( ) Reference radius for minimum boom angle capacities (shown in parentheses) are in feet.

301F12.EPS

Figure 12 ◆ Hydraulic boom load/capacity chart.

BOOM LENGTH (CHART HEADING)

LOAD RATINGS COLUMN

CAPACITY IN ANY QUADRANT

LOADED BOOM ANGLE COLUMN

CAPACITY IN EITHER END QUADRANT ONLY

COUNTERWEIGHT CONFIGURATIONS (LWR = BOTH LOWER CRAWLER FRAME COUNTERWEIGHTS)

Main Boom Capacities* – 110 ft — Open Throat Boom

Load Radius (ft)	Boom Angle (deg)	360° Rotation 135K + LWR (lb)	360° Rotation 135K + 0 (lb)	360° Rotation 83K + 0 (lb)	360° Rotation 31K + 0 (lb)	Over End Blocked 135K + LWR(lb)	Over End Blocked 135K + 0 (lb)
18.5	82.0	346,500	346,500	329,000	283,900	346,500	346,500
19	81.7	344,400	344,400	320,000	276,000	344,400	344,400
20	81.2	341,100	341,100	303,900	246,900	341,100	341,100
25	78.6	276,100	276,100	233,200	158,900	276,100	276,100
30	75.9	229,100	226,500	171,200	116,000	229,100	229,100
35	73.2	192,600	178,500	134,500	90,500	195,200	195,200
40	70.4	158,500	146,800	110,300	73,700	169,700	169,700
50	64.8	116,200	107,400	80,100	52,900	133,700	133,000
60	58.9	90,900	83,900	62,200	40,400	109,700	102,800
70	52.6	74,100	68,300	50,200	32,100	90,700	83,100
80	45.7	62,100	57,200	41,600	26,100	75,800	69,300
90	37.9	53,100	48,800	35,200	21,700	64,700	59,100
100	28.3	46,100	42,200	30,200	18,100	56,100	51,100
110	13.8	40,400	36,900	26,100	15,200	49,200	44,700

***Ratings are not more than 75% of tipping loads**

301F13.EPS

Figure 13 ◆ Lattice boom crawler crane load/capacity chart.

Instructor's Notes:

The operator must first know what boom length the crane is operating at, determine the radius, and then read across to determine the gross capacity, depending on the counterweight configuration. Some manufacturers may also include the **boom elevation point** in their charts for reference.

Remember that the load/capacity charts shown in this module are examples from certain manufacturers. The standards do not require or specify that this information be presented in a particular format. Load/capacity chart layouts are as diverse as the number of manufacturers that build cranes. They often vary greatly between different crane models from the same manufacturer. Some have illustrations and counterweight requirements at the top; some express their load radius or boom lengths in feet and/or meters (*Figures 14, 15,* and *16*), and some list their capacities in kilograms and pounds (*Figures 14* and *16*). This variety presents a challenge to operators; they must pay close attention to load/capacity charts and their introductory information when switching between cranes to avoid making mistakes on load chart calculations and interpretations.

It is imperative for operators to remember that if a load radius, boom length, or boom angle falls between values listed on the load/capacity chart, there is no **interpolation** of values allowed. When either boom length or radius, or both, are between values listed, the smallest load shown at either the next larger radius or next longer boom length shall be used.

Explain that the format for load charts is not standard and varies greatly.

Show Transparencies 7 and 8 (Figures 14 and 16). Discuss the differences in format for various load charts. Explain how to find information on each type. Explain that operators must pay close attention to each chart.

Give some example problems and have trainees use Figures 14 through 16 to determine crane capacity.

Define interpolation, and explain that it is not allowed when using load charts.

Outriggers Fully Extended

25' - 43' *(7.62m - 13.11m)* Offset Telescoping Fly + 70.25' *(21.41m)* Main Boom with Manual Retracted

	25' (7.62m) Fly						43' (13.11m) Fly						
	2° OFFSET		15° OFFSET		30° OFFSET		2° OFFSET		15° OFFSET		30° OFFSET		
LOAD RADIUS (ft/m)	LOADED BOOM ANGLE (deg.)	LOAD (lbs/kgs)	LOADED BOOM ANGLE (deg.)	LOAD (lbs/kgs)	LOADED BOOM ANGLE (deg.)	LOAD (lbs/kgs)	LOADED BOOM ANGLE (deg.)	LOAD (lbs/kgs)	LOADED BOOM ANGLE (deg.)	LOAD (lbs/kgs)	LOADED BOOM ANGLE (deg.)	LOAD (lbs/kgs)	LOAD RADIUS (ft/m)
20 *6.10*	78.5	12,700 *5,761*											20 *6.10*
25 *7.62*	75.5	11,500 *5,216*	79.0	8,800 *3,992*			77.5	4,900 *2,223*					25 *7.62*
30 *9.14*	72.5	10,600 *4,808*	76.0	8,200 *3,720*	79.0	6,100 *2,767*	75.5	4,700 *2,132*					30 *9.14*
35 *10.67*	69.5	9,600 *4,355*	73.0	7,500 *3,402*	76.0	5,700 *2,586*	73.0	4,500 *2,041*	78.0	4,000 *1,614*			35 *10.67*
40 *12.19*	66.0	8,700 *3,946*	69.5	6,900 *3,130*	72.5	5,400 *2,449*	70.0	4,300 *1,950*	75.5	3,900 *1,769*			40 *12.19*
45 *13.72*	63.0	8,000 *3,629*	66.0	6,400 *2,903*	69.0	5,100 *2,313*	67.5	4,100 *1,860*	72.5	3,700 *1,678*	77.5	3,000 *1,361*	45 *13.72*
50 *15.24*	59.5	7,400 *3,357*	62.5	5,900 *2,676*	65.5	4,800 *2,177*	65.0	3,900 *1,769*	70.0	3,500 *1,588*	74.5	2,900 *1,315*	50 *15.24*
55 *16.76*	56.0	6,500 *2,948*	59.0	5,300 *2,404*	62.0	4,500 *2,041*	62.0	3,800 *1,724*	67.0	3,300 *1,497*	71.5	2,700 *1,225*	55 *16.76*
60 *18.29*	52.0	5,400 *2,449*	55.0	4,900 *2,223*	58.0	4,200 *1,905*	59.5	3,700 *1,678*	64.0	3,100 *1,406*	68.5	2,600 *1,179*	60 *18.29*
65 *19.81*	48.0	4,500 *2,041*	51.0	4,500 *2,041*	53.5	3,900 *1,769*	56.5	3,500 *1,588*	61.0	2,900 *1,315*	65.5	2,500 *1,134*	65 *19.81*
70 *21.34*	43.5	3,700 *1,678*	46.5	3,900 *1,769*	49.0	3,600 *1,633*	53.0	3,200 *1,452*	58.0	2,800 *1,270*	62.5	2,400 *1,089*	70 *21.34*
75 *22.86*	38.5	3,000 *1,361*	41.5	3,200 *1,452*	44.0	3,400 *1,542*	50.5	3,000 *1,361*	54.5	2,600 *1,179*	59.0	2,300 *1,043*	75 *22.86*
80 *24.38*	33.0	2,500 *1,134*	36.0	2,600 *1,179*	37.5	2,700 *1,225*	46.5	2,800 *1,270*	51.0	2,500 *1,134*	55.0	2,200 *998*	80 *24.38*
85 *25.91*	26.5	2,000 *907*	29.0	2,100 *953*	30.0	2,100 *953*	43.0	2,700 *1,225*	47.5	2,400 *1,089*	51.0	2,100 *953*	85 *25.91*
90 *27.43*	17.5	1,600 *726*	19.5	1,600 *726*	17.0	1,600 *726*	38.5	2,300 *1,043*	43.0	2,300 *1,043*	46.5	2,000 *907*	90 *27.43*
95 *28.96*							34.0	2,000 *907*	38.5	2,200 *998*	41.5	2,000 *907*	95 *28.96*
100 *30.48*							28.5	1,600 *726*	33.0	1,800 *816*	35.5	1,900 *862*	100 *30.48*
105 *32.00*									25.5	1,400 *635*	26.0	1,400 *635*	105 *32.00*

Warning: Do not lower 25' offset telescoping fly in working position below 12° unless main boom is 68' or less, since loss of stability will occur causing a tipping condition.

Warning: Do not lower 43' offset telescoping fly in working position below 20° unless main boom is 65' or less, since loss of stability will occur causing a tipping condition.

© 1992 LinkBelt Construction Company

301F14.EPS

Figure 14 ◈ Load/capacity chart in English and metric units with left- and right-hand load radius columns.

LOAD CHARTS 1.11

CAUTION

Interpolation (mathematical approximation) is not permitted in determining load chart capacities.

Lifting capacities main boom
Tragfähigkeiten Hauptausleger
Capacités de levage flèche principale

140 t + 40 t ZB | 7,25 m | 360° | **DIN/ISO**

	Radius Ausladung Porteé	Main boom · Hauptausleger · Flèche principale											
	m	18,0	24,0	30,0	36,0	42,0	48,0	54,0	60,0	66,0	72,0	78,0	84,0
	m	t	t	t	t	t	t	t	t	t	t	t	t
	4,5	300,0*	295,0*	-	-	-	-	-	-	-	-	-	-
	5	300,0*	295,0*	289,0*	-	-	-	-	-	-	-	-	-
	5,5	300,0*	295,0*	289,0*	274,0*	245,0	-	-	-	-	-	-	-
	6	300,0*	295,0*	289,0*	274,0*	245,0	215,0	-	-	-	-	-	-
	7	288,0*	295,0*	280,0*	274,0*	245,0	215,0	171,0	143,0	116,0	97,5	-	-
	8	271,0*	276,0*	271,0*	255,0*	245,0	215,0	171,0	143,0	116,0	97,5	82,5	69,5
	9	251,0*	247,0	249,0	235,0	225,0	202,0	171,0	143,0	116,0	97,5	82,5	69,5
	10	209,0	209,0	209,0	208,0	205,0	189,0	165,0	139,0	116,0	97,5	82,5	69,5
	12	157,0	156,0	156,0	155,0	155,0	154,0	151,0	133,0	115,0	97,0	82,0	69,5
	14	125,0	124,0	124,0	123,0	122,0	122,0	122,0	121,0	113,0	96,5	81,5	69,0
	16	103,0	103,0	102,0	101,0	101,0	100,0	100,0	100,0	100,0	96,0	81,5	68,5
	18	88,5	87,5	87,0	86,0	85,5	85,0	84,5	84,5	84,0	83,5	81,0	68,5
	20	-	76,0	75,0	74,5	73,5	73,0	73,0	72,5	72,5	72,0	71,0	66,5
SH	22	-	67,0	66,0	65,5	64,5	64,0	63,5	63,5	63,0	62,5	62,0	61,5
	24	-	-	59,0	58,0	57,5	56,5	56,5	56,0	56,0	55,0	54,5	54,0
	26	-	-	53,0	52,0	51,5	50,5	50,5	50,0	49,9	49,4	48,7	48,0
	28	-	-	48,3	47,4	46,5	45,9	45,5	45,1	44,9	44,3	43,6	42,9
	30	-	-	-	43,2	42,3	41,7	41,2	40,8	40,6	40,0	39,4	38,6
	34	-	-	-	-	35,7	35,0	34,4	34,0	33,8	33,2	32,5	31,8
	38	-	-	-	-	30,7	29,9	29,3	28,8	28,5	27,9	27,2	26,5
	42	-	-	-	-	-	25,9	25,3	24,7	24,4	23,8	23,1	22,4
	46	-	-	-	-	-	-	22,1	21,5	21,1	20,5	19,7	18,8
	50	-	-	-	-	-	-	-	18,8	18,4	17,6	16,8	15,9
	54	-	-	-	-	-	-	-	16,5	16,0	15,2	14,4	13,5
	58	-	-	-	-	-	-	-	-	14,0	13,2	12,3	11,4
	62	-	-	-	-	-	-	-	-	-	11,5	10,6	9,6
	66	-	-	-	-	-	-	-	-	-	-	9,1	8,1
	70	-	-	-	-	-	-	-	-	-	-	7,8	6,8
	74	-	-	-	-	-	-	-	-	-	-	-	5,7

301F15.EPS

Figure 15 ◆ Load/capacity chart in metric units only with left-hand load radius column.

Instructor's Notes:

TEREX LIFTING

AMERICAN HC 60

Hydraulic Crawler Crane 46 HI Boom

With 46HI Angle Boom, 4 Sheave T ip, 39,000 Pound Counterweight

Boom Length	Radius (Feet)	Boom Angle (Degrees)	Side Frames Extended (Pounds)	From Boom Pt. to Ground (Feet)
40' (12.2M)	10	80.3	120,000*	45
	12	77.4	120,000*	44
	15	73.0	93,390	44
	20	65.3	59,020	42
	25	57.1	42,780	39
	30	48.1	33,370	35
	35	37.5	27,190	30
	40	23.4	22,840	21
50' (15.2M)	12	80.0	120,000*	55
	15	76.4	93,310	54
	20	70.5	58,910	53
	25	64.2	42,640	50
	30	57.7	33,230	48
	35	50.6	27,040	44
	40	42.7	22,670	39
	45	33.5	19,460	33
	50	20.9	16,960	23
60' (18.3M)	13	80.7	117,460*	65
	15	78.7	93,230	64
	20	73.8	58,800	63
	25	68.8	42,510	61
	30	63.6	33,100	59
	35	58.1	26,890	56
	40	52.3	22,520	53
	45	46.0	19,330	49
	50	38.9	16,820	43
	55	30.5	14,820	36
	60	19.0	13,190	[illegible]

Boom Length	Radius (Feet)	Boom Angle (Degrees)	Side Frames Extended (Pounds)	From Boom Pt. to Ground (Feet)
90' (con't)	45	62.4	18,820	85
	50	58.7	16,290	82
	55	54.9	14,290	79
	60	50.9	12,650	75
	65	46.7	11,300	71
	70	42.2	10,160	66
	75	37.2	9,180	60
	80	31.5	8,330	52
	85	24.8	7,610	43
	90	15.5	6,970	29
100' (30.5M)	19	81.0	62,850	104
	20	80.4	58,180	104
	25	77.5	41,800	103
	30	74.5	32,400	102
	35	71.5	26,170	100
	40	68.5	21,770	98
	45	65.3	18,630	96
	50	62.2	16,110	94
	55	58.9	14,100	91
	60	55.4	12,460	88
	65	51.9	11,110	84
	70	48.2	9,970	80
	75	44.2	8,980	75
	80	39.9	8,140	[illegible]
	85	35.2	[illegible]	[illegible]
	90	[illegible]	[illegible]	[illegible]

Boom Length	Radius (Feet)	Boom Angle (Degrees)	Side Frames Extended (Pounds)	From Boom Pt. to Ground (Feet)
120' (con't)	90	43.7	6,350	88
	95	40.2	5,770	83
	100	36.3	5,260	76
	105	32.0	4,800	69
	110	27.2	4,380	60
	115	21.4	4,020	49
	120	13.4	3,680	33
130' (39.6M)	24	80.8	43,780	134
	25	80.4	41,220	134
	30	78.1	31,830	133
	35	75.9	25,570	131
	40	73.6	21,160	130
	45	71.3	18,050	129
	50	68.9	15,520	127
	55	66.6	13,500	125
	60	64.1	11,870	122
	65	61.7	10,500	120
	70	59.1	9,350	117
	75	56.5	8,370	114
	80	53.8	7,520	110
	85	51.0	6,780	107
	[illegible]	[illegible]	[illegible]	102

301F16.EPS

Figure 16 ◆ Load/capacity chart in English and metric units with left-hand boom length column.

Some hydraulic telescoping cranes are equipped with dual boom mode capabilities, known as A and B (*Figures 17* and *18*). They are engaged or disengaged by means of a switch. Boom mode A could extend all sections of the boom equally, whereas boom mode B could just extend the first, second, or third section. The different configurations will result in different lifting capacities, which, in turn, necessitate different load/capacity charts for each mode.

6.2.0 Quadrants of Operation

As a crane swings to various quadrants, the distance between the load's position and the tipping axis changes dramatically. This is because cranes are typically built to be longer than they are wide to allow them to fit on a roadway or trailer for transportation from job to job. This rectangular shape tends to result in more stability of the crane from front to rear than from side to side.

Explain that some cranes have dual boom mode capabilities, and discuss how that affects capacity.

Give some example problems, and have trainees use Figures 17 and 18 to determine crane capacity.

Demonstrate how to identify the boom angle, boom length, and load radius on a load/capacity chart.

Have trainees practice identifying the boom angle, boom length, and load radius on a load/capacity chart. Note the proficiency of each trainee. This laboratory corresponds to Performance Task 1.

Explain how the quadrant of operation affects the tipping axis and stability.

LOAD CHARTS 1.13

35 - 138 ft. (10.8 - 42.0 m) | 0 lbs. (0 kg) | 100% | 360°

85% Domestic (Pounds)

(Feet)	35	61	74	87	99	112	138
10	139,500 (65.5)	42,900 (77.5)	*32,100 (80)				
12	110,000 (62)	42,000 (75.5)	32,100 (78.5)	*31,850 (80)			
15	92,450 (56)	36,550 (72.5)	32,100 (76)	31,850 (78.5)	*21,350 (80)		
20	64,600 (44.5)	29,400 (67.5)	31,350 (71.5)	28,850 (75)	21,350 (77.5)	19,000 (79.5)	
25	48,350 (29.5)	24,350 (62)	26,450 (67.5)	25,050 (71.5)	20,850 (74.5)	18,150 (77)	*19,000 (80)
30		20,500 (56.5)	22,300 (63)	21,550 (68)	18,650 (71.5)	17,300 (74.5)	18,300 (78.5)
35		17,450 (50)	19,100 (58.5)	18,500 (64)	16,900 (68.5)	16,450 (71.5)	17,650 (76.5)
40		15,050 (43)	16,500 (53.5)	16,000 (60)	15,300 (65)	15,650 (69)	17,000 (74)
45		13,100 (35)	14,450 (48.5)	14,000 (56)	13,650 (61.5)	14,150 (66)	16,350 (72)
50		11,450 (24.5)	12,750 (42.5)	12,350 (52)	12,100 (58.5)	12,700 (63)	15,450 (69.5)
60			9,160 (28)	9,710 (42.5)	9,580 (51)	10,150 (57)	10,400 (65)
70				5,990 (30)	7,430 (42.5)	8,220 (50)	7,140 (60)
80					4,720 (32)	5,790 (42.5)	4,820 (55)
90					2,550 (15.5)	3,700 (33.5)	3,100 (49.5)
100						1,990 (21)	1,760 (43)
Minimum boom angle (deg.) for indicated length (no load)						0	40

Maximum boom length (ft.) at 0 degree boom angle (no load) 112

NOTE: ( ) Boom angles are in degrees.

*This capacity is based on maximum boom angle.

Boom Angle	35	61	74	87	99	112
0	26,400 (28.2)	10,150 (53.8)	6,240 (66.6)	3,330 (79.4)	2,130 (92.2)	1,260 (105)

NOTE: ( ) Reference radii are in feet.

A6-829-014471

T1 T2 T3 T4 % **MODE A**

© Grove Europe Limited Form number TMS//TTS870

301F17.EPS

Figure 17 ◈ Hydraulic crane load/capacity chart for mode A boom configuration.

Instructor's Notes:

35 - 138 ft. (10.8 - 42.0 m) | 0 lbs. (0 kg) | 100% | 360°

85% Domestic (Pounds)

(Feet)	35	55	61	74	87	99	112	125	138
10	139,500 (65.5)	79,100 (76)	78,450 (77.5)	*57,050 (80)					
12	110,000 (62)	79,100 (73.5)	77,500 (75.5)	57,050 (78.5)	*43,300 (80)				
15	92,450 (56)	79,100 (70)	69,850 (72.5)	51,650 (76)	43,300 (78.5)	*32,100 (80)			
20	64,600 (44.5)	64,250 (64.5)	59,850 (67.5)	44,350 (71.5)	39,550 (75)	32,100 (77.5)	30,050 (79.5)	*20,150 (80)	
25	48,350 (29.5)	48,050 (58)	48,450 (62)	38,750 (67.5)	33,800 (71.5)	32,100 (74.5)	30,050 (77)	20,150 (79)	*19,000 (80)
30		34,400 (51)	34,050 (56.5)	34,050 (63)	29,200 (68)	30,200 (71.5)	27,350 (74.5)	19,100 (76.5)	18,300 (78.5)
35		25,150 (43.5)	24,500 (50)	25,200 (58.5)	25,250 (64)	26,600 (68.5)	24,300 (71.5)	18,100 (74)	17,650 (76.5)
40		18,650 (34.5)	18,400 (43)	19,300 (53.5)	18,750 (60)	20,050 (65)	21,350 (69)	17,250 (72)	17,000 (74)
45		14,150 (21.5)	14,150 (35)	14,400 (48.5)	14,100 (56)	15,350 (61.5)	16,550 (66)	16,450 (69)	16,350 (72)
50			11,050 (24.5)	10,750 (42.5)	10,700 (52)	11,850 (58.5)	13,050 (63)	14,250 (66.5)	15,450 (69.5)
60				5,810 (28)	6,040 (42.5)	7,110 (51)	8,210 (57)	9,310 (61.5)	10,400 (65)
70					2,970 (30)	3,990 (42.5)	5,030 (50)	6,080 (56)	7,140 (60)
80						1,780 (32)	2,780 (42.5)	3,800 (49.5)	4,820 (55)
90							1,120 (33.5)	2,100 (43)	3,100 (49.5)
100									1,760 (43)
Minimum boom angle (deg.) for indicated length (no load)					20	25	33	37	40
Maximum boom length (ft.) at 0 degree boom angle (no load)							74		

NOTE: ( ) Boom angles are in degrees.

*This capacity is based on maximum boom angle.

Boom Angle	35	55	61	74
0	26,400 (28.2)	12,500 (47.4)	9,190 (53.8)	3,540 (66.6)

NOTE: ( ) Reference radii are in feet.

A6-829-014539

T1 T2 T3 T4 %

MODE B

© Grove Europe Limited Form number TMS//TTS870

301F18.EPS

Figure 18 ◆ Hydraulic crane load/capacity chart for mode B boom configuration.

LOAD CHARTS 1.15

Show Transparency 9 (Figure 19). Discuss the different quadrants of operation.

Explain that there may be separate load charts for different quadrants of operation. Discuss how to identify this information on the load charts.

The operational quadrants are termed over-the-front, over-the-rear, over-the-side, and 360 degrees (*Figure 19*).

Crane manufacturers produce separate load/capacity charts for certain quadrants because capacities will vary. For example, notice the capacity differences in the load/capacity charts in *Figure 20* for a hydraulic boom crane.

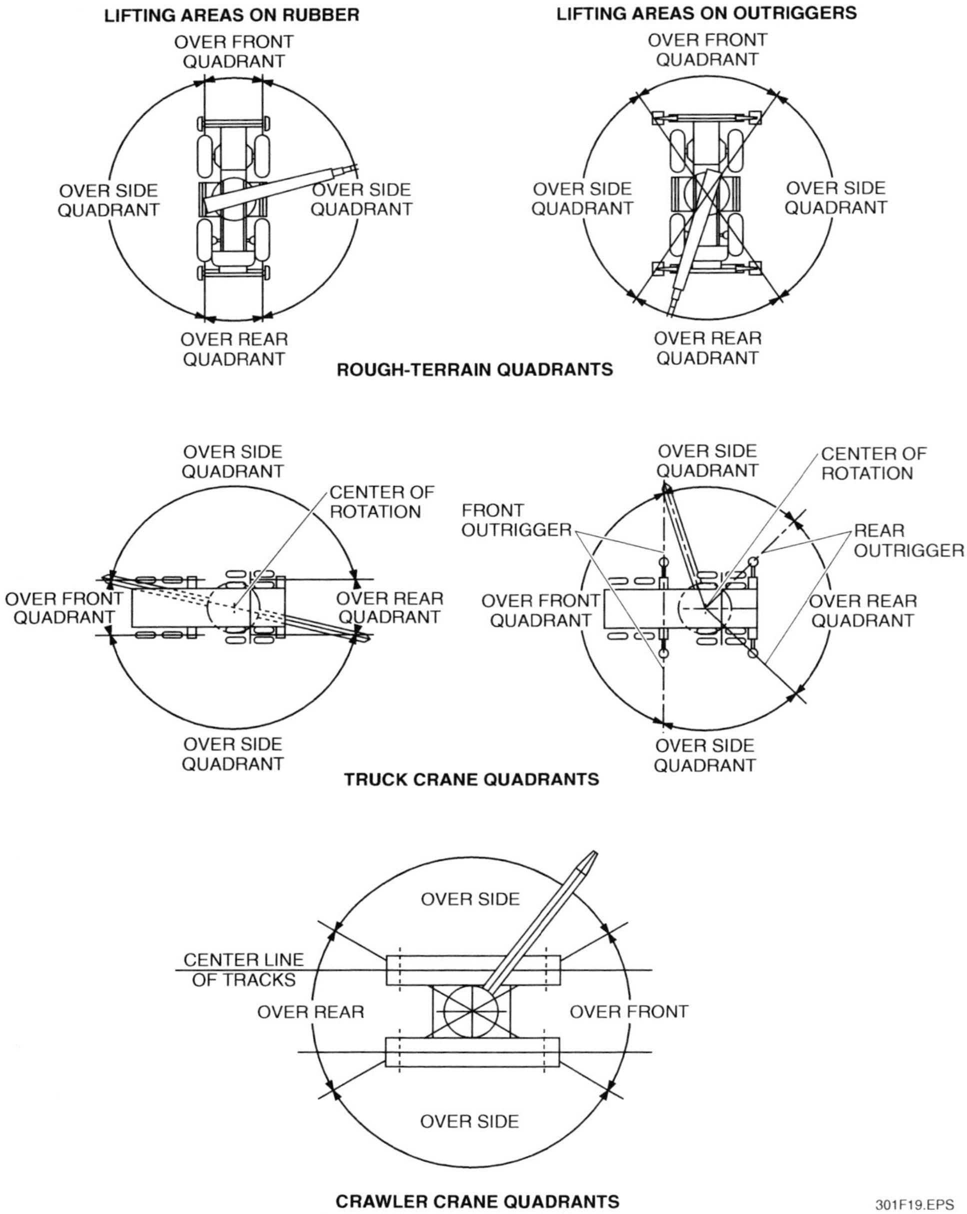

Figure 19 ◆ Quadrant definition drawings for rough-terrain, truck, and crawler cranes.

Instructor's Notes:

36-70 ft. | 12,150 lbs | Stationary | 360° (QUADRANT)

Pounds — #9005

Feet	Main Boom Length in Feet 36	40	50	*60	70
10	45,300 (69)	39,700 (72)			
12	41,750 (65.5)	39,700 (68.5)	29,600 (73.5)		
15	29,350 (59.5)	26,450 (63.5)	26,450 (70)	20,900 (74)	
20	17,800 (49)	17,650 (55)	17,050 (63.5)	16,250 (69)	16,250 (72)
25	11,750 (36)	11,700 (45)	11,350 (56.5)	10,850 (63.5)	10,850 (68)
30		8,040 (31.5)	7,820 (48.5)	7,470 (57.5)	7,470 (63)
35			5,400 (40)	5,120 (51.5)	5,120 (58)
40			3,660 (28)	3,200 (45)	3,430 (53)
45				1,000 (37)	2,150 (47.5)
50					1,150 (41.0)

Lifting Capacities at Zero Degree Boom Angle On Rubber - Stationary 360

Boom Angle	Main Boom Length in Feet 36	40	50
0°	8,180 (29.7)	5,890 (34.2)	2,170 (44.2)

Note: ( ) Referance radii in feat. A6-829-101048A

#LMI operating code. Refer to LMI manual for instructions.

*60 ft. boom length is with inner-mid extended and outer-mid & fly retracted.

36-70 ft. | 12,150 lbs | Stationary | Defined Arc Over Front (QUADRANT)

Pounds — #9005

Feet	Main Boom Length in Feet 36	40	50	*60	70
10	45,300 (69)	42,850 (72)	29,600 (76)		
12	43,650 (65.5)	41,350 (68.5)	29,600 (73.5)		
15	38,300 (59.5)	36,300 (63.5)	29,600 (70)	20,900 (74)	17,300 (76.5)
20	31,150 (49)	29,550 (55)	25.900 (63.5)	20,900 (69)	17.300 (72)
25	24,100 (36)	24,150 (45)	21,800 (56.5)	18,800 (63.5)	17,300 (68)
30		17,400 (31.5)	17,200 (48.5)	15,300 (57.5)	15,300 (63)
35			12,800 (40)	12,500 (51.5)	11,000 (58)
40			9,720 (28)	9,390 (45)	9,390 (53)
45				7,090 (37)	7,090 (47.5)
50				5,310 (26.5)	5,310 (41)
55					3,870 (33.5)
60					3,170 (24)

Lifting Capacities at Zero Degree Boom Angle On Rubber - Defined Arc Over Front

Boom Angle	Main Boom Length in Feet 36	40	50	*60	70
0°	17,600 (29.7)	13,600 (34.2)	7,750 (44.2)	4,010 (54.6)	2,670 (64.2)

Note: ( ) Reference radii in feet. A6-829-101047

#LMI operating code. Refer to LMI manual for instructions.

*60 ft. boom length is with inner-mid extended and outer-mid & fly retracted.

301F20.EPS

Figure 20 ◆ Capacity differences.

With a boom length of 36 feet, the gross capacity remains the same in both configurations at an operating radius of 10 feet. However, at all other boom lengths and operating radii, the capacity for 360 degrees is less than for over-the-front operation. The differences in capacities can vary significantly at different radii as the boom length increases.

6.3.0 Configuration of the Crane Base

The crane can be set up on rubber, outriggers, with crawlers extended or retracted, or on stabilizers. Each configuration affects capacity, which translates into a different load/capacity chart.

6.3.1 On Rubber

On-rubber load/capacity charts are shown in *Figure 21*. Depending on the manufacturer, the required tire pressures may be defined in notes for the charts or in a separate table.

There are several requirements for on-rubber lifting. Capacities for on-rubber lifting do not exceed 75 percent of tipping loads as determined by test in accordance with *ASME/SAE J765*. This percentage is strictly an engineering stipulation used by the manufacturer in creating load/capacity chart capacities. The operator does not have a 25-percent cushion to work with. The load chart capacities must not be exceeded.

Ask a trainee to name different crane base configurations that affect capacity.

Discuss requirements for on-rubber lifting.

Explain how to read a load chart for on-rubber lifting.

Demonstrate how to identify requirements of the on-rubber load/capacity chart.

Remind trainees that even though capacities are only a percentage of tipping loads, they should never be exceeded.

Have trainees practice identifying requirements of the on-rubber load/capacity chart. Note the proficiency of each trainee. This laboratory corresponds to Performance Task 2.

LOAD CHARTS 1.17

36-70 ft. | 12,150 lbs | Stationary | 360°

Pounds

#9005

Feet	Main Boom Length in Feet 36	40	50	*60	70
10	45,300 (69)	39,700 (72)			
12	41,750 (65.5)	39,700 (68.5)	29,600 (73.5)		
15	29,350 (59.5)	26,450 (63.5)	26,450 (70)	20,900 (74)	
20	17,800 (49)	17,650 (55)	17,050 (63.5)	16,250 (69)	16,250 (72)
25	11,750 (36)	11,700 (45)	11,350 (56.5)	10,850 (63.5)	10,850 (68)
30		8,040 (31.5)	7,820 (48.5)	7,470 (57.5)	7,470 (63)
35			5,400 (40)	5,120 (51.5)	5,120 (58)
40			3,660 (28)	3,200 (45)	3,430 (53)
45				1,000 (37)	2,150 (47.5)
50					1,150 (41.0)

Lifting Capacities at Zero Degree Boom Angle On Rubber - Stationary 360

Boom Angle	Main Boom Length in Feet 36	40	50
0°	8,180 (29.7)	5,890 (34.2)	2,170 (44.2)

Note: ( ) Reference radii in feet. A6-829-101048A

#LMI operating code. Refer to LMI manual for instructions.
*60 ft. boom length is with inner-mid extended and outer-mid & fly retracted.

36-70 ft. | 12,150 lbs | Stationary | Defined Arc Over Front

Pounds

#9005

Feet	Main Boom Length in Feet 36	40	50	*60	70
10	45,300 (69)	42,850 (72)	29,600 (76)		
12	43,650 (65.5)	41,350 (68.5)	29,600 (73.5)		
15	38,300 (59.5)	36,300 (63.5)	29,600 (70)	20,900 (74)	17,300 (76.5)
20	31,150 (49)	29,550 (55)	25,900 (63.5)	20,900 (69)	17,300 (72)
25	24,100 (36)	24,150 (45)	21,800 (56.5)	18,800 (63.5)	17,300 (68)
30		17,400 (31.5)	17,200 (48.5)	15,300 (57.5)	15,300 (63)
35			12,800 (40)	12,500 (51.5)	11,000 (58)
40			9,720 (28)	9,390 (45)	9,390 (53)
45				7,090 (37)	7,090 (47.5)
50				5,310 (26.5)	5,310 (41)
55					3,870 (33.5)
60					3,170 (24)

Lifting Capacities at Zero Degree Boom Angle On Rubber - Defined Arc Over Front

Boom Angle	Main Boom Length in Feet 36	40	50	*60	70
0°	17,600 (29.7)	13,600 (34.2)	7,750 (44.2)	4,010 (54.6)	2,670 (64.2)

Note: ( ) Reference radii in feet. A6-829-101047

#LMI operating code. Refer to LMI manual for instructions.
*60 ft. boom length is with inner-mid extended and outer-mid & fly retracted.

36-70 ft. | 12,150 lbs | Pick & Carry Up to 2.5 mph | Boom Center Over Front

Pounds

#9006

Feet	Main Boom Length in Feet 36	40	50	*60	70
10	42,500 (69)	37,600 (72)	30,700 (76)		
12	42,500 (65.5)	37,600 (68.5)	30,700 (73.5)	23,450 (77)	
15	37,600 (59.5)	37,600 (63.5)	30,700 (70)	19,700 (74)	16,250 (76.5)
20	32,950 (49)	32,900 (55)	30,700 (63.5)	19,700 (69)	16,250 (72)
25	24,100 (36)	24,150 (45)	23,850 (56.5)	19,700 (63.5)	16,250 (68)
30		17,400 (31.5)	17,200 (48.5)	16,800 (57.5)	16,250 (63)
35			12,800 (40)	12,500 (51.5)	12,500 (58)
40			9,720 (28)	9,390 (45)	9,390 (53)
45				7,090 (37)	7,090 (47.5)
50				5,310 (26.5)	5,310 (41)
55					3,870 (33.5)
60					3,090 (24)

Lifting Capacities at Zero Degree Boom Angle On Rubber - Pick & Carry

Boom Angle	Main Boom Length in Feet 36	40	50	*60	70
0°	17,600 (29.7)	13,600 (34.2)	7,750 (44.2)	4,010 (54.6)	2,670 (64.2)

Note: ( ) Reference radii in feet. A6-829-101049

#LMI operating code. Refer to LMI manual for instructions.
*60 ft. boom length is with inner-mid extended and outer-mid & fly retracted.

NOTES:

1. Capacities are in pounds and do not exceed 75% of tipping loads as determined by test in accordance with SAE J765.
2. Capacities are applicable to machines equipped with 29.6x25 (28 ply) tires at 65 psi cold inflation pressure.
3. Defined Arc – Over front includes 6˚ on either side of longitudinal centerline of machine (ref. drawing C6-829-003529).
4. Capacities appearing above the bold line are based on structural strength and tipping should not be relied upon as a capacity limitation.
5. Capacities are applicable only with machine on firm level surface.
6. On rubber lifting with boom extensions not permitted.
7. For pick and carry operation, boom must be centered over front of machine, mechanical swing lock engaged and load restrained from swinging. When handling loads in the structural range with capacities close to maximum ratings, travel should be reduced to creep speeds.
8. Axle lockouts must be functioning when lifting on rubber.
9. All lifting depends on proper tire inflation, capacity and condition. Capacities must be reduced for lower tire inflation pressures. See lifting capacity chart for tire used. Damaged tires are hazardous to safe operation of crane.
10. Creep – Not over 200 ft. of movement in any 30 minute period and not exceeding 1 mph.

301F21.EPS

Figure 21 ◆ On-rubber load/capacity chart.

Instructor's Notes:

6.3.2 On Outriggers

Capacities for on-outrigger lifting do not exceed 85 percent of tipping loads. As mentioned with on-rubber lifting, this is an engineering stipulation. It does not mean the operator has a 15-percent cushion. Depending on the design of the crane, there may be load/capacity charts for outriggers at no extension, intermediate extension, and full extension (*Figures* 22, 23, and 24). At no extension, capacities do not exceed 75 percent of tipping. In this situation, the crane may have less capacity but more stability than when on rubber.

6.3.3 On Crawlers

Crawlers can be extended or retracted for lifting operations. Capacities do not exceed 75 percent of tipping on crawler cranes. Some cranes, however, may have outrigger extensions along with crawlers where capacities can be 85 percent of tipping. The operator should consult the operator's manual and follow the manufacturer's recommendations when lifting.

Discuss the limitations of on-outrigger lifting.

Explain how to read a load chart for on-outrigger lifting.

Demonstrate how to identify requirements of the on-outrigger load/capacity chart.

Have trainees practice identifying requirements of the on-outrigger load/capacity chart. Note the proficiency of each trainee. This laboratory corresponds to Performance Task 3.

Discuss the limitations on capacity for on-crawler lifting.

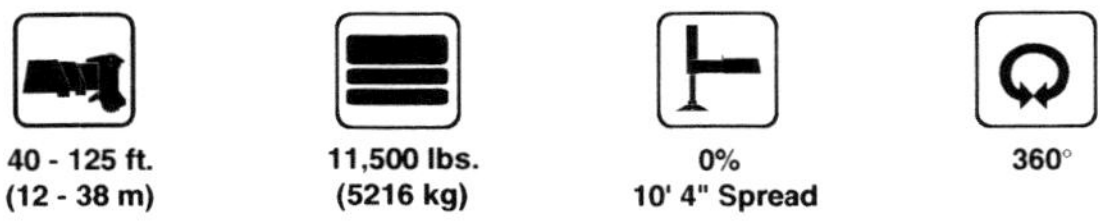

Pounds

Feet	40	45	55	65	75	85	95	105	115	125
10	91,900 (70)	86,850 (72.5)								
12	70,100 (67)	66,600 (70)	61,200 (74)							
15	50,500 (61.5)	48,150 (65.5)	44,550 (70.5)	41,550 (74)						
20	31,200 (52.5)	30,950 (58)	28,900 (65)	26,950 (69)	26,600 (72.5)	26,600 (75)				
25	20,850 (41.5)	20,550 (49.5)	19,900 (58.5)	18,500 (64.5)	18,500 (68.5)	18,800 (71.5)	18,950 (73.5)	19,050 (75.5)		
30	14,650 (26)	14,350 (39.5)	13,850 (52)	12,950 (59)	13,850 (64)	13,650 (67.5)	13,950 (70.5)	14,200 (72.5)	14,200 (75)	*14,200 (76.5)
35		10,200 (26)	9,750 (44.5)	9,090 (53.5)	9,370 (59.5)	9,970 (64)	10,350 (67)	10,700 (69.5)	10,750 (72)	10,850 (74)
40			6,810 (35.5)	6,190 (47.5)	6,550 (54.5)	7,200 (60)	7,660 (63.5)	8,070 (66.5)	8,200 (69.5)	8,350 (71.5)
45			4,600 (23)	3,950 (40.5)	4,350 (49.5)	5,040 (55.5)	5,540 (60)	5,990 (63.5)	6,170 (66.5)	6,360 (69)
50				2,170 (32.5)	2,610 (44)	3,320 (51.5)	3,850 (56.5)	4,330 (60.5)	4,540 (64)	4,750 (66.5)
60							1,300 (48.5)	1,820 (53.5)	2,070 (58)	2,320 (61.5)
Minimum boom angle (deg.) for indicated length (no load)										51
Maximum boom length (ft.) at 0 degree boom angle (no load)										65

NOTE: ( ) Boom angles are in degrees.
*Based on maximum obtainable boom angle.

A6-829-011927A

Boom Angle	40	45	55
0°	12,550 (32.3)	8,440 (37.8)	3,610 (47.8)

NOTE: ( ) Reference radii are in feet.

A6-829-012282

301F22.EPS

Figure 22 ◆ Capacities for on outrigger at no extension (0 percent).

LOAD CHARTS 1.19

40 - 125 ft. (12 - 38 m)	11,500 lbs. (5216 kg)	50% 17' 4" Spread	360°

Feet	40	45	55	65	75	85	95	105	115	125
						Pounds				
10	116,000 (70)	105,000 (72.5)								
12	102,500 (67)	102,000 (70)	94,600 (74)							
15	86,800 (61.5)	86,350 (65.5)	85,800 (70.5)	71,050 (74)						
20	67,250 (52.5)	64,850 (58)	60,550 (65)	56,950 (69)	55,250 (72.5)	48,150 (75)				
25	45,750 (41.5)	44,950 (49.5)	42,050 (58.5)	39,700 (64.5)	39,100 (68.5)	38,900 (71.5)	38,000 (73.5)	33,350 (75.5)		
30	32,500 (26)	32,050 (39.5)	31,000 (52)	29,250 (59)	29,050 (64)	29,250 (67.5)	29,300 (70.5)	29,250 (72.5)	24,550 (75)	*23,700 (76.5)
35		23,900 (26)	23,350 (44.5)	22,250 (53.5)	22,300 (59.5)	22,700 (64)	22,900 (67)	23,050 (69.5)	21,700 (72)	21,900 (74)
40			17,650 (35.5)	17,200 (47.5)	17,400 (54.5)	17,900 (60)	18,250 (63.5)	18,550 (66.5)	18,550 (69.5)	18,550 (71.5)
45			13,400 (23)	13,000 (40.5)	13,600 (49.5)	14,300 (55.5)	14,700 (60)	15,050 (63.5)	15,150 (66.5)	15,250 (69)
50				9,850 (32.5)	10,450 (44)	11,450 (51.5)	11,900 (56.5)	12,350 (60.5)	12,450 (64)	12,600 (66.5)
60					5,970 (30)	6,940 (41.5)	7,840 (48.5)	8,310 (53.5)	8,500 (58)	8,700 (61.5)
70						4,020 (28.5)	4,840 (39)	5,410 (46)	5,700 (51.5)	5,930 (55.5)
80							2,630 (27)	3,170 (37)	3,620 (44.5)	3,870 (49.5)
90								1,490 (25.5)	2,000 (36)	2,280 (42.5)
100										1,020 (34.5)
Minimum boom angle (deg.) for indicated length (no load)										27.5
Maximum boom length (ft.) at 0 degree boom angle (no load)										105

NOTE: ( ) Boom angles are in degrees.
*Based on maximum obtainable boom angle.

A6-829-011802A

Boom Angle	40	45	55	65	75	85	95
0°	22,800 (32.3)	18,250 (37.8)	11,500 (47.8)	6,220 (57.8)	3,740 (67.8)	2,350 (77.8)	1,300 (87.8)

NOTE: ( ) Reference radii are in feet.

A6-829-012282

301F23.EPS

Figure 23 ◈ Capacities for on outrigger at intermediate extension (50 percent).

Instructor's Notes:

40 - 125 ft. (12 - 38 m)	11,500 lbs. (5216 kg)	100%	360°

Pounds

Feet	40	45	55	65	75	85	95	105	115	125
10	130,000 (70)	105,000 (72.5)								
12	111,000 (67)	105,000 (70)	94,600 (74)							
15	91,450 (61.5)	91,000 (65.5)	88,250 (70.5)	71,050 (74)						
20	69,550 (52.5)	69,050 (58)	68,400 (65)	60,400 (69)	55,250 (72.5)	48,150 (75)				
25	55,050 (41.5)	54,600 (49.5)	53,950 (58.5)	53,450 (64.5)	47,950 (68.5)	41,700 (71.5)	38,000 (73.5)	33,350 (75.5)		
30	42,950 (26)	42,450 (39.5)	41,700 (52)	41,200 (59)	41,950 (64)	36,700 (67.5)	33,300 (70.5)	30,750 (72.5)	24,550 (75)	*23,700 (76.5)
35		33,700 (26)	33,000 (44.5)	32,500 (53.5)	33,250 (59.5)	32,600 (64)	29,550 (67)	27,300 (69.5)	21,700 (72)	21,900 (74)
40			26,650 (35.5)	26,150 (47.5)	26,900 (54.5)	27,850 (60)	26,450 (63.5)	24,450 (66.5)	19,350 (69.5)	20,300 (71.5)
45			21,750 (23)	21,300 (40.5)	22,050 (49.5)	23,000 (55.5)	23,700 (60)	22,000 (63.5)	17,450 (66.5)	18,800 (69)
50				17,500 (32.5)	18,250 (44)	19,150 (51.5)	19,900 (56.5)	19,850 (60.5)	15,800 (64)	17,050 (66.5)
60					12,400 (30)	13,250 (41.5)	14,100 (48.5)	14,650 (53.5)	13,250 (58)	14,150 (61.5)
70						9,190 (28.5)	9,910 (39)	10,400 (46)	10,850 (51.5)	11,350 (55.5)
80							6,930 (27)	6,740 (37)	7,850 (44.5)	8,290 (49.5)
90								5,170 (25.5)	5,600 (36)	6,010 (42.5)
100									3,880 (25)	4,250 (34.5)
110										2,840 (24)
Minimum boom angle (deg.) for indicated length (no load)										0
Maximum boom length (ft.) at 0 degree boom angle (no load)										125

NOTE: ( ) Boom angles are in degrees.
*Based on maximum obtainable boom angle.

A6-829-011801A

Boom Angle	40	45	55	65	75	85	95	105	115
0°	22,800 (32.3)	18,250 (37.8)	12,200 (47.8)	7,990 (57.8)	5,720 (67.8)	4,320 (77.8)	3,210 (87.8)	2,380 (97.8)	1,570 (107.8)

NOTE: ( ) Reference radii are in feet.

A6-829-012282

301F24.EPS

Figure 24 ◈ Capacities for on outrigger at full extension (100 percent).

Show Transparency 10 (Figure 25). Describe a tower attachment.

Remind trainees to follow the manufacturer's recommendations for lifting with a tower attachment.

6.4.0 Tower and Ring Attachments

A tower attachment is used as an alternative to a freestanding tower crane where a lower erection height allows its use (*Figure 25*). It is also used to increase the load radius where reach is the primary operational concern. As with all attachments, the operator should follow the crane manufacturer's and/or attachment manufacturer's recommendations for lifting.

LOAD CHARTS 1.21

Discuss a ring attachment.

Have trainees review Sections 6.5.0–6.6.2.

Ensure that you have everything required for teaching this session.

Discuss counterweight configurations. Explain that load capacities are listed with specific counterweights in place.

Demonstrate how to identify load/capacity charts that are used in different configurations.

Have trainees practice identifying load/capacity charts that are used in different configurations. Note the proficiency of each trainee. This laboratory corresponds to Performance Task 4.

A ring attachment is used to greatly increase the crane's capacity through configuration changes and weight transfer to the ring foundation (*Figure 26*).

6.5.0 Counterweight Configurations

Each crane has counterweight specifications. Counterweights give extra stability to the crane during assembly and lifting operations. They can be front, rear, side, or upper counterweights. Manufacturers specify in the operational notes or with specific load/capacity charts that the capacities listed are with counterweights in place. The counterweight sections must be combined with each other only in the way specified in the load charts. If a counterweight is extendable, it must be extended according to the manufacturer's recommendations. *Figures 27* through *29* show examples of various counterweight configurations on different models of cranes.

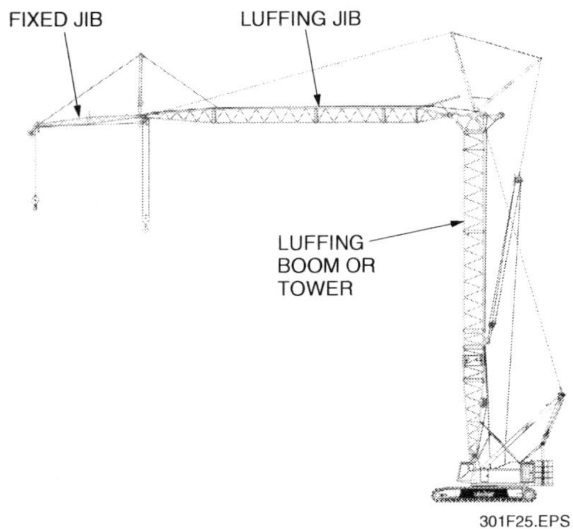

Figure 25 ◆ Crane with tower attachment.

Figure 27 ◆ Crawler crane counterweights.

Figure 26 ◆ 200-ton ringer crane.

Instructor's Notes:

301F28.EPS

Figure 28 ◆ Ringer crane counterweights.

301F29.EPS

Figure 29 ◆ Counterweights on an auxiliary carrier.

LOAD CHARTS 1.23

Explain that allowances or deductions must be made to the overall capacity of a crane.

Show Transparency 11 (Figure 30). Describe the various deductions, and explain how to use a deduction chart.

Describe a jib chart, and explain how to read it.

Give some example problems, and have trainees use Figures 30 and 31 to determine capacity.

6.6.0 Deduction Charts

In addition to the actual load weight, allowance must be made for additional weight. For example, the weight of the load block, headache ball, wire rope, parts of hoist line, and any boom extensions may require deduction from the gross capacity of the crane. These are all items that the crane will interpret as load. They are listed in charts supplied by the manufacturer (*Figure 30*).

6.6.1 Jib Charts

Charts provided by a manufacturer, such as the one shown in *Figure 31*, show the capacities in pounds of the luffing- or fixed-jib boom extensions at different degrees of offset. The capacity of a crane equipped with a jib boom extension is affected by the following:

- Load radius
- Length of the main boom
- Main boom angle
- Jib length
- Jib angle to the horizontal
- Jib angle with respect to the boom
- Weight of the rigging and hooks on both the jib and main boom
- Mathematical determination of whether the capacity of the crane is based on these factors:
  - Strength of the jib
  - Strength of the main boom
  - Stability of the crane

When a jib and/or boom extension is stowed on the heel section of the main boom, the stowed weight must be deducted from the gross capacity listed on the load/capacity chart in order to calculate the crane's net capacity.

When the jib/extension is mounted on the boom tip, the effective weight must be deducted from the gross capacity. This deduction is made only when the hoisting is done from the main boom. If you are lifting from an erected jib, do not count the weight of the jib as part of the load. It has already been compensated for in the jib's specific capacity chart. Effective weight is calculated by the manufacturer as the weight which, when applied at the boom tip, will have the same effect on the crane as the boom extension and/or jib.

CAPACITY DEDUCTIONS FOR AUXILIARY LOAD HANDLING EQUIPMENT

Load Handling Equipment:	**(lbs.)**
Auxiliary Head Attached	100
25-ton quick reeve 3 sheave hook block (see hook block for actual weight)	670
40-ton quick reeve 4 sheave hook block (see hook block for actual weight)	780
8.5-ton hook ball (see hook ball for actual weight)	360
Lifting From Main Boom With:	**(lbs.)**
28.5 ft. or 51 ft. fly stowed on base (see operation note 4)	0
28.5 ft. offset fly erected but not used	2600
51 ft. offset fly erected but not used	4800
Lifting From 28.5 ft. Offset Fly With:	
22.5 ft. fly tip erected but not used	**PROHIBITED**
22.5 ft. fly tip stowed on 28.5 ft. offset fly	**PROHIBITED**
Note: Capacity deductions are for Link-Belt supplied equipment only.	

301F30.EPS

Figure 30 ◆ Deductions chart.

Instructor's Notes:

6.6.2 Parts of Line

Before any lift is made, the operator must ensure that the crane has sufficient net capacity to lift the load and that it is also reeved with enough parts of line (*Figure 32*) to lift the load without breaking the hoist rope.

There are several methods used to determine the parts of line for a lift. Some crane manufacturers provide this information along with the load/capacity charts (*Figure 33*). This is the easiest of the methods. This method is also accurate if all the material used is recommended by the manufacturer.

On some crane charts, the safe working load of the hoist rope may be reduced for multiple-part reeving due to block- and boom-point sheave friction. On other cranes, the determination of the parts of line may be in direct proportion to the number of layers on the winch drum. The greater the number of layers on the drum, the less line pull available to lift the load.

Show Transparency 12 (Figure 32). Explain that the operator must ensure that the block is reeved properly so that the hoist rope will not be overloaded.

Show Transparency 13 (Figure 33). Describe a wire rope capacity chart. Explain that this is the easiest method if all of the material used is recommended by the manufacturer.

Discuss deductions and limitations to the safe working load for different parts of line.

Demonstrate how to identify parts of line and counterweight considerations on a load/capacity chart.

Have trainees practice identifying parts of line and counterweight considerations on a load/capacity chart. Note the proficiency of each trainee. This laboratory corresponds to Performance Task 5.

33-56 ft. luffing folding boom extension (mode B) (fixed offset angles)

37.3-141.7 ft. | 33 - 56 ft. | 22,000 lbs | 100% 34'-6" Spread | 360°

Pounds

Feet	33 ft. LENGTH 5° OFFSET #0091	33 ft. LENGTH 20° OFFSET #0091	33 ft. LENGTH 40° OFFSET #0091	56 ft. LENGTH 5° OFFSET #0092	56 ft. LENGTH 20° OFFSET #0092	56 ft. LENGTH 40° OFFSET #0092
40	*13,700 (78)					
45	13,700 (77)					
50	13,700 (75)	13,700 (77.5)		*8,200 (78)		
55	13,700 (73.5)	13,700 (75.5)	*11,000 (78)	8,200 (77.5)		
60	13,700 (71.5)	13,700 (74)	11,000 (76)	8,200 (76)		
65	13,700 (70)	12,850 (72)	10,950 (74.5)	8,200 (74.5)	8,200 (77.5)	
70	12,500 (68)	12,000 (70)	10,350 (72.5)	8,200 (73)	8,200 (76)	
75	11,350 (66)	11,200 (68)	9,830 (70.5)	8,200 (71.5)	8,100 (74)	6,400 (77.5)
80	9,730 (64.5)	10,450 (66.5)	9,330 (68.5)	8,200 (69.5)	7,600 (72.5)	6,400 (76)
85	8,300 (62.5)	8,980 (64.5)	8,860 (66.5)	8,200 (68)	7,150 (71)	6,230 (74)
90	7,060 (60.5)	7,660 (62.5)	8,210 (64.5)	7,740 (66.5)	6,730 (69)	5,920 (72.5)
95	5,960 (58.5)	6,500 (60.5)	6,980 (62)	7,130 (64.5)	6,350 (67.5)	5,640 (70.5)
100	4,990 (56.5)	5,470 (58)	5,880 (60)	6,130 (63)	6,000 (65.5)	5,380 (68.5)
105	4,120 (54)	4,560 (56)	4,900 (58)	5,230 (61)	5,690 (64)	5,140 (67)
110	3,340 (52)	3,730 (54)	4,020 (55.5)	4,430 (59.5)	5,290 (62)	4,900 (65)
115	2,640 (49.5)	2,990 (51.5)	3,230 (53)	3,700 (57.5)	4,490 (60)	4,690 (63)
120	2,000 (47.5)	2,320 (49)	2,510 (50.5)	3,040 (55.5)	3,760 (58.5)	4,470 (61)
125	1,420 (45)	1,700 (46.5)	1,850 (47.5)	2,440 (53.5)	3,100 (56.5)	3,710 (58.5)
130		1,140 (44)	1,250 (45)	1,900 (51.5)	2,500 (54.5)	3,030 (56.5)
135				1,390 (49.5)	1,940 (52)	2,390 (54)
140					1,420 (50)	1,810 (52)
145						1,270 (49)
Minimum boom angle (°) for indicated length (no load)	42	43	43	48	48	47
Maximum boom length (ft.) at 0° boom angle (no load)		89.8			76.7	

NOTE: ( ) Boom angles are in degrees.

A6-829-103522

#LMI operating code. Refer to LMI manual for operating instructions.
*This capacity is based upon maximum boom angle.

NOTES:

1. All capacities above the bold line are based on structural strength of boom extension and do not exceed 85% of tipping loads, in accordance with SAE J-765.
2. The 33 ft. luffing folding boom extension may be used for single or double line lifting service. The 56 ft. luffing folding boom extension may be used for single line lifting service only.
   WARNING: Lifting with the 33 ft. extension base, with the 23 ft. extension fly either erected or folded along side of extension base, is strictly prohibited.
3. WARNING: Operation of this machine with heavier loads than the capacities listed is strictly prohibited. Machine tipping with boom extension occurs rapidly and without advance warning.
4. Boom angle is the angle above or below horizontal of the longitudinal axis of the boom base section after lifting rated load.
5. Capacities listed are with outriggers properly extended and vertical jacks set only.
6. For main boom lengths less than 141.7 ft. with the boom extension erected, the rated loads are determined by boom angle. Use only the column which corresponds to the boom extension length and offset for which the machine is set up. For boom angles not shown, use rating of the next lower boom angle.
7. When lifting over the main boom nose with 33 ft. or 56 ft. extension erected, the outriggers must be fully extended or 50% extended (17.3 ft. spread).

301F31.EPS

Figure 31 ◈ Offsetable jib/boom extension chart.

LOAD CHARTS 1.25

Show Transparency 14 (Table 1). Explain how to determine parts of line using the table.

Demonstrate how to calculate parts of line using a formula. Work the example on the board.

Have trainees practice calculating parts of line. See the suggested exercise for Performance Task 6 located in the Test Booklet. Note the proficiency of each trainee. This laboratory corresponds to Performance Task 6.

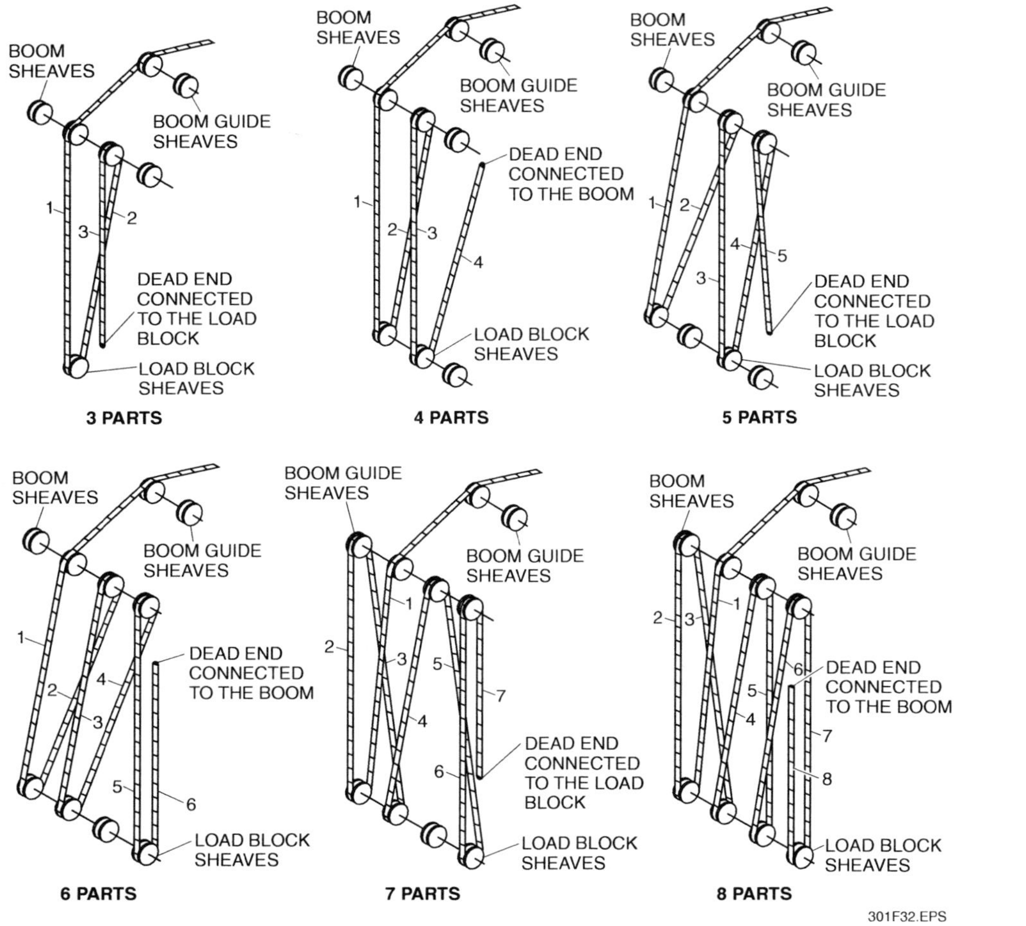

Figure 32 ◈ Reeving for various parts of line.

WIRE ROPE CAPACITY

Maximum Lifting Capacities Based On Wire Rope Strength			
Parts of Line	5/8" Type RB	5/8" Type ZB	Notes
1	9,080	11,080	Capacities shown are in pounds and working loads must not exceed the ratings on the capacity charts in the Crane Rating Manual.
2	18,160	22,160	
3	27,240	33,240	
4	36,320	44,320	
5	45,400	55,400	
6	54,480	66,480	Study Operator's Manual for wire rope inspection procedures and single part of line applications.
7	63,560	77,560	
8	72,640	88,640	
9	81,720		

LBCE	DESCRIPTION
TYPE RB	18 × 19 Rotation Resistant – Compact Strand, High Strength Preformed, Right Regular Lay
TYPE ZB	36 × 7 Rotation Resistant – Extra Improved Plow Steel – Right Regular Lay

301F33.EPS

Figure 33 ◈ Wire rope capacity chart.

You can use the following formula to calculate the parts of line:

$$\text{Parts of line} = \frac{\text{suspended weight}}{\text{safe working load (SWL) of hoist rope}}$$

Where:

Suspended weight = load weight + weight of hook block + weight of slings and rigging hardware

SWL = diameter squared × 8

Example:

- Size of wire rope: 1¼ inches
- Load weight: 50 tons
- Type of sheaves: anti-friction bearing

SWL = diameter squared × 8 = 1.25 × 1.25 × 8 = 12.5

Total weight of load = 50

50 tons ÷ 12.5 = 4 (ratio)

Instructor's Notes:

Looking at *Table 1*, find the column under Ratio B, anti-friction bearing sheaves. Note that 4.71 is the next highest number over the calculated number of 4. The number of line parts column gives 5 parts of line.

Table 1 Ratio Factors

Ratio A Bronze Bushed Sheaves	Ratio B Anti-friction Bearing Sheaves	Number Of Line Parts
0.96	0.98	1
1.87	1.94	2
2.75	2.88	3
3.59	3.81	4
4.39	4.71	5
5.16	5.60	6
5.90	6.47	7
6.60	7.32	8
7.27	8.16	9
7.91	8.98	10
8.52	9.79	11
9.11	10.60	12
9.68	11.40	13
10.20	12.10	14
10.70	12.90	15
11.20	13.60	16
11.70	14.30	17
12.20	15.00	18
12.60	15.70	19
13.00	16.40	20

301T01.EPS

7.0.0 ◆ CALCULATING CRANE CAPACITY

Providing information on particular lifting configurations and presenting load/capacity charts applicable to those configurations is useless unless the trainee can perform the math required to find capacities. For this reason, this section will demonstrate, step by step, how to find the gross and net capacities for a crane.

Crane manufacturers provide operational notes (notes for lifting capacities or other information) with the load charts for each crane model. Refer to *Figure 31* for an example. These notes stress the important facts of load/capacity charts and lifting covered in this module. It is important that the operator read and understand the notes because they contain stipulations regarding the operation of the crane.

Example One:

When lifting from the main boom, find the net capacity of a Grove TM-1500 truck-mounted hydraulic crane (*Figure 34*) under the following conditions:

- Outriggers properly set and level
- Over-the-rear lift
- Power-pinned fly retracted
- 33- to 58-ft extension stowed
- Boom length of 52 ft
- Load radius of 23 ft
- 150-ton, 8-sheave block
- Auxiliary boom head attached
- 10-ton headache ball reeved over auxiliary boom head
- Rigging and other accessories of 195 pounds

The net capacity is 123,984 lbs. How was the net capacity determined? Refer to the load/capacity charts in *Figure 35* and the weight reduction charts in *Figure 36*.

WARNING!
Operators must follow load/capacity chart information according to manufacturer's specifications.

301F34.EPS

Figure 34 ◆ Grove TM-1500 crane.

Have trainees review Sections 7.0.0–8.0.0.

Ensure that you have everything required for teaching this session.

See the Teaching Tips for Section 7.0.0 at the end of this module.

Explain that a crane operator must be able to calculate the capacity from the information on load charts.

Refer to Figure 31. Discuss the importance of the notes that accompany a load chart.

Show Transparencies 15 and 16 (Figures 35 and 36). Discuss the considerations and deductions in the example.

Reiterate the importance of following the manufacturer's specifications for load chart information.

LOAD CHARTS 1.27

Rated Lifting Capacities in Pounds
46 Ft – 173 Ft Boom on Outriggers - Over Rear

Radius in Feet	#0001 Main Boom Length in Feet (Power Pinned Fly Retracted)									#0002 Power Pin. Fly Ext. & 141 ft.
	46	58	70	82	94	106	118	130	141	173
10	300,000 (74.5)									
12	280,000 (72)	143,500 (76)	142,000 (79)							
15	235,000 (67.5)	143,500 (72.5)	141,500 (76.5)	130,000 (78.5)						
20	173,500 (60.5)	143,500 (67.5)	123,500 (72)	112,000 (75)	102,000 (77.5)	90,300 (79.5)				
25	136,500 (52)	131,500 (61.5)	110,500 (67.5)	98,650 (71)	89,250 (74)	78,550 (76.5)	73,700 (78.5)	69,300 (80)		
30	106,000 (43)	106,000 (55.5)	98,000 (63)	88,350 (67.5)	78,750 (71)	69,250 (73.5)	65,100 (76)	61,000 (77.5)	60,000 (79.5)	
35	84,700 (30.5)	84,700 (49)	84,700 (58)	80,150 (63.5)	69,000 (67.5)	60,750 (70.5)	57,150 (73)	54,000 (75.5)	52,150 (77.5)	
40		70,500 (41)	70,500 (52.5)	70,500 (59.5)	61,300 (64)	54,000 (67.5)	50,600 (70.5)	48,300 (73)	45,850 (75)	38,000 (79)
45		58,850 (32)	58,850 (47)	58,850 (55)	55,000 (60.5)	48,500 (64.5)	46,200 (68)	43,050 (71)	40,400 (73)	35,750 (77)
50		49,600 (17.5)	49,600 (40.5)	49,600 (50.5)	48,750 (57)	43,050 (61.5)	40,700 (65)	38,250 (68.5)	35,750 (71)	32,100 (75.5)
60			36,200 (22.5)	36,200 (39.5)	36,200 (48.5)	34,300 (55)	33,600 (59.5)	30,750 (63.5)	28,500 (66.5)	26,350 (72)
70				26,050 (25)	26,050 (39.5)	26,050 (47.5)	26,050 (53)	24,750 (58)	23,100 (61.5)	22,000 (68.5)
80					18,850 (27)	18,850 (39)	18,850 (46.5)	18,850 (52.5)	18,700 (56.5)	18,500 (64.5)
90						13,500 (28)	13,500 (38.5)	13,500 (46.5)	13,500 (51.5)	15,250 (60.5)
100							9,390 (29)	9,390 (38)	9,390 (45.5)	12,600 (56.5)
110							6,080 (12.5)	6,080 (30.5)	6,080 (39)	10,100 (52)
120								3,390 (17.5)	3,390 (31)	7,530 (47.5)
130									1,150 (19.5)	5,390 (42.5)
140										3,610 (36.5)
150										2,100 (30)
Minimum boom angle (deg.) for indicated length (no load)									10	19
Maximum boom length (ft.) at 0 deg. boom angle (no load)									140	167

Note: ( ) Boom angles are in degrees.
#LMI operating code. Refer to LMI manual for instructions.

301F35.EPS

Figure 35 ◆ TM-1500 load/capacity chart.

Instructor's Notes:

WEIGHT REDUCTIONS FOR LOAD HANDLING DEVICES

33 ft. Extension	
*Stowed -	892 lbs.
*Erected -	5,704 lbs.

33 ft. - 58 ft. Extension	
*Stowed -	1,246 lbs.
*Erected (Ret.) -	8,412 lbs.
*Erected (Ext.) -	11,406 lbs.

46 ft. - 173 ft. Boom with	
*46 ft. Jib Erected -	9,613 lbs.
*60 ft. Jib Erected -	14,571 lbs.
*74 ft. Jib Erected -	20,443 lbs.
*88 ft. Jib Erected -	27,199 lbs.
*Fixed Jib Accessories -	327 lbs.

*Reduction of main boom capacities

HOOKBLOCKS:	
30 Ton, 1 Sheave	1,022 lbs.
150 Ton, 8 Sheave	5,254 lbs.
Auxiliary Boom Head	261 lbs.
10 Ton Headache Ball	560 lbs.
15 Ton Headache Ball	803 lbs.

301F36.EPS

Figure 36 ◈ TM-1500 weight reduction charts

Read down the Radius column to 23 ft (*Figure 35*). This number is not listed, so go to the next longer load radius, which is 25 ft. Read across the Length row to 52 ft. It is not listed, so go to the boom length with the next lower gross capacity, which is 58 ft. The gross capacity where these two intersect is 131,500 lbs. Look up the deduction weights in the deduction charts (*Figure 36*), and subtract these from the gross capacity.

131,500 lbs (gross capacity)
– 1,246 lbs (extension stowed)
– 5,254 lbs (block)
– 261 lbs (aux. boom head)
– 560 lbs (headache ball)
– 195 lbs (rigging)
123,984 lbs (net capacity)

Example Two:

Find the net capacity of the Manitowoc 4100W-2 in the following configuration:

- Crane counterweights of 146,400 lbs installed
- 22C boom, 230-ft main boom no offset
- 125 ft of load radius
- 2-sheave boom point attached
- 15-ton hook block 20 ft below boom point
- 100-ton block reeved with 4 parts of line
- Crawlers extended with two 12,000-lb side counterweights installed
- Rigging of 345 lbs
- Lifting from the main hoist
- Two 30,000-lb carbody counterweights installed

The net capacity is 17,382 lbs. How was the net capacity determined? Refer to the Manitowoc 4100W-2 load/capacity charts (*Figure 37*).

Read the section of the load/capacity chart that applies to 230 ft of boom length. Read down the Operating Radius column. The radius of 125 ft has a gross capacity rating of 24,100 lbs. Deductions for rope and equipment below the boom or attached to the boom must be made for a lattice boom crane per the notes or information furnished with the load charts.

Weight of rope:

Boom tip height (boom point elevation) = 202.6

Whip line = 20' × 2.34 lbs = 46.8 lbs

Load line = 202.6' × 2.34 lbs = 474.1 lbs × 4 lines = 1,896.4 lbs

Total line weight = 1,896.4 lbs + 46.8 lbs = 1,943.2 lbs (rounded to 1943 lbs)

24,100 lbs (gross capacity)
– 1,500 lbs (two-sheave point)
– 865 lbs (15-ton hook block)
– 2,065 lbs (100-ton block)
– 345 lbs (rigging)
– 1,943 lbs (rope)
17,382 lbs (net capacity)

Show Transparencies 15 through 17 (Figures 35 through 37). Explain the steps for determining the net capacity of the cranes using the load charts.

Demonstrate how to calculate crane capacity using a load/capacity chart.

Have trainees practice calculating crane capacity using a load/capacity chart. See the suggested exercise for Performance Task 7 located in the Test Booklet. Note the proficiency of each trainee. This laboratory corresponds to Performance Task 7.

LOAD CHARTS 1.29

MANITOWOC 4100 W-2

LIFTCRANE CAPACITIES

BOOM NO. 22C WITH OPEN THROAT TOP
146,400 LB. CRANE COUNTERWEIGHT
60,000 LB. CARBODY COUNTERWEIGHT
26'6" CRAWLERS EXTENDED

WARNING: ***This chart will apply only when two 12,000 lb. side ctwts. and two 30,000 lb. carbody ctwts. bear MEC registered Serial Numbers.***

LIFTING CAPACITIES: Capacities for various boom lengths and operating radii may be based on percent of tipping, strength of structural components, operating speeds and other factors.

Capacities are for freely suspended loads and do not exceed 75% of a static tipping load. Capacities based on structural competence are shown by shaded areas.

Capacities are shown in pounds. Deduct 1200 pounds from capacities listed when single sheave upper boom point is attached and 1500 pounds when two sheave upper boom point is attached. To comply with B30.5 requirements, upper boompoint cannot be used on the 260 ft. boom. Weight of jib, (see chart A), all load blocks, hooks, weight ball, slings, hoist lines beneath boom and jib point sheaves, etc., is considered part of the main boom load. Boom is not to be lowered beyond radii where combined weights are greater than rated capacity. Where no capacity is shown, operation is not intended or approved.

OPERATING CONDITIONS: Machine to operate in a level position on a firm surface with crawlers fully extended and gantry in working position and be rigged in accordance with and under conditions referred to in rigging drawing No. 190693 and load line specification chard No. 6592-A.

HOIST REEVING FOR MAIN LOAD BLOCK						
No.Parts Of Line	1	2	3	4	5	6
Max Load - Lbs.	32,500	65,000	97,500	130,000	162,500	195,000
No. Parts of Line	7	8	9	10	11	12
Max. Load - Lbs.	227,500	260,000	292,500	325,000	357,500	400,000
No. Parts of Line	13					
Max. Load - Lbs.	430,000					

Crane operator judgement must be used to allow for dynamic load effects of swinging, hoisting or lowering, travel, as well as adverse operating conditions & physical machine depreciation.

OPERATOR RADIUS: Operating is the horizontal distance from the axis of rotation to the center of vertical hoist line or load block with the load freely suspended. Add 14" to boom point radius for radius of sheave when using single part hoist line.

Boom angle is the angle between horizontal and centerline of boom butt and inserts and is an indication of operating radius. In all cases, operating radius shall govern capacity.

BOOM POINT ELEVATION: Boom point elevation, in feet, is the vertical distance from ground level to centerline of boom point shaft.

MACHINE EQUIPMENT: Machine equipped with 26'6" extendible crawlers, 48" treads, 17' retractable gantry, 12 part boom hoist reeving, four 1 3/8" boom pendants, 1st ctwt. 41,900 lbs., 2nd ctwt. 41,500 lbs., 3rd ctwt. 39,000 lbs., two 12,000 lbs. side ctwt's. and two 30,000 lbs. carbody ctwt's.

LOAD AND WHIP LINE SPECIFICATIONS	
LOAD LINE:	1-1/8" - 6 x 31 Warrington-Seale, Extra Improved Plow Steel, Regular Lay, IWRC. Minimum Breaking Strength 65 Ton. (Approx.Weight Per Ft. in Lbs. 2.34)
WHIP LINE:	1-1/8" - Warrington-Seale, Improved Plow Steel, Regular Lay, IWRC. Minimum Breaking Strength 56.5 Ton. Maximum Load - 28,300 Lbs. Per Line. (Approx. Weight Per Ft. in Lbs. 2.34)

MAXIMUM BOOM AND JIB LENGTH LIFTED UNASSISTED				DEDUCT FROM CAPACITIES WHEN JIB IS ATTACHED	
OVER FRONT OF BLOCKED CRAWLERS		OVER SIDE OF EXTENDED CRAWLERS			
BOOM LENGTH	JIB NO. 123	BOOM LENGTH	JIB NO. 123	JIB LENGTH	JIB NO. 123
260'	--	260'	--	30'	3,000 lbs.
250'	--	250'	--	40'	3,600 lbs.
240'	40'	240'	40'	50'	4,200 lbs.
230'	60'	230'	60'	60'	4,900 lbs.
Load block, hook and weight ball on ground to start.					

FOR JIB CAPACITIES, CONSULT JIB CHART.

BOOM LGTH FEET	OPER. RAD. FEET	BOOM ANG. DEG.	BOOM POINT ELEV.	CAPACITY: CRAWLERS EXTENDED
70	16.5	79.7	75.9	460,000
	17	79.3	75.8	400,000
	18	78.5	75.6	380,100
	19	77.6	75.4	363,000
	20	76.8	75.1	347,300
	22	75.1	74.6	319,600
	24	73.4	74.1	293,400
	26	71.7	73.5	266,100
	28	69.9	72.8	237,500
	30	68.2	72.0	214,300
	32	66.4	71.2	195,100
	34	64.6	70.2	178,900
	36	62.8	69.3	165,200
	38	60.9	68.2	153,300
	40	59.1	67.0	143,000
	45	54.1	63.7	122,100
	50	48.9	59.8	106,300
	55	43.2	54.9	93,900
	60	36.9	49.0	84,000
	65	29.4	41.3	75,800
	70	19.5	30.3	63,900

BOOM LGTH FEET	OPER. RAD. FEET	BOOM ANG. DEG.	BOOM POINT ELEV.	CAPACITY: CRAWLERS EXTENDED
80	17	80.6	85.9	392,800
	18	79.9	85.8	378,900
	19	79.2	85.6	361,800
	20	78.5	85.4	346,100
	22	77.0	84.9	318,400
	24	75.5	84.5	292,500
	26	74.0	83.9	265,600
	28	72.5	83.3	237,000
	30	71.0	82.7	213,800
	32	69.5	81.9	194,600
	34	68.0	81.2	178,500
	36	66.4	80.3	164,700
	38	64.8	79.4	152,800
	40	63.3	78.4	142,400
	45	59.2	75.7	121,500
	50	54.9	72.5	105,700
	55	50.4	68.6	93,300
	60	45.6	64.1	83,400
	65	40.3	58.8	75,200
	70	34.4	52.2	68,300
	75	27.4	43.9	52,500
	80	18.2	32.0	53,900

BOOM LGTH FEET	OPER. RAD. FEET	BOOM ANG. DEG.	BOOM POINT ELEV.	CAPACITY: CRAWLERS EXTENDED
90	18	81.1	95.9	355,400
	19	80.4	95.7	346,900
	20	79.8	95.6	336,900
	22	78.5	95.2	317,400
	24	77.2	94.7	291,700
	26	75.9	94.3	264,800
	28	74.5	93.7	236,600
	30	73.2	93.2	213,400
	32	71.9	92.5	194,200
	34	70.5	91.9	178,000
	36	69.2	91.1	164,200
	38	67.8	90.3	152,300
	40	66.4	89.5	142,000
	45	62.9	87.1	121,100
	50	59.3	84.4	105,200
	55	55.5	81.2	92,800
	60	51.5	77.5	82,900
	65	47.3	73.2	74,700
	70	42.8	68.2	67,800
	75	37.9	62.3	62,000
	80	32.4	55.2	57,000
	85	25.8	46.2	52,600
	90	17.1	33.5	45,900

BOOM LGTH FEET	OPER. RAD. FEET	BOOM ANG. DEG.	BOOM POINT ELEV.	CAPACITY: CRAWLERS EXTENDED
100	19	81.4	105.9	332,900
	20	80.8	105.7	327,100
	22	79.6	105.4	316,200
	24	78.5	105.0	290,800
	26	77.3	104.5	263,900
	28	76.1	104.1	236,200
	30	74.9	103.6	212,900
	32	73.7	103.0	193,700
	34	72.5	102.4	177,500
	36	71.3	101.7	163,700
	38	70.1	101.0	151,800
	40	68.9	100.3	141,400
	45	65.8	98.2	120,500
	50	62.6	95.8	104,700
	55	59.3	93.0	92,300
	60	55.9	89.8	82,300
	65	52.4	86.2	74,100
	70	48.7	82.1	67,200
	75	44.8	77.4	61,400
	80	40.5	72.0	56,400
	85	35.9	65.6	52,000
	90	30.7	58.0	48,200
	95	24.5	48.5	44,900
	100	16.3	35.0	39,300

CAUTION! CHECK AMOUNT OF COUNTERWEIGHT ON MACHINE BEFORE USE OF THIS CHART

© Manitowoc 1997

301F37A.EPS

Figure 37 ◈ Manitowoc 4100 W-2 load/capacity chart. (1 of 4)

Instructor's Notes:

LIFTCRANE CAPACITIES – 4100W SERIES 2 BOOM NO 22C WITH OPEN-THROAT TOP, CONTINUED

BOOM LGTH FEET	OPER. RAD. FEET	BOOM ANG. DEG.	BOOM POINT ELEV.	CAPACITY: CRAWLERS EXTENDED
110	22	82.3	115.7	240,500
	24	81.3	115.4	232,100
	26	80.2	115.0	224,400
	28	79.1	114.6	217,300
	30	78.1	114.1	210,700
	32	77.0	113.6	193,300
	34	75.9	113.1	177,100
	36	74.8	112.5	163,300
	38	73.7	111.8	151,400
	40	72.6	111.2	141,000
	45	69.9	109.3	120,100
	50	67.0	107.2	104,200
	55	64.1	104.7	91,800
	60	61.2	101.9	81,800
	65	58.1	98.8	73,600
	70	54.9	95.3	66,800
	75	51.6	91.3	60,900
	80	48.1	86.8	55,900
	85	44.4	81.7	51,600
	90	40.4	75.9	47,800
	95	36.1	69.2	44,400
	100	31.1	61.1	41,400
	105	25.3	51.1	38,700
	110	17.6	37.2	33,900
120	22	83.0	125.9	228,700
	24	82.0	125.5	220,500
	26	81.0	125.2	213,000
	28	80.0	124.8	206,100
	30	79.1	124.4	199,700
	32	78.1	123.9	192,800
	34	77.1	123.4	176,600
	36	76.1	122.9	162,800
	38	75.1	122.3	150,900
	40	74.1	121.7	140,500
	45	71.6	120.0	119,500
	50	69.1	118.1	103,700
	55	66.5	115.9	91,300
	60	63.8	113.4	81,300
	65	61.1	110.6	73,100
	70	58.3	107.5	66,200
	75	55.4	104.0	60,400
	80	52.4	100.1	55,300
	85	49.2	95.8	51,000
	90	45.9	91.0	47,200
	95	42.4	85.5	43,800
	100	38.6	79.4	40,800
	105	34.4	72.2	38,100
	110	29.7	63.7	35,700
	115	24.1	53.2	33,500
130	24	82.6	135.7	212,700
	26	81.7	135.3	205,300
	28	80.8	135.0	198,600
	30	79.9	134.6	192,300
	32	79.0	134.2	186,500
	34	78.1	133.7	176,300
	36	77.2	133.2	162,500
	38	76.3	132.7	150,600
	40	75.4	132.1	140,200
	45	73.1	130.6	119,300
	50	70.8	128.8	103,400
	55	68.4	126.8	91,000
	60	66.0	124.6	81,000
	65	63.5	122.0	72,800
	70	61.0	119.2	65,900
	75	58.4	116.1	60,000
	80	55.7	112.7	55,000
	85	53.0	108.9	50,700
	90	50.1	104.8	46,800
	95	47.1	100.1	43,500
	100	43.9	95.0	40,500
	105	40.6	89.2	37,800
	110	36.9	82.6	35,300
	115	33.0	75.1	33,100
	120	28.5	66.1	31,100
	125	23.1	55.1	29,300
140	26	82.3	145.5	198,400
	28	81.5	145.1	191,800
	30	80.7	144.8	185,600
	32	79.8	144.4	179,900
	34	79.0	144.0	174,700
	36	78.1	143.5	162,000
	38	77.3	143.0	150,100
	40	76.5	142.5	139,700
	45	74.3	141.1	118,700
	50	72.2	139.4	102,800
	55	70.0	137.6	90,400
	60	67.8	135.5	80,400
	65	65.6	133.2	72,200
	70	63.3	130.7	65,300
	75	60.9	127.9	59,500
	80	58.5	124.8	54,400
	85	56.1	121.4	50,100
	90	53.5	117.7	46,200
	95	50.9	113.7	42,900
	100	48.1	109.2	39,900
140	105	45.3	104.2	37,200
	110	42.2	98.8	34,700
	115	39.0	92.7	32,500
	120	35.5	85.7	30,500
	125	31.7	77.8	28,700
	130	27.4	68.5	27,000
	135	22.2	57.0	25,500
150	28	82.1	155.3	185,600
	30	81.3	154.9	179,600
	32	80.5	154.6	174,000
	34	79.7	154.2	168,800
	36	79.0	153.8	161,600
	38	78.2	153.3	149,600
	40	77.4	152.8	139,200
	45	75.4	151.5	118,200
	50	73.4	150.0	102,400
	55	71.4	148.3	89,900
	60	69.4	146.4	79,900
	65	67.3	144.3	71,700
	70	65.2	141.9	64,800
	75	63.0	139.4	59,000
	80	60.9	136.6	53,900
	85	58.6	133.5	49,600
	90	56.3	130.1	45,700
	95	54.0	126.5	42,400
	100	51.5	122.5	39,400
	105	49.0	118.2	36,700
	110	46.4	113.4	34,200
	115	43.6	108.2	32,000
	120	40.7	102.4	30,000
	125	37.6	96.0	28,200
	130	34.2	88.8	26,500
	135	30.5	80.5	25,000
	140	26.4	70.8	23,500
	145	21.4	58.8	22,200
160	28	82.6	165.4	179,800
	30	81.8	165.1	173,900
	32	81.1	164.7	168,400
	34	80.4	164.4	163,300
	36	79.7	164.0	158,600
	38	78.9	163.6	149,100
	40	78.2	163.1	138,700
	45	76.3	161.9	117,700
	50	74.5	160.5	101,800
	55	72.6	158.9	89,300
	60	70.7	157.1	79,300
	65	68.8	155.1	71,100
	70	66.8	153.0	64,200
	75	64.9	150.6	58,400
	80	62.9	148.0	53,300
	85	60.8	145.2	49,000
	90	58.7	142.2	45,100
	95	56.6	138.8	41,800
	100	54.4	135.2	38,700
	105	52.1	131.4	36,000
	110	49.7	127.1	33,600
	115	47.3	122.5	31,400
	120	44.8	117.5	29,400
	125	42.1	112.0	27,600
	130	39.3	105.9	25,900
	135	36.3	99.2	24,300
	140	33.1	91.7	22,900
	145	29.5	83.0	21,600
	150	25.5	72.9	20,400
	155	20.7	60.5	19,200
170	30	82.3	175.2	169,000
	32	81.6	174.9	163,600
	34	81.0	174.5	158,700
	36	80.3	174.2	154,000
	38	79.6	173.8	148,800
	40	78.9	173.3	138,400
	45	77.2	172.2	117,400
	50	75.4	170.9	101,500
	55	73.7	169.4	89,100
	60	71.9	167.7	79,000
	65	70.1	165.9	70,800
	70	68.3	163.9	63,900
	75	66.4	161.7	58,100
	80	64.6	159.3	53,000
	85	62.7	156.7	48,700
	90	60.8	153.9	44,900
	95	58.8	150.8	41,500
	100	56.8	147.5	38,500
	105	54.7	144.0	35,800
	110	52.6	140.2	33,300
	115	50.4	136.0	31,100
	120	48.1	131.5	29,100
	125	45.8	126.7	27,300
	130	43.3	121.4	25,600
	135	40.8	115.7	24,100
	140	38.1	109.3	22,600
	145	35.2	102.3	21,300
	150	32.0	94.5	20,100
	155	28.6	85.5	18,900
	160	24.7	75.1	17,800
	165	20.0	62.2	16,400
180	32	82.1	185.0	158,700
	34	81.5	184.7	154,000
	36	80.8	184.3	149,500
	38	80.2	184.0	145,200
	40	79.5	183.6	137,900
	45	77.9	182.5	116,900
	50	76.3	181.2	101,000
	55	74.6	179.8	88,500
	60	72.9	178.3	78,500
	65	71.3	176.6	70,300
	70	69.6	174.7	63,400
	75	67.8	172.6	57,500
	80	66.1	170.4	52,500
	85	64.3	168.0	48,100
	90	62.5	165.4	44,300
	95	60.7	162.5	40,900
	100	58.9	159.5	37,900
	105	57.0	156.2	35,200
	110	55.0	152.7	32,700
	115	53.0	149.0	30,500
	120	51.0	144.9	28,500
	125	48.9	140.5	26,700
	130	46.7	135.8	25,000
	135	44.4	130.7	23,500
	140	42.0	125.2	22,000
	145	39.6	119.2	20,700
	150	36.9	112.6	19,500
	155	34.1	105.3	18,300
	160	31.1	97.2	17,200
	165	27.7	87.9	16,200
	170	23.9	77.1	15,300
	175	19.4	63.8	13,500
190	32	82.5	195.1	153,100
	34	81.9	194.8	148,800
	36	81.3	194.5	144,800
	38	80.7	194.1	140,800
	40	80.1	193.8	137,000
	45	78.5	192.7	116,400
	50	77.0	191.5	100,500
	55	75.4	190.2	88,000
	60	73.9	188.8	78,000
	65	72.3	187.1	69,700
	70	70.7	185.4	62,800
	75	69.1	183.4	57,000
	80	67.4	181.4	52,000
	85	65.8	179.1	47,600
	90	64.1	176.6	43,800
	95	62.4	174.0	40,400
	100	60.7	171.2	37,400
	105	58.9	168.2	34,700
	110	57.1	164.9	32,200
	115	55.3	161.4	30,000
	120	53.4	157.7	28,000
	125	51.5	153.7	26,200
	130	49.5	149.5	24,500
	135	47.5	144.9	22,900
	140	45.3	140.0	21,500
	145	43.1	134.6	20,200
	150	40.8	128.9	18,900
	155	38.4	122.6	17,800
	160	35.9	115.8	16,700
	165	33.2	108.3	15,700
	170	30.2	99.8	14,800
	175	27.0	90.3	13,900
	180	23.3	79.1	13,100
	185	18.9	65.4	11,000
200	34	82.3	204.9	143,100
	36	81.7	204.6	139,500
	38	81.2	204.3	136,000
	40	80.6	203.9	132,500
	45	79.1	203.0	115,900
	50	77.7	201.8	100,000
	55	76.2	200.6	87,500
	60	74.7	199.2	77,400
	65	73.2	197.7	69,200
	70	71.7	196.0	62,300
	75	70.2	194.2	56,400
	80	68.6	192.2	51,400
	85	67.1	190.1	47,000
	90	65.5	187.8	43,200
	95	63.9	185.3	39,800
	100	62.3	182.7	36,800
	105	60.6	179.8	34,100
	110	59.0	176.8	31,600
	115	57.3	173.6	29,400
	120	55.5	170.2	27,400
	125	53.0	166.5	25,600
	130	51.9	162.6	23,900
	135	50.1	158.4	22,300
	140	48.1	153.9	20,900
	145	46.2	149.1	19,600
	150	44.1	144.0	18,300
	155	42.0	138.4	17,200
	160	39.7	132.5	16,100
	165	37.4	126.0	15,100
	170	34.9	118.9	14,200
200	175	32.3	111.1	13,300
	180	29.4	102.4	12,400
	185	26.2	92.5	11,700
	190	22.6	81.0	10,900
	195	18.4	67.0	8,500
210	36	82.1	214.7	134,200
	38	81.6	214.4	131,000
	40	81.0	214.1	127,900
	45	79.6	213.2	115,500
	50	78.3	212.1	99,600
	55	76.9	210.9	87,100
	60	75.4	209.6	77,100
	65	74.0	208.1	68,800
	70	72.6	206.6	61,900
	75	71.2	204.8	56,100
	80	69.7	203.0	51,000
	85	68.2	201.0	46,700
	90	66.7	198.8	42,800
	95	65.2	196.5	39,400
	100	63.7	194.0	36,400
	105	62.2	191.3	33,700
	110	60.6	188.5	31,300
	115	59.0	185.5	29,100
	120	57.4	182.3	27,100
	125	55.7	178.9	25,200
	130	54.1	175.2	23,500
	135	52.3	171.4	22,000
	140	50.6	167.3	20,600
	145	48.8	162.9	19,200
	150	46.9	158.2	18,000
	155	45.0	153.2	16,800
	160	43.0	147.9	15,800
	165	40.9	142.1	14,800
	170	38.7	135.9	13,800
	175	36.5	129.2	12,900
	180	34.0	121.9	12,100
	185	31.5	113.9	11,300
	190	28.7	104.9	10,600
	195	25.6	94.7	9,900
	200	22.1	82.9	8,800
	205	17.9	68.5	6,500
220	38	82.0	224.6	125,700
	40	81.4	224.2	122,800
	45	80.1	223.3	115,000
	50	78.8	222.3	99,100
	55	77.5	221.2	86,600
	60	76.1	220.0	76,600
	65	74.8	218.6	68,300
	70	73.4	217.1	61,400
	75	72.0	215.4	55,500
	80	70.7	213.7	50,500
	85	69.3	211.8	46,100
	90	67.9	209.7	42,300
	95	66.4	207.5	38,900
	100	65.0	205.2	35,800
	105	63.6	202.7	33,100
	110	62.1	200.0	30,700
	115	60.6	197.2	28,500
	120	59.1	194.2	26,500
	125	57.5	191.0	24,600
	130	55.9	187.6	23,000
	135	54.3	184.0	21,400
	140	52.7	180.2	20,000
	145	51.0	176.1	18,600
	150	49.3	171.8	17,400
	155	47.6	167.3	16,300
	160	45.7	162.4	15,200
	165	43.9	157.2	14,200
	170	41.9	151.7	13,200
	175	39.9	145.7	12,300
	180	37.8	139.3	11,500
	185	35.6	132.4	10,700
	190	33.2	124.8	10,000
	195	30.7	116.6	9,300
	200	28.0	107.3	8,600
	205	25.0	96.9	8,000
	210	21.5	84.8	6,500

CAPACITIES CONTINUED ON NEXT PAGE

These load charts are intended for instructional purposes only. They were derived from manufacturer sales information which may not be complete or machine specific. Not responsible for typographical errors.

© MANITOWOC 1997

301F37B.EPS

Figure 37 ◈ Manitowoc 4100 W-2 load/capacity chart. (2 of 4)

LOAD CHARTS 1.31

LIFTCRANE CAPACITIES – 4100W SERIES 2 BOOM NO 22C WITH OPEN-THROAT TOP, CONTINUED

SEE CONDITIONS ON FRONT PAGE

BOOM LGTH FEET	OPER. RAD. FEET	BOOM ANG. DEG.	BOOM POINT ELEV.	CAPACITY: CRAWLERS EXTENDED
230	38	81.5	234.5	132,900
	40	81.0	234.2	125,200
	45	79.7	233.3	114,500
	50	78.5	232.4	98,600
	55	77.2	231.3	86,100
	60	75.9	230.1	76,100
	65	74.6	228.8	67,800
	70	73.3	227.3	60,900
	75	72.0	225.8	55,000
	80	70.7	224.1	50,000
	85	69.4	222.3	45,600
	90	68.0	220.3	41,700
	95	66.7	218.2	38,400
	100	65.3	216.0	35,300
	105	64.0	213.6	32,600
	110	62.6	211.1	30,200
	115	61.1	208.4	28,000
	120	59.7	205.6	26,000
	125	58.3	202.6	24,100
	130	56.8	199.4	22,400
	135	55.3	196.0	20,900
	140	53.8	192.5	19,400
	145	52.2	188.7	18,100
	150	50.6	184.7	16,900
	155	49.0	180.5	15,700
	160	47.3	176.0	14,700
	165	45.6	171.3	13,600
	170	43.8	166.2	12,700
	175	42.0	160.8	11,800
	180	40.1	155.1	11,000
	185	38.1	148.9	10,200
	190	36.0	142.3	9,400
	195	33.9	135.1	8,700
	200	31.6	127.3	8,100
	205	29.1	118.8	7,400
	210	26.4	109.3	6,800
240	40	81.4	244.3	123,400
	45	80.2	243.5	112,600
	50	78.9	242.6	98,100
	55	77.7	241.5	85,600
	60	76.5	240.4	75,600
	65	75.3	239.1	67,300
	70	74.0	237.7	60,400
	75	72.8	236.3	54,500
	80	71.5	234.6	49,500
	85	70.3	232.9	45,100
	90	69.0	231.1	41,200
	95	67.7	229.1	37,800
	100	66.4	227.0	34,800
	105	65.1	224.7	32,100
	110	63.8	222.3	29,700
	115	62.5	219.8	27,400
	120	61.1	217.1	25,400
	125	59.7	214.3	23,600
	130	58.3	211.3	21,900
	135	56.9	208.1	20,300
	140	55.5	204.7	18,900
	145	54.0	201.2	17,600
	150	52.5	197.5	16,300
	155	51.0	193.5	15,200
	160	49.5	189.4	14,100
	165	47.9	185.0	13,100
	170	46.2	180.3	12,200
	175	44.6	175.4	11,300
	180	42.8	170.2	10,400
	185	41.0	164.6	9,600
	190	39.2	158.7	8,900
	195	37.3	152.3	8,200
	200	35.2	145.5	7,500
	205	33.1	138.1	6,700
	210	30.9	130.1	6,000

BOOM LGTH FEET	OPER. RAD. FEET	BOOM ANG. DEG.	BOOM POINT ELEV.	CAPACITY: CRAWLERS EXTENDED
250	45	80.6	253.6	106,600
	50	79.4	252.7	97,800
	55	78.2	251.7	85,200
	60	77.1	250.6	75,200
	65	75.9	249.4	66,900
	70	74.7	248.1	60,000
	75	73.5	246.7	54,100
	80	72.3	245.2	49,100
	85	71.1	243.5	44,700
	90	69.9	241.7	40,800
	95	68.7	239.8	37,400
	100	67.4	237.8	34,400
	105	66.2	235.7	31,700
	110	64.9	233.4	29,300
	115	63.6	231.0	27,100
	120	62.4	228.5	25,000
	125	61.1	225.8	23,200
	130	59.7	222.9	21,500
	135	58.4	219.9	20,000
	140	57.0	216.8	18,500
	145	55.7	213.4	17,200
	150	54.3	209.9	16,000
	155	52.8	206.2	14,800
	160	51.4	202.4	13,700
	165	49.9	198.3	12,700
	170	48.4	193.9	11,800
	175	46.8	189.4	10,900
	180	45.3	184.5	10,100
	185	43.6	179.4	9,300
	190	41.9	174.0	8,500
	195	40.2	168.3	7,800
	200	38.4	162.2	7,000
	205	36.5	155.7	6,200
	210	34.5	148.6	5,500
260	45	80.9	263.7	104,800
	50	79.8	262.9	95,700
	55	78.7	261.9	84,700
	60	77.6	260.9	74,700
	65	76.4	259.7	66,400
	70	75.3	258.5	59,500
	75	74.2	257.1	53,600
	80	73.0	255.6	48,500
	85	71.8	254.1	44,100
	90	70.7	252.4	40,300
	95	69.5	250.6	36,900
	100	68.3	248.6	33,900
	105	67.1	246.6	31,100
	110	65.9	244.4	28,700
	115	64.7	242.1	26,500
	120	63.5	239.7	24,500
	125	62.3	237.1	22,600
	130	61.0	234.4	20,900
	135	59.7	231.6	19,400
	140	58.5	228.6	17,900
	145	57.2	225.4	16,600
	150	55.8	222.1	15,400
	155	54.5	218.7	14,200
	160	53.1	215.0	13,100
	165	51.7	211.2	12,100
	170	50.3	207.1	11,200
	175	48.9	202.9	10,300
	180	47.4	198.4	9,500
	185	45.9	193.7	8,700
	190	44.3	188.7	7,900
	195	42.7	183.4	7,000
	200	41.1	177.8	6,200
	205	39.4	171.9	5,400
	210	37.6	165.6	4,700

© Manitowoc 1997

301F37C.EPS

Figure 37 ◆ Manitowoc 4100 W-2 load/capacity chart. (3 of 4)

Instructor's Notes:

WEIGHTS 4100W Cont'd

DESCRIPTION	APPROX. WEIGHT (IN LBS.)
JIB NO. 123	
Jib Top - 15' (w/Jib point)	695
Jib Butt - 15'	690
Jib Insert - 10'	340
Basic Pendant - 33' 3-3/4" (2 Req'd)	115 each
Pendant - 10' (2 Per insert)	65 each
Jib Backstay Pendant	155 each
Jib Strut - 12'- 6"	365
COMPONENTS	
Hook Rollers (6) - w/shafts	1,020
Light Plant - 6.5KW - w/mounting platform	1,390
Catwalk - left and right side w/rails	1,320
Lagging - 27-5/8" dia. plain	1,410
Boom Hoist Rope - 12 Part - 760' of 7/8" - 6 x 26	1,080
Wire Rope Guide Assembly - Lower	325
Wire Rope Guide Assembly - Upper	510
Rope Guide Roller Assembly	55 each
2-Part Gantry w/Telescopic Backhitch	7,805
Equalizer	2,000
Hoist Line - 1-1/8" - 6 x 31	2.34 lbs./ft.
Whip Line - 1-1/8" - 6 x 31	2.34 lbs./ft.
15-Ton Hook and Weight Ball	865
100-Ton Hook Block Assembly	2,065
200-Ton Hook Block Assembly	4,900
230-Ton Hook Block Assembly	5,375
Boom Stop - Telescopic Air Cushioned	675
Dragline Fairlead - Revolving	1,910
Dragline Fairlead - Hinged	9,330

NOTE: The above weights may fluctuate up or down 5% due to manufacturing tolerances.

© Manitowoc 1997

301F37D.EPS

Figure 37 ◈ Manitowoc 4100 W-2 load/capacity chart. (4 of 4)

8.0.0 ◈ RANGE DIAGRAM

A range diagram (*Figure 38*) is a side view of a crane with its full range of boom lengths, jibs, and boom extensions. It serves three important functions:

- Determines the configuration of the crane for a particular job. The operator can determine the boom length, boom angle, boom point height or hook elevation, jib length, and jib offset required to suit the lift conditions.
- Identifies the minimum allowable clearances between the load blocks and boom tip.
- Determines the main boom angle of telescopic booms when the boom is only partially extended and only the load radius is known.

> **NOTE**
> Boom angles shown in a range diagram are for booms when no load is being lifted. Boom deflection with subsequent load radius and boom angle change must be accounted for when a load is applied.

Show Transparency 18 (Figure 38). Explain how to read a range diagram to determine the configuration of a crane for a particular job.

Explain how to read the range diagram to identify the minimum clearances.

Explain how to determine the main boom angle of telescopic booms.

LOAD CHARTS 1.33

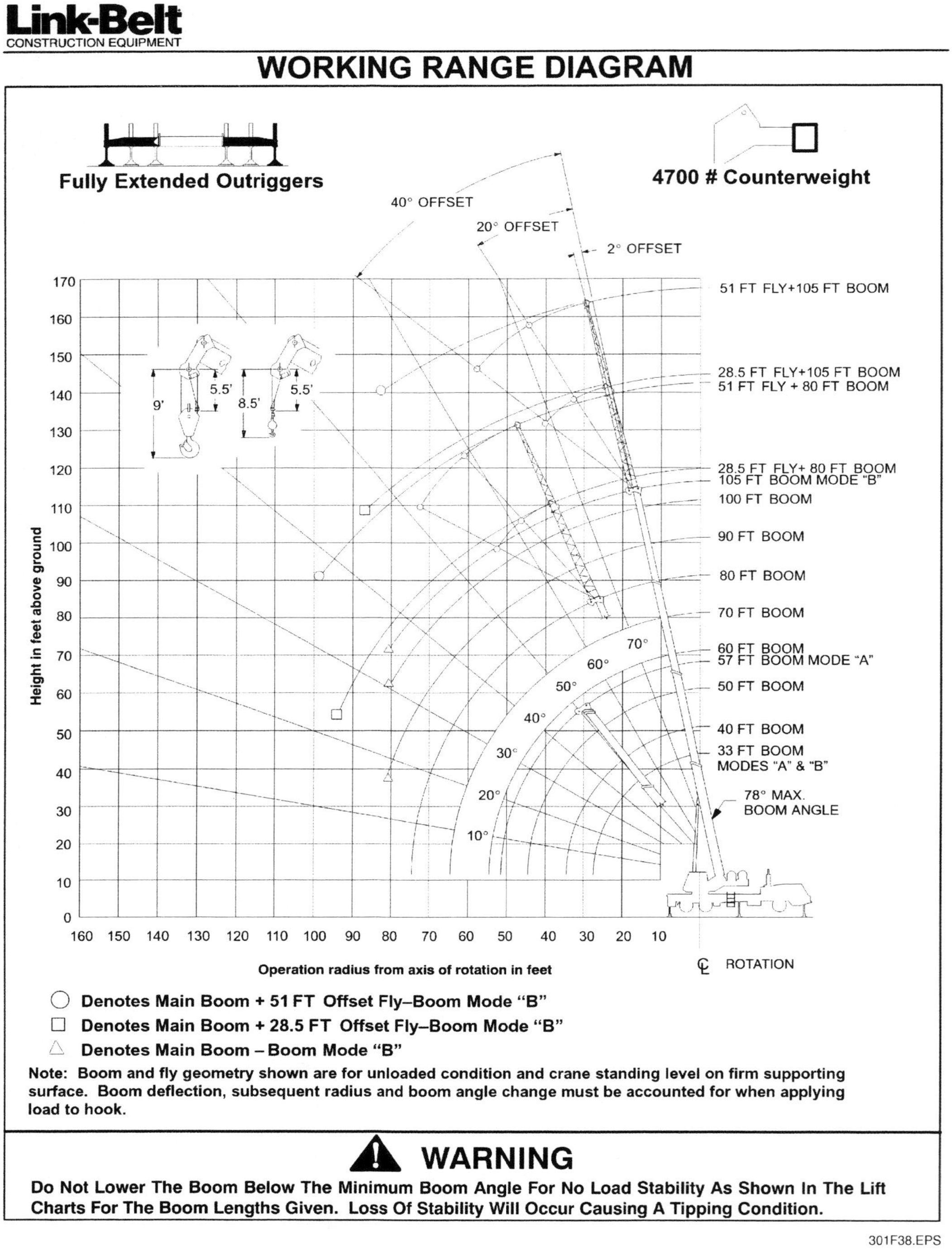

Figure 38 ◈ Range diagram.

Instructor's Notes:

Summary

It is imperative that crane operators understand how to read and interpret the various load/capacity charts that are included with a particular crane. The information used to determine whether a load and the lifting attachments are safe to lift is provided by the specifications in the charts. Manufacturers establish the gross capacities listed on the charts in ideal lifting conditions. The purpose of the charts is to provide the operator with a guideline on how much they can lift in a particular situation so that accidents can be avoided.

Review Questions

1. A crane's leverage is determined by _____.

a. manufacturer recommendations
b. the crane weight times the horizontal distance from the crane's center of gravity to the tipping axis
c. the crane weight times the horizontal distance from the load's center of gravity to the tipping axis
d. operator interpolation

2. The least capacity of a typical truck-mounted crane without a forward outrigger occurs when the tipping axis is _____.

a. above the crane
b. over the rear of the crane
c. over the side of the crane
d. over the front of the crane

3. Rough-terrain cranes are designed to be _____.

a. operated within five percent of level on a wide variety of supporting surfaces
b. transported from job site to job site with major component removal
c. operated from a standing position behind the controls
d. operated within one percent of level on a firm, supporting surface

4. Site preparation and crane erection take the most time for _____ cranes.

a. industrial hydraulic
b. all-terrain
c. truck-mounted telescopic
d. lattice boom

5. Boom length does not include the length of the jib.

a. True
b. False

6. The maximum horizontal distance from a projection of the axis of rotation to the center of the vertical hoist line or tackle with load applied for the given boom length, angle, and operating quadrant is the _____.

a. boom angle
b. boom length
c. boom point elevation
d. load operating radius

7. According to *ASME B30.5*, a load rating chart must include _____.

a. capacities for adverse operating conditions
b. capacities in both kilograms and pounds
c. counterweight requirements at the top of each chart sheet
d. work areas and load ratings where no loads are permitted to be handled

8. A crane can _____.

a. lift any load in any condition
b. be overloaded before any signs of tipping are evident
c. be operated with a bent chord in the lattice boom
d. overcome an overload situation by keeping the load still for 5 minutes and then proceeding with the operation

9. On lattice boom cranes, the _____ is measured from the center line of the boom to the horizontal.

a. boom length
b. boom angle
c. counterweight
d. range diagram

10. If the boom length falls between values listed on the load chart, the operator should use the capacity rating for the _____.

a. smallest load shown at either the next larger radius or next longer or shorter boom length
b. next shorter boom length
c. next longer load radius to compensate for the boom length
d. next shorter boom angle to compensate for the boom length

Summarize the major concepts presented in the module.

Have trainees complete the Review Questions, and go over the answers prior to administering the Module Examination.

Administer the Module Examination. Record the results on Craft Training Report Form 200, and submit the results to the Training Program Sponsor.

Administer the Performance Test, and fill out Performance Profile Sheets for each trainee. If desired, trainee proficiency noted during laboratory sessions may be used to complete the Performance Test. Record the results on Craft Training Report Form 200, and submit the results to the Training Program

LOAD CHARTS 1.35

Sponsor.

11. The boom point elevation is the distance from _____.
    a. the ground to the center of the boom point sheaves
    b. a projection of the axis of rotation to the center of the vertical hoist line
    c. the boom foot pins to the center of the boom point sheaves
    d. the bottom of the base section to the center line of the boom

12. _____ are different boom configurations that are activated and deactivated by means of a switch.
    a. Boom modes
    b. Boom ranges
    c. Electrical extensions
    d. Mechanical extensions

13. Interpolation is permitted in determining load capacities.
    a. True
    b. False

14. As a crane swings to various quadrants, the distance that the load is positioned from the tipping axis _____.
    a. doesn't change
    b. results in no stability change
    c. may change dramatically
    d. can be corrected with proper interpolation

15. Cranes are typically built to be _____.
    a. as wide as they are long
    b. wider than they are long
    c. longer than they are wide
    d. wider in the back than in the front

16. Crane manufacturers produce a separate load/capacity chart for each operational quadrant.
    a. True
    b. False

17. Crane capacities must be reduced for lower tire inflation pressure.
    a. True
    b. False

18. For on-rubber lifting, capacities do not exceed _____ of tipping.
    a. 70 percent
    b. 75 percent
    c. 85 percent
    d. 95 percent

19. The _____ gives extra stability to the crane during lifting operations.
    a. counterweight
    b. auxiliary boom sheave
    c. hoist rope
    d. hook block

20. The weights of the load block, headache ball, and wire rope of a crane are _____.
    a. weights that need to be added to the gross capacity
    b. capacity deductions that need to be subtracted from the gross capacity
    c. already compensated for in the capacity chart
    d. deducted from gross capacity only when hoisting is done from the main boom

21. Jib charts show the capacities of the jib _____.
    a. with rigging and load attachments
    b. at different degrees of offset
    c. in the stowed position
    d. with rigging attached

22. When lifting from the main boom with the jib extension mounted on the boom tip, the effective jib weight must be _____ the gross capacity.
    a. added to
    b. multiplied by
    c. deducted from
    d. divided by

23. To find the parts of line, the operator _____.
    a. adds the suspended weight and the safe working load of the hoist rope
    b. divides the suspended weight by the safe working load of the hoist rope
    c. subtracts the safe working load of the hoist rope from the suspended weight
    d. divides the load weight by the safe working load of the hoist rope

Instructor's Notes:

24. Suspended weight includes all of the following *except* _____.

a. hook block weight
b. sling and rigging hardware
c. hoist rope weight
d. load weight

25. Referring to the Grove RT522 load/capacity charts in *Appendix A*, find the net capacity of the Grove RT522 configured as follows:

- On outriggers, over-the-front lift
- 23- to 38-ft telescoping jib stowed on boom
- 40-ft main boom length
- 23-ft load radius
- 15-ton, 2-sheave block on main boom
- Auxiliary boom head attached
- 5-ton headache ball on auxiliary boom head
- Rigging and accessories of 65 lbs
- Lifting from the main boom

a. 16,522 lbs
b. 16,731 lbs
c. 16,918 lbs
d. 17,141 lbs

26. Referring to the Grove RT522 load/capacity charts in *Appendix A*, find the net load that can be handled safely with the RT522 configured as follows:

- On rubber, 360 degrees, stationary
- 20.5 × 25 tires
- 23-foot A-frame jib stowed on boom
- 22-foot load radius
- 12-ton, 1-sheave (12½" OD) block on main boom
- Rigging and accessories of 70 pounds

a. 3,211 lbs
b. 3,594 lbs
c. 3,728 lbs
d. 3,914 lbs

27. Referring to the Grove RT522 load/capacity charts in *Appendix A*, find the net capacity of the Grove RT522 when configured as follows:

- On outriggers fully extended, 360 degrees
- 46-ft boom length
- 15-ft load radius
- 22-ton, 3-sheave hook block
- Rigging and accessories of 45 pounds

a. 29,755 lbs
b. 29,555 lbs
c. 29,315 lbs
d. 29,050 lbs

28. Referring to the Manitowoc 4100W-2 load/capacity chart in *Appendix B*, find the gross capacity of the Manitowoc 4100W-2 in the following configuration:

- Crane counterweights of 146,400 lbs installed
- 22C open-throat top boom
- 180 ft of main boom
- Boom angle of 60 degrees
- Crawlers extended with two 12,000-lb side counterweights installed
- Two 30,000-lb carbody counterweights installed

a. 37,900 lbs
b. 40,900 lbs
c. 45,300 lbs
d. 52,100 lbs

29. Referring to the Manitowoc 4100W-2 load/capacity charts in *Appendix B*, find the net capacity in the following configuration:

- Crane counterweights of 146,000 lbs installed
- 22C boom, 190 ft of main boom, no offset
- 130 ft of load radius
- 2-sheave boom point attached
- 15-ton hook block 20 ft below boom point
- 100-ton block reeved with 4 parts of line
- Crawlers extended with two 12,000-lb side counterweights installed
- Rigging of 345 lbs
- Lifting from the main hoist
- Two 30,000-lb carbody counterweights installed

a. 17,213 lbs
b. 17,729 lbs
c. 17,811 lbs
d. 18,292 lbs

30. The diagram that is used to identify the minimum allowable clearances between the load blocks and boom tip is the _____.

a. jib chart
b. range diagram
c. reeving diagram
d. counterweight illustration

LOAD CHARTS 1.37

Instructor's Notes:

GLOSSARY

Trade Terms Introduced in This Module

ANSI/SAE J765: American Society of Mechanical Engineers/Society of Automotive Engineers consensus standard known as the *Crane Load Stability Test Code*.

ANSI/SAE J987: American Society of Mechanical Engineers/Society of Automotive Engineers consensus standard known as the *Crane Structures – Method of Test*.

ANSI/SAE J1028: American Society of Mechanical Engineers/Society of Automotive Engineers consensus standard known as the *Mobile Crane Working Area Definitions*.

Boom angle: The angle above or below horizontal of the longitudinal axis of the base boom section. On hydraulic cranes, it is measured between the bottom of the base section and the horizontal with the boom loaded. On lattice boom cranes, it is measured from the center line of the boom to the horizontal.

Boom length: Measured from the boom foot pins to the center of the boom point sheaves. It does not include the length of the jib.

Boom point elevation: Distance from the ground to the center of the boom point sheaves.

Gross capacity: Values listed in load charts.

Interpolation: Method of estimating or finding an intermediate value or term in a sequence or between two known values.

Load operating radius: The maximum horizontal distance from a projection of the axis of rotation to the center of the vertical hoist line or tackle with load applied for the given boom length, angle, and operating quadrant. Also called load radius or operating radius.

Net capacity: Gross capacity minus deductions.

Tipping axis (point): The fulcrum point on which a lever is balanced.

Grove RT522 Crane Load Charts

RANGE DIAGRAM

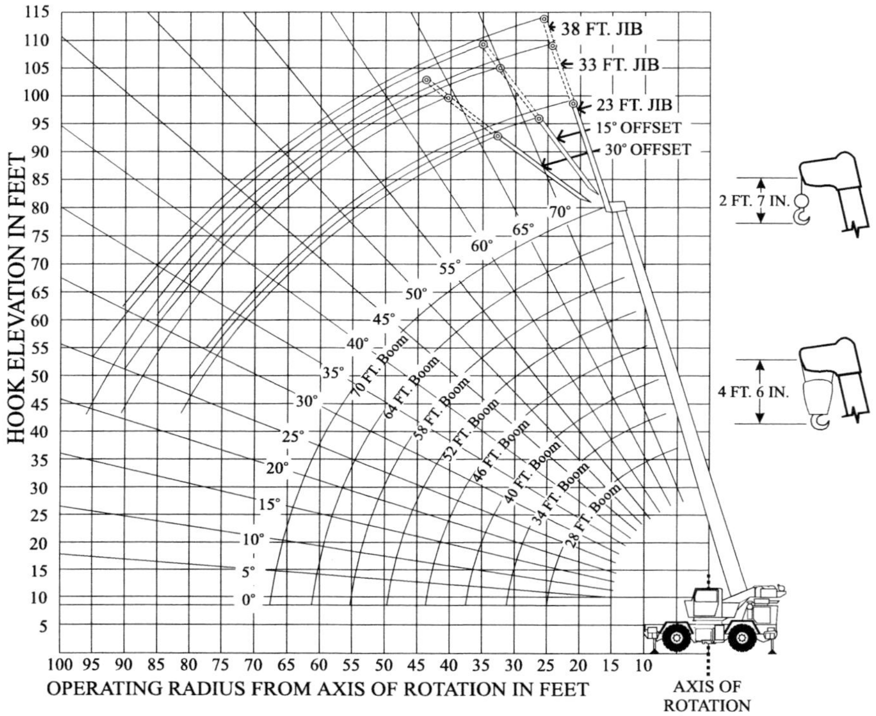

HOOK BLOCKS

22 Ton, 3 Sheave	490 lbs.
15 Ton, 2 Sheave	300 lbs.
12 Ton, 1 Sheave (15-7/8" OD)	400 lbs.
12 Ton, 1 Sheave (12-1/8" OD)	285 lbs.
Auxiliary Boom Head	100 lbs.
5-Ton Headache Ball	150 lbs.

NOTE: All load handling devices and boom attachments are considered part of the load and suitable allowances MUST BE MADE for their combined weights. Weights are for Grove furnished equipment.

© 1985 Grove Manufacturing Co.

WEIGHT REDUCTIONS FOR LOAD HANDLING DEVICES

23 FT. Jib WITH 28-70 FT. BOOM	
* Stowed	381 LBS.
* Erected	1,950 LBS.

** Reduction of main boom capacities.*

23-38 FT. TELE. Jib WITH 28-70 FT. BOOM	
* Stowed	604 LBS.
* Erected (Retracted)	3,659 LBS.
* Erected (Extended)	4,583 LBS.

** Reduction of main boom capacities.*

301A01.EPS

Instructor's Notes:

On Outriggers Fully Extended - 360°

Radius In Feet	Boom Length in Feet							
	28	34	40	46	52	58	64	70
10	44,000 (64)	36,000 (69)	36,000 (73)					
12	40,000 (59.5)	36,000 (65.5)	36,000 (70)	35,000 (73)				
15	31,000 (51.5)	31,000 (59.5)	30,700 (65)	29,850 (69)	29,150 (72)	28,600 (74.5)		
20	23,200 (36.5)	23,200 (49)	23,200 (57)	23,200 (62)	23,000 (66)	22,600 (69.5)	22,150 (72)	20,500 (74)
25	17,950 (6)	17,950 (36)	17,950 (47.5)	17,950 (54.5)	17,950 (60)	17,950 (64)	17,950 (67)	17,650 (69.5)
30		13,470 (15.5)	13,470 (36.5)	13,470 (46.5)	13,470 (53)	13,470 (58)	13,470 (62)	13,470 (65)
35			10,220 (20)	10,220 (36.5)	10,220 (45.5)	10,220 (51.5)	10,220 (56.5)	10,220 (60)
40	See Warning Note 16			8,010 (23)	8,010 (36.5)	8,010 (45)	8,010 (50.5)	8,010 (55)
45					6,530 (25)	6,530 (37)	6,530 (44.5)	6,530 (49.5)
50						5,430 (26.5)	5,430 (37)	5,430 (43.5)
55						4,440 (3.5)	4,440 (28)	4,440 (37)
60							3,620 (13)	3,620 (28.5)
65								2,980 (15.5)
Min. boom angle (deg.) for indicated length (No Load)								0
Max. boom length (Ft.) at 0 degree boom angle (No Load)								70.0

Note: Boom Angles are in degrees.

On Outriggers Fully Extended - Over Front

Radius In Feet	Boom Length in Feet							
	28	34	40	46	52	58	64	70
10	44,000 (64)	36,000 (69)	36,000 (73)					
12	40,000 (59.5)	36,000 (65.5)	36,000 (70)	35,000 (73)				
15	31,000 (51.5)	31,000 (59.5)	30,700 (65)	29,850 (69)	29,150 (72)	28,600 (74.5)		
20	23,200 (36.5)	23,200 (49)	23,200 (57)	23,200 (62)	23,000 (66)	22,600 (69.5)	22,150 (72)	20,500 (74)
25	17,950 (6)	17,950 (36)	17,950 (47.5)	17,950 (54.5)	17,950 (60)	17,950 (64)	17,950 (67)	17,650 (69.5)
30		15,350 (15.5)	15,350 (36.5)	15,350 (46.5)	15,350 (53)	15,150 (58)	14,950 (62)	14,750 (65)
35			11,900 (20)	11,900 (36.5)	11,900 (45.5)	11,900 (51.5)	11,900 (56.5)	11,900 (60)
40	See Warning Note 16			9,410 (23)	9,410 (36.5)	9,410 (45)	9,410 (50.5)	9,410 (55)
45					7,720 (25)	7,720 (37)	7,720 (44.5)	7,720 (49.5)
50						6,410 (26.5)	6,410 (37)	6,410 (43.5)
55						5,410 (3.5)	5,410 (28)	5,410 (37)
60							4,530 (13)	4,530 (28.5)
65								3,780 (15.5)
Min. boom angle (deg.) for indicated length (No Load)								0
Max. boom length (Ft.) at 0 degree boom angle (No Load)								70.0

Note: *Boom Angles are in degrees.*

© 1985 Grove Manufacturing Co.

301A02.EPS

LOAD CHARTS 1.41

14.00 x 24 Tires

Radius In Feet	Stationary Capacity Defined Arc (3) Over Front	Stationary Capacity 360° Arc	Pick & Carry Cap. Up to 2.5 MPH Boom Centered (7) Over Front
10	25,600 (a)	21,100 (a)	21,610 (a)
12	21,500 (a)	15,780 (a)	18,520 (a)
15	17,770 (a)	10,430 (b)	15,050 (a)
20	12,840 (b)	5,950 (c)	12,060 (a)
25	8,480 (b)	3,570 (d)	8,480 (b)
30	6,230 (c)	2,320 (e)	6,230 (c)
35	4,650 (d)	1,470 (f)	4,650 (d)
40	3,520 (e)	910 (f)	3,000 (d)
45	2,660 (f)		2,270 (e)
50	2,020 (g)		1,680 (f)
55	1,520 (g)		1,220 (f)
60	1,130 (h)		
65	880 (h)		

16.00 x 25 Tires

Radius In Feet	Stationary Capacity Defined Arc (3) Over Front	Stationary Capacity 360° Arc	Pick & Carry Cap. Up to 2.5 MPH Boom Centered (7) Over Front
10	29,150 (a)	19,640 (a)	30,460 (a)
12	24,030 (a)	14,600 (a)	26,320 (a)
15	20,150 (a)	10,390 (b)	21,670 (a)
20	13,350 (b)	5,980 (c)	13,350 (a)
25	8,610 (b)	3,500 (d)	8,290 (b)
30	6,290 (c)	2,160 (e)	6,290 (c)
35	4,650 (d)	1,430 (f)	4,650 (d)
40	3,620 (e)	810 (f)	3,620 (d)
45	2,800 (f)		2,800 (e)
50	2,120 (g)		2,120 (f)
55	1,630 (g)		1,630 (f)
60	1,240 (h)		
65	960 (h)		

Maximum Permissible Boom Length:	
(a) 28 Ft.	(e) 52 Ft.
(b) 34	(f) 58
(c) 40	(g) 64
(d) 46	(h) 70

		Main Boom 70 Ft.	Main Boom w/23 Ft. Jib
Front (No Load)	Min. Boom Angle (deg.) for indicated Boom Length	0	0
	Max. Boom Length (Ft.) at 0 degree Boom Angle	70	93
360° (No Load)	Min. Boom Angle (deg.) for indicated Boom Length	42	51
	Max. Boom Length (Ft.) at 0 degree Boom Angle	52	57

20.5 x 25 Tires

Radius In Feet	Stationary Capacity Defined Arc Over Front (3)	Stationary Capacity 360° Arc	Pick & Carry Cap. Up to 2.5 MPH Boom Centered Over Front (7)
10	28,380 (a)	24,450 (a)	26,020 (a)
12	23,800 (a)	17,190 (a)	22,420 (a)
15	19,100 (a)	11,640 (b)	18,380 (a)
20	13,940 (b)	7,180 (c)	13,940 (a)
25	9,160 (b)	4,330 (c)	9,160 (b)
30	6,640 (c)	2,930 (d)	4,830 (c)
35	5,100 (d)	2,000 (e)	3,790 (d)
40	4,020 (e)	1,370 (f)	3,020 (d)
45	3,140 (f)	880 (g)	2,310 (e)
50	2,430 (g)		1,710 (f)
55	1,860 (g)		1,210 (f)
60	1,440 (h)		
65	1,100 (h)		

NOTES FOR RUBBER CAPACITIES

1. Capacities do not exceed 75% of tipping loads as determined by test in accordance with SAE J-765.
2. Capacities are applicable to machines equipped with:

	Cold Inflation	2.5 MPH
14.00 x 24 (20 ply)	115 PSI	110 PSI
16.00 x 25 (20 ply)	95 PSI	80 PSI
20.5 x 25 (20 ply)	80 PSI	65 PSI

3. Defined Arc - Over front includes " 6° on either side of longitudinal centerline of machine.
4. Capacities appearing above bold line are based on structural strength and tipping should not be relied upon as a capacity limitation.
5. Capacities are applicable only with machine on a firm level surface.
6. On-rubber lifting with jib not permitted.
7. For pick and carry operation, boom must be centered over front of machine and mechanical swing lock engaged. When handling loads in the structural range with capacities close to maximum ratings, travel should be reduced to creep speed.
8. Axle lockouts must be functioning before lifting on rubber. (Check automatic lockout system for proper functioning: Refer to "Operation and Maintenance Manual" for description of a proper functioning axle lockout system).
9. All lifting depends on proper tire inflation, capacity and condition. Capacities must be reduced for lower tire inflation pressures. See lifting capacity chart

© 1985 Grove Manufacturing Co.

301A03.EPS

Instructor's Notes:

APPENDIX B

Manitowoc 4100W-2 Crane Load Charts

MANITOWOC 4100 W-2

LIFTCRANE CAPACITIES

BOOM NO. 22C WITH OPEN THROAT TOP
146,400 LB. CRANE COUNTERWEIGHT
60,000 LB. CARBODY COUNTERWEIGHT
26'6" CRAWLERS EXTENDED

WARNING: This chart will apply only when two 12,000 lb. side ctwts. and two 30,000 lb. carbody ctwts. bear MEC registered Serial Numbers.

LIFTING CAPACITIES: Capacities for various boom lengths and operating radii may be based on percent of tipping, strength of structural components, operating speeds and other factors.

Capacities are for freely suspended loads and do not exceed 75% of a static tipping load. Capacities based on structural competence are shown by shaded areas.

Capacities are shown in pounds. Deduct 1200 pounds from capacities listed when single sheave upper boom point is attached and 1500 pounds when two sheave upper boom point is attached. To comply with B30.5 requirements, upper boompoint cannot be used on the 260 ft. boom. Weight of jib, (see chart A), all load blocks, hooks, weight ball, slings, hoist lines beneath boom and jib point sheaves, etc., is considered part of the main boom load. Boom is not to be lowered beyond radii where combined weights are greater than rated capacity. Where no capacity is shown, operation is not intended or approved.

OPERATING CONDITIONS: Machine to operate in a level position on a firm surface with crawlers fully extended and gantry in working position and be rigged in accordance with and under conditions referred to in rigging drawing No. 190693 and load line specification chard No. 6592-A.

Crane operator judgement must be used to allow for dynamic load effects of swinging, hoisting or lowering, travel, as well as adverse operating conditions & physical machine depreciation.

OPERATOR RADIUS: Operating is the horizontal distance from the axis of rotation to the center of vertical hoist line or load block with the load freely suspended. Add 14" to boom point radius for radius of sheave when using single part hoist line.

Boom angle is the angle between horizontal and centerline of boom butt and inserts and is an indication of operating radius. In all cases, operating radius shall govern capacity.

BOOM POINT ELEVATION: Boom point elevation, in feet, is the vertical distance from ground level to centerline of boom point shaft.

MACHINE EQUIPMENT: Machine equipped with 26'6" extendible crawlers, 48" treads, 17' retractable gantry, 12 part boom hoist reeving, four 1 3/8" boom pendants, 1st ctwt. 41,900 lbs., 2nd ctwt. 41,500 lbs., 3rd ctwt. 39,000 lbs., two 12,000 lbs. side ctwt's. and two 30,000 lbs. carbody ctwt's.

LOAD AND WHIP LINE SPECIFICATIONS	
LOAD LINE:	1-1/8" - 6 x 31 Warrington-Seale, Extra Improved Plow Steel, Regular Lay, IWRC. Minimum Breaking Strength 65 Ton. (Approx. Weight Per Ft. in Lbs. 2.34)
WHIP LINE:	1-1/8" - Warrington-Seale, Improved Plow Steel, Regular Lay, IWRC. Minimum Breaking Strength 56.5 Ton. Maximum Load - 28,300 Lbs. Per Line. (Approx. Weight Per Ft. in Lbs. 2.34)

HOIST REEVING FOR MAIN LOAD BLOCK						
No. Parts Of Line	1	2	3	4	5	6
Max Load - Lbs.	32,500	65,000	97,500	130,000	162,500	195,000
No. Parts of Line	7	8	9	10	11	12
Max. Load - Lbs.	227,500	260,000	292,500	325,000	357,500	400,000
No. Parts of Line	13					
Max. Load - Lbs.	430,000					

MAXIMUM BOOM AND JIB LENGTH LIFTED UNASSISTED				DEDUCT FROM CAPACITIES WHEN JIB IS ATTACHED	
OVER FRONT OF BLOCKED CRAWLERS		OVER SIDE OF EXTENDED CRAWLERS			
BOOM LENGTH	JIB NO. 123	BOOM LENGTH	JIB NO. 123	JIB LENGTH	JIB NO. 123
260'	--	260'	--	30'	3,000 lbs.
250'	--	250'	--	40'	3,600 lbs.
240'	40'	240'	40'	50'	4,200 lbs.
230'	60'	230'	60'	60'	4,900 lbs.
Load block, hook and weight ball on ground to start.					

FOR JIB CAPACITIES, CONSULT JIB CHART.

BOOM LGTH FEET	OPER. RAD. FEET	BOOM ANG. DEG.	BOOM POINT ELEV.	CAPACITY: CRAWLERS EXTENDED
70	16.5	79.7	75.9	460,000
	17	79.3	75.8	400,000
	18	78.5	75.6	380,100
	19	77.6	75.4	363,000
	20	76.8	75.1	347,300
	22	75.1	74.6	319,600
	24	73.4	74.1	293,400
	26	71.7	73.5	266,100
	28	69.9	72.8	237,500
	30	68.2	72.0	214,300
	32	66.4	71.2	195,100
	34	64.6	70.2	178,900
	36	62.8	69.3	165,200
	38	60.9	68.2	153,300
	40	59.1	67.0	143,000
	45	54.1	63.7	122,100
	50	48.9	59.8	106,300
	55	43.2	54.9	93,900
	60	36.9	49.0	84,000
	65	29.4	41.3	75,800
	70	19.5	30.3	63,900

BOOM LGTH FEET	OPER. RAD. FEET	BOOM ANG. DEG.	BOOM POINT ELEV.	CAPACITY: CRAWLERS EXTENDED
80	17	80.6	85.9	392,800
	18	79.9	85.8	378,900
	19	79.2	85.6	361,800
	20	78.5	85.4	346,100
	22	77.0	84.9	318,400
	24	75.5	84.5	292,500
	26	74.0	83.9	265,600
	28	72.5	83.3	237,000
	30	71.0	82.7	213,800
	32	69.5	81.9	194,600
	34	68.0	81.2	178,500
	36	66.4	80.3	164,700
	38	64.8	79.4	152,800
	40	63.3	78.4	142,400
	45	59.2	75.7	121,500
	50	54.9	72.5	105,700
	55	50.4	68.6	93,300
	60	45.6	64.1	83,400
	65	40.3	58.8	75,200
	70	34.4	52.2	68,300
	75	27.4	43.9	52,500
	80	18.2	32.0	53,900

BOOM LGTH FEET	OPER. RAD. FEET	BOOM ANG. DEG.	BOOM POINT ELEV.	CAPACITY: CRAWLERS EXTENDED
90	18	81.1	95.9	355,400
	19	80.4	95.7	346,900
	20	79.8	95.6	336,900
	22	78.5	95.2	317,400
	24	77.2	94.7	291,700
	26	75.9	94.3	264,800
	28	74.5	93.7	236,600
	30	73.2	93.2	213,400
	32	71.9	92.5	194,200
	34	70.5	91.9	178,000
	36	69.2	91.1	164,200
	38	67.8	90.3	152,300
	40	66.4	89.5	142,000
	45	62.9	87.1	121,100
	50	59.3	84.4	105,200
	55	55.5	81.2	92,800
	60	51.5	77.5	82,900
	65	47.3	73.2	74,700
	70	42.8	68.2	67,800
	75	37.9	62.3	62,000
	80	32.4	55.2	57,000
	85	25.8	46.2	52,600
	90	17.1	33.5	45,900

BOOM LGTH FEET	OPER. RAD. FEET	BOOM ANG. DEG.	BOOM POINT ELEV.	CAPACITY: CRAWLERS EXTENDED
100	19	81.4	105.9	332,900
	20	80.8	105.7	327,100
	22	79.6	105.4	316,200
	24	78.5	105.0	290,800
	26	77.3	104.5	263,900
	28	76.1	104.1	236,200
	30	74.9	103.6	212,900
	32	73.7	103.0	193,700
	34	72.5	102.4	177,500
	36	71.3	101.7	163,700
	38	70.1	101.0	151,800
	40	68.9	100.3	141,400
	45	65.8	98.2	120,500
	50	62.6	95.8	104,700
	55	59.3	93.0	92,300
	60	55.9	89.8	82,300
	65	52.4	86.2	74,100
	70	48.7	82.1	67,200
	75	44.8	77.4	61,400
	80	40.5	72.0	56,400
	85	35.9	65.6	52,000
	90	30.7	58.0	48,200
	95	24.5	48.5	44,900
	100	16.3	35.0	39,300

CAUTION! CHECK AMOUNT OF COUNTERWEIGHT ON MACHINE BEFORE USE OF THIS CHART

© Manitowoc 1997

301A04.EPS

LOAD CHARTS 1.43

LIFTCRANE CAPACITIES – 4100W SERIES 2 BOOM NO 22C WITH OPEN-THROAT TOP, CONTINUED

BOOM LGTH FEET	OPER. RAD. FEET	BOOM ANG. DEG.	BOOM POINT ELEV.	CAPACITY: CRAWLERS EXTENDED
110	22	82.3	115.7	240,500
	24	81.3	115.4	232,100
	26	80.2	115.0	224,400
	28	79.1	114.6	217,300
	30	78.1	114.1	210,700
	32	77.0	113.6	193,300
	34	75.9	113.1	177,100
	36	74.8	112.5	163,300
	38	73.7	111.8	151,400
	40	72.6	111.2	141,000
	45	69.9	109.3	120,100
	50	67.0	107.2	104,200
	55	64.1	104.7	91,800
	60	61.2	101.9	81,800
	65	58.1	98.8	73,600
	70	54.9	95.3	66,800
	75	51.6	91.3	60,900
	80	48.1	86.8	55,900
	85	44.4	81.7	51,600
	90	40.4	75.9	47,800
	95	36.1	69.2	44,400
	100	31.1	61.1	41,400
	105	25.3	51.1	38,700
	110	17.6	37.2	33,900
120	22	83.0	125.9	228,700
	24	82.0	125.5	220,500
	26	81.0	125.2	213,000
	28	80.0	124.8	206,100
	30	79.1	124.4	199,700
	32	78.1	123.9	192,800
	34	77.1	123.4	176,600
	36	76.1	122.9	162,800
	38	75.1	122.3	150,900
	40	74.1	121.7	140,500
	45	71.6	120.0	119,500
	50	69.1	118.1	103,700
	55	66.5	115.9	91,300
	60	63.8	113.4	81,300
	65	61.1	110.6	73,100
	70	58.3	107.5	66,200
	75	55.4	104.0	60,400
	80	52.4	100.1	55,300
	85	49.2	95.8	51,000
	90	45.9	91.0	47,200
	95	42.4	85.5	43,800
	100	38.6	79.4	40,800
	105	34.4	72.2	38,100
	110	29.7	63.7	35,700
	115	24.1	53.2	33,500
130	24	82.6	135.7	212,700
	26	81.7	135.3	205,300
	28	80.8	135.0	198,600
	30	79.9	134.6	192,300
	32	79.0	134.2	186,500
	34	78.1	133.7	176,300
	36	77.2	133.2	162,500
	38	76.3	132.7	150,600
	40	75.4	132.1	140,200
	45	73.1	130.6	119,300
	50	70.8	128.8	103,400
	55	68.4	126.8	91,000
	60	66.0	124.6	81,000
	65	63.5	122.0	72,800
	70	61.0	119.2	65,900
	75	58.4	116.1	60,000
	80	55.7	112.7	55,000
	85	53.0	108.9	50,700
	90	50.1	104.8	46,800
	95	47.1	100.1	43,500
	100	43.9	95.0	40,500
	105	40.6	89.2	37,800
	110	36.9	82.6	35,300
	115	33.0	75.1	33,100
	120	28.5	66.1	31,100
	125	23.1	55.1	29,300
140	26	82.3	145.5	198,400
	28	81.5	145.1	191,800
	30	80.7	144.8	185,600
	32	79.8	144.4	179,900
	34	79.0	144.0	174,700
	36	78.1	143.5	162,000
	38	77.3	143.0	150,100
	40	76.5	142.5	139,700
	45	74.3	141.1	118,700
	50	72.2	139.4	102,800
	55	70.0	137.6	90,400
	60	67.8	135.5	80,400
	65	65.6	133.2	72,200
	70	63.3	130.7	65,300
	75	60.9	127.9	59,500
	80	58.5	124.8	54,400
	85	56.1	121.4	50,100
	90	53.5	117.7	46,200
	95	50.9	113.7	42,900
	100	48.1	109.2	39,900
140	105	45.3	104.2	37,200
	110	42.2	98.8	34,700
	115	39.0	92.7	32,500
	120	35.5	85.7	30,500
	125	31.7	77.8	28,700
	130	27.4	68.5	27,000
	135	22.2	57.0	25,500
150	28	82.1	155.3	185,600
	30	81.3	154.9	179,600
	32	80.5	154.6	174,000
	34	79.7	154.2	168,800
	36	79.0	153.8	161,600
	38	78.2	153.3	149,600
	40	77.4	152.8	139,200
	45	75.4	151.5	118,200
	50	73.4	150.0	102,400
	55	71.4	148.3	89,900
	60	69.4	146.4	79,900
	65	67.3	144.3	71,700
	70	65.2	141.9	64,800
	75	63.0	139.4	59,000
	80	60.9	136.6	53,900
	85	58.6	133.5	49,600
	90	56.3	130.1	45,700
	95	54.0	126.5	42,400
	100	51.5	122.5	39,400
	105	49.0	118.2	36,700
	110	46.4	113.4	34,200
	115	43.6	108.2	32,000
	120	40.7	102.4	30,000
	125	37.6	96.0	28,200
	130	34.2	88.8	26,500
	135	30.5	80.5	25,000
	140	26.4	70.8	23,500
	145	21.4	58.8	22,200
160	28	82.6	165.4	179,800
	30	81.8	165.1	173,900
	32	81.1	164.7	168,400
	34	80.4	164.4	163,300
	36	79.7	164.0	158,600
	38	78.9	163.6	149,100
	40	78.2	163.1	138,700
	45	76.3	161.9	117,700
	50	74.5	160.5	101,800
	55	72.6	158.9	89,300
	60	70.7	157.1	79,300
	65	68.8	155.1	71,100
	70	66.8	153.0	64,200
	75	64.9	150.6	58,400
	80	62.9	148.0	53,300
	85	60.8	145.2	49,000
	90	58.7	142.2	45,100
	95	56.6	138.8	41,800
	100	54.4	135.2	38,700
	105	52.1	131.4	36,000
	110	49.7	127.1	33,600
	115	47.3	122.5	31,400
	120	44.8	117.5	29,400
	125	42.1	112.0	27,600
	130	39.3	105.9	25,900
	135	36.3	99.2	24,300
	140	33.1	91.7	22,900
	145	29.5	83.0	21,600
	150	25.5	72.9	20,400
	155	20.7	60.5	19,200
170	30	82.3	175.2	169,000
	32	81.6	174.9	163,600
	34	81.0	174.5	158,700
	36	80.3	174.2	154,000
	38	79.6	173.8	148,800
	40	78.9	173.3	138,400
	45	77.2	172.2	117,400
	50	75.4	170.9	101,500
	55	73.7	169.4	89,100
	60	71.9	167.7	79,000
	65	70.1	165.9	70,800
	70	68.3	163.9	63,900
	75	66.4	161.7	58,100
	80	64.6	159.3	53,000
	85	62.7	156.7	48,700
	90	60.8	153.9	44,900
	95	58.8	150.8	41,500
	100	56.8	147.5	38,500
	105	54.7	144.0	35,800
	110	52.6	140.2	33,300
	115	50.4	136.0	31,100
	120	48.1	131.5	29,100
	125	45.8	126.7	27,300
	130	43.3	121.4	25,600
	135	40.8	115.7	24,100
	140	38.1	109.3	22,600
	145	35.2	102.3	21,300
	150	32.0	94.5	20,100
	155	28.6	85.5	18,900
	160	24.7	75.1	17,800
	165	20.0	62.2	16,400
180	32	82.1	185.0	158,700
	34	81.5	184.7	154,000
	36	80.8	184.3	149,500
	38	80.2	184.0	145,200
	40	79.5	183.6	137,900
	45	77.9	182.5	116,900
	50	76.3	181.2	101,000
	55	74.6	179.8	88,500
	60	72.9	178.3	78,500
	65	71.3	176.6	70,300
	70	69.6	174.7	63,400
	75	67.8	172.6	57,500
	80	66.1	170.4	52,500
	85	64.3	168.0	48,100
	90	62.5	165.4	44,300
	95	60.7	162.5	40,900
	100	58.9	159.5	37,900
	105	57.0	156.2	35,200
	110	55.0	152.7	32,700
	115	53.0	149.0	30,500
	120	51.0	144.9	28,500
	125	48.9	140.5	26,700
	130	46.7	135.8	25,000
	135	44.4	130.7	23,500
	140	42.0	125.2	22,000
	145	39.6	119.2	20,700
	150	36.9	112.6	19,500
	155	34.1	105.3	18,300
	160	31.1	97.2	17,200
	165	27.7	87.9	16,200
	170	23.9	77.1	15,300
	175	19.4	63.8	13,500
190	32	82.5	195.1	153,100
	34	81.9	194.8	148,800
	36	81.3	194.5	144,800
	38	80.7	194.1	140,800
	40	80.1	193.8	137,000
	45	78.5	192.7	116,400
	50	77.0	191.5	100,500
	55	75.4	190.2	88,000
	60	73.9	188.8	78,000
	65	72.3	187.1	69,700
	70	70.7	185.4	62,800
	75	69.1	183.4	57,000
	80	67.4	181.4	52,000
	85	65.8	179.1	47,600
	90	64.1	176.6	43,800
	95	62.4	174.0	40,400
	100	60.7	171.2	37,400
	105	58.9	168.2	34,700
	110	57.1	164.9	32,200
	115	55.3	161.4	30,000
	120	53.4	157.7	28,000
	125	51.5	153.7	26,200
	130	49.5	149.5	24,500
	135	47.5	144.9	22,900
	140	45.3	140.0	21,500
	145	43.1	134.6	20,200
	150	40.8	128.9	18,900
	155	38.4	122.6	17,800
	160	35.9	115.8	16,700
	165	33.2	108.3	15,700
	170	30.2	99.8	14,800
	175	27.0	90.3	13,900
	180	23.3	79.1	13,100
	185	18.9	65.4	11,000
200	34	82.3	204.9	143,100
	36	81.7	204.6	139,500
	38	81.2	204.3	136,000
	40	80.6	203.9	132,500
	45	79.1	203.0	115,900
	50	77.7	201.8	100,000
	55	76.2	200.6	87,500
	60	74.7	199.2	77,400
	65	73.2	197.7	69,200
	70	71.7	196.0	62,300
	75	70.2	194.2	56,400
	80	68.6	192.2	51,400
	85	67.1	190.1	47,000
	90	65.5	187.8	43,200
	95	63.9	185.3	39,800
	100	62.3	182.7	36,800
	105	60.6	179.8	34,100
	110	59.0	176.8	31,600
	115	57.3	173.6	29,400
	120	55.5	170.2	27,400
	125	53.0	166.5	25,600
	130	51.9	162.6	23,900
	135	50.1	158.4	22,300
	140	48.1	153.9	20,900
	145	46.2	149.1	19,600
	150	44.1	144.0	18,300
	155	42.0	138.4	17,200
	160	39.7	132.5	16,100
	165	37.4	126.0	15,100
	170	34.9	118.9	14,200
200	175	32.3	111.1	13,300
	180	29.4	102.4	12,400
	185	26.2	92.5	11,700
	190	22.6	81.0	10,900
	195	18.4	67.0	8,500
210	36	82.1	214.7	134,200
	38	81.6	214.4	131,000
	40	81.0	214.1	127,900
	45	79.6	213.2	115,500
	50	78.3	212.1	99,600
	55	76.9	210.9	87,100
	60	75.4	209.6	77,100
	65	74.0	208.1	68,800
	70	72.6	206.6	61,900
	75	71.2	204.8	56,100
	80	69.7	203.0	51,000
	85	68.2	201.0	46,700
	90	66.7	198.8	42,800
	95	65.2	196.5	39,400
	100	63.7	194.0	36,400
	105	62.2	191.3	33,700
	110	60.6	188.5	31,300
	115	59.0	185.5	29,100
	120	57.4	182.3	27,100
	125	55.7	178.9	25,200
	130	54.1	175.2	23,500
	135	52.3	171.4	22,000
	140	50.6	167.3	20,600
	145	48.8	162.9	19,200
	150	46.9	158.2	18,000
	155	45.0	153.2	16,800
	160	43.0	147.9	15,800
	165	40.9	142.1	14,800
	170	38.7	135.9	13,800
	175	36.5	129.2	12,900
	180	34.0	121.9	12,100
	185	31.5	113.9	11,300
	190	28.7	104.9	10,600
	195	25.6	94.7	9,900
	200	22.1	82.9	8,800
	205	17.9	68.5	6,500
220	38	82.0	224.6	125,700
	40	81.4	224.2	122,800
	45	80.1	223.3	115,000
	50	78.8	222.3	99,100
	55	77.5	221.2	86,600
	60	76.1	220.0	76,600
	65	74.8	218.6	68,300
	70	73.4	217.1	61,400
	75	72.0	215.4	55,500
	80	70.7	213.7	50,500
	85	69.3	211.8	46,100
	90	67.9	209.7	42,300
	95	66.4	207.5	38,900
	100	65.0	205.2	35,800
	105	63.6	202.7	33,100
	110	62.1	200.0	30,700
	115	60.6	197.2	28,500
	120	59.1	194.2	26,500
	125	57.5	191.0	24,600
	130	55.9	187.6	23,000
	135	54.3	184.0	21,400
	140	52.7	180.2	20,000
	145	51.0	176.1	18,600
	150	49.3	171.8	17,400
	155	47.6	167.3	16,300
	160	45.7	162.4	15,200
	165	43.9	157.2	14,200
	170	41.9	151.7	13,200
	175	39.9	145.7	12,300
	180	37.8	139.3	11,500
	185	35.6	132.4	10,700
	190	33.2	124.8	10,000
	195	30.7	116.6	9,300
	200	28.0	107.3	8,600
	205	25.0	96.9	8,000
	210	21.5	84.8	6,500

CAPACITIES CONTINUED ON NEXT PAGE

These load charts are intended for instructional purposes only. They were derived from manufacturer sales information which may not be complete or machine specific. Not responsible for typographical errors.

© MANITOWOC 1997

301A05.EPS

Instructor's Notes:

LIFTCRANE CAPACITIES – 4100W SERIES 2 BOOM NO 22C WITH OPEN-THROAT TOP, CONTINUED

SEE CONDITIONS ON FRONT PAGE

BOOM LGTH FEET	OPER. RAD. FEET	BOOM ANG. DEG.	BOOM POINT ELEV.	CAPACITY: CRAWLERS EXTENDED
230	38	81.5	234.5	132,900
	40	81.0	234.2	125,200
	45	79.7	233.3	114,500
	50	78.5	232.4	98,600
	55	77.2	231.3	86,100
	60	75.9	230.1	76,100
	65	74.6	228.8	67,800
	70	73.3	227.3	60,900
	75	72.0	225.8	55,000
	80	70.7	224.1	50,000
	85	69.4	222.3	45,600
	90	68.0	220.3	41,700
	95	66.7	218.2	38,400
	100	65.3	216.0	35,300
	105	64.0	213.6	32,600
	110	62.6	211.1	30,200
	115	61.1	208.4	28,000
	120	59.7	205.6	26,000
	125	58.3	202.6	24,100
	130	56.8	199.4	22,400
	135	55.3	196.0	20,900
	140	53.8	192.5	19,400
	145	52.2	188.7	18,100
	150	50.6	184.7	16,900
	155	49.0	180.5	15,700
	160	47.3	176.0	14,700
	165	45.6	171.3	13,600
	170	43.8	166.2	12,700
	175	42.0	160.8	11,800
	180	40.1	155.1	11,000
	185	38.1	148.9	10,200
	190	36.0	142.3	9,400
	195	33.9	135.1	8,700
	200	31.6	127.3	8,100
	205	29.1	118.8	7,400
	210	26.4	109.3	6,800
240	40	81.4	244.3	123,400
	45	80.2	243.5	112,600
	50	78.9	242.6	98,100
	55	77.7	241.5	85,600
	60	76.5	240.4	75,600
	65	75.3	239.1	67,300
	70	74.0	237.7	60,400
	75	72.8	236.3	54,500
	80	71.5	234.6	49,500
	85	70.3	232.9	45,100
	90	69.0	231.1	41,200
	95	67.7	229.1	37,800
	100	66.4	227.0	34,800
	105	65.1	224.7	32,100
	110	63.8	222.3	29,700
	115	62.5	219.8	27,400
	120	61.1	217.1	25,400
	125	59.7	214.3	23,600
	130	58.3	211.3	21,900
	135	56.9	208.1	20,300
	140	55.5	204.7	18,900
	145	54.0	201.2	17,600
	150	52.5	197.5	16,300
	155	51.0	193.5	15,200
	160	49.5	189.4	14,100
	165	47.9	185.0	13,100
	170	46.2	180.3	12,200
	175	44.6	175.4	11,300
	180	42.8	170.2	10,400
	185	41.0	164.6	9,600
	190	39.2	158.7	8,900
	195	37.3	152.3	8,200
	200	35.2	145.5	7,500
	205	33.1	138.1	6,700
	210	30.9	130.1	6,000

BOOM LGTH FEET	OPER. RAD. FEET	BOOM ANG. DEG.	BOOM POINT ELEV.	CAPACITY: CRAWLERS EXTENDED
250	45	80.6	253.6	106,600
	50	79.4	252.7	97,800
	55	78.2	251.7	85,200
	60	77.1	250.6	75,200
	65	75.9	249.4	66,900
	70	74.7	248.1	60,000
	75	73.5	246.7	54,100
	80	72.3	245.2	49,100
	85	71.1	243.5	44,700
	90	69.9	241.7	40,800
	95	68.7	239.8	37,400
	100	67.4	237.8	34,400
	105	66.2	235.7	31,700
	110	64.9	233.4	29,300
	115	63.6	231.0	27,100
	120	62.4	228.5	25,000
	125	61.1	225.8	23,200
	130	59.7	222.9	21,500
	135	58.4	219.9	20,000
	140	57.0	216.8	18,500
	145	55.7	213.4	17,200
	150	54.3	209.9	16,000
	155	52.8	206.2	14,800
	160	51.4	202.4	13,700
	165	49.9	198.3	12,700
	170	48.4	193.9	11,800
	175	46.8	189.4	10,900
	180	45.3	184.5	10,100
	185	43.6	179.4	9,300
	190	41.9	174.0	8,500
	195	40.2	168.3	7,800
	200	38.4	162.2	7,000
	205	36.5	155.7	6,200
	210	34.5	148.6	5,500
260	45	80.9	263.7	104,800
	50	79.8	262.9	95,700
	55	78.7	261.9	84,700
	60	77.6	260.9	74,700
	65	76.4	259.7	66,400
	70	75.3	258.5	59,500
	75	74.2	257.1	53,600
	80	73.0	255.6	48,500
	85	71.8	254.1	44,100
	90	70.7	252.4	40,300
	95	69.5	250.6	36,900
	100	68.3	248.6	33,900
	105	67.1	246.6	31,100
	110	65.9	244.4	28,700
	115	64.7	242.1	26,500
	120	63.5	239.7	24,500
	125	62.3	237.1	22,600
	130	61.0	234.4	20,900
	135	59.7	231.6	19,400
	140	58.5	228.6	17,900
	145	57.2	225.4	16,600
	150	55.8	222.1	15,400
	155	54.5	218.7	14,200
	160	53.1	215.0	13,100
	165	51.7	211.2	12,100
	170	50.3	207.1	11,200
	175	48.9	202.9	10,300
	180	47.4	198.4	9,500
	185	45.9	193.7	8,700
	190	44.3	188.7	7,900
	195	42.7	183.4	7,000
	200	41.1	177.8	6,200
	205	39.4	171.9	5,400
	210	37.6	165.6	4,700

© Manitowoc 1997

301A06.EPS

LOAD CHARTS 1.45

DESCRIPTION	APPROX. WEIGHT (IN LBS.)
JIB NO. 123	
Jib Top - 15' (w/Jib point)	695
Jib Butt - 15'	690
Jib Insert - 10'	340
Basic Pendant - 33' 3-3/4" (2 Req'd)	115 each
Pendant - 10' (2 Per insert)	65 each
Jib Backstay Pendant	155 each
Jib Strut - 12'- 6"	365
COMPONENTS	
Hook Rollers (6) - w/shafts	1,020
Light Plant - 6.5KW - w/mounting platform	1,390
Catwalk - left and right side w/rails	1,320
Lagging - 27-5/8" dia. plain	1,410
Boom Hoist Rope - 12 Part - 760' of 7/8" - 6 x 26	1,080
Wire Rope Guide Assembly - Lower	325
Wire Rope Guide Assembly - Upper	510
Rope Guide Roller Assembly	55 each
2-Part Gantry w/Telescopic Backhitch	7,805
Equalizer	2,000
Hoist Line - 1-1/8" - 6 x 31	2.34 lbs./ft.
Whip Line - 1-1/8" - 6 x 31	2.34 lbs./ft.
15-Ton Hook and Weight Ball	865
100-Ton Hook Block Assembly	2,065
200-Ton Hook Block Assembly	4,900
230-Ton Hook Block Assembly	5,375
Boom Stop - Telescopic Air Cushioned	675
Dragline Fairlead - Revolving	1,910
Dragline Fairlead - Hinged	9,330

NOTE: The above weights may fluctuate up or down 5% due to manufacturing tolerances.

© Manitowoc 1997

301A07.EPS

Instructor's Notes:

Figure Credits

Grove Crane	301F03, 301F04, 301F05, 301F11B, 301F17, 301F18, 301F20–301F24, 301F31, 301F34–301F36, Appendix A
Link-Belt Construction Equipment Company	301F06, 301F08, 301F09, 301F11A (top), 301F12–301F14, 301F25, 301F27, 301F30, 301F33, 301F38
Topaz Publications, Inc.	301F07
Terex	301F11A (bottom), 301F15, 301F16
Essex Crane Rental Corp., www.essexcrane.com	301F26
The Crosby Group, Inc.	301T01
Manitowoc Crane Group	301F37, Appendix B

LOAD CHARTS 1.47

MODULE 21301-05 — TEACHING TIPS

The following are suggested activities or instructional methods to help you teach the material in this module.

General

When you call on someone to answer a question, the rest of the class relaxes or even tunes out because they expect that the question and answer will take place only between you and the trainee you called on. Instead, use this technique to involve more trainees in answering questions and to keep them on their toes.

1. Ask trainees to define a term or explain a concept.
2. After one trainee has answered, ask a trainee seated nearby if the answer is right. Then ask whether a trainee in the back of the room agrees.
3. Ask trainees to explain why they think an answer is right or wrong.
4. Use the session to clear up incorrect ideas and encourage trainees to learn from their mistakes.

Section 4.0.0 *Types of Cranes*

Mobile crane operators must be familiar with different types of cranes. Trainees will need appropriate personal protective equipment, pencils, and paper. Arrange for an opportunity to observe several types of cranes in operation; alternatively, have a manufacturer's representative give a presentation on different types of cranes. Allow 30 to 60 minutes for this exercise.

1. Before the tour discuss the different types of cranes and their configurations with trainees.
2. Identify the different types of cranes and attachments. Have trainees take notes and write down questions during the tour.
3. After the tour answer any questions the trainees may have.

Section 5.0.0 *Load Chart*

This Quick Quiz will familiarize trainees with the terms and definitions that are commonly used in describing load charts. You will need photocopies of the quiz on the following page. Trainees will need pencils. If you allow trainees to use the Trainee Module, decrease the amount of time you give them to complete the quiz.

1. Make a photocopy of the quiz for each trainee.
2. Give trainees between 5 and 10 minutes to complete the quiz.
3. Go over the answers to the quiz.
4. Ask trainees if they have questions.

Answers to Quick Quiz

1. b
2. f
3. d
4. e
5. a
6. c

Quick Quiz *Terms and Definitions*

For each description listed, identify the term that the text best describes. Write the corresponding letter in the blank provided.

_____ **1.** Values listed in load charts

_____ **2.** Value listed in the load chart minus deductions

_____ **3.** Items such as the load block, headache ball, and boom extension that the crane will interpret as load

_____ **4.** Method of estimating an intermediate value between two known values

_____ **5.** The fulcrum point upon which a lever is balanced

_____ **6.** The angle above or below horizontal of the longitudinal axis of the base boom section

a. Tipping point
b. Gross capacity
c. Boom angle
d. Deductions
e. Interpolation
f. Net capacity

Copyright © 2005 National Center for Construction Education and Research. Permission is granted to reproduce this page provided that copies are for local use only and that each copy contains this notice.

Section 7.0.0 *Lift Calculations*

A crane operator must know how to properly calculate crane capacity for a particular lift. Trainees will need pencils and paper. The following video is available from the Maryland or North Carolina Department of Labor. Check your state Department of Labor to see if it is available for loan. Allow 20 to 30 minutes for this exercise.

Lift Calculations, NUS Training Corporation (1994) 12:16 minutes. Demonstrates and explains how to use a load chart to do lift calculations.

1. Before the presentation discuss lift calculations with trainees.
2. Show the video.
3. After the presentation answer any questions the trainees may have.

Section 7.0.0 *Calculating Capacity*

Calculating crane capacity is a critical aspect of an operator's job. The load moment indicator, covered in a previous module, uses the crane's sensors and a computer program to calculate the capacity. Trainees will need appropriate personal protective equipment, pencils, and paper. Arrange for an opportunity to work with a crane that has a load moment indicator. Allow 30 to 60 minutes for this exercise.

1. Explain how to calculate capacity by hand using charts. Explain how the load moment indicator uses the crane's sensors and a computer program to determine capacity.
2. For an example load, have trainees calculate the crane capacity by hand.
3. Have the crane pick up the example load. Compare the readout in the load moment

MODULE 21301-05 — ANSWERS TO REVIEW QUESTIONS

Answer	Section
1. b	2.0.0
2. d	2.0.0; Figure 1
3. d	4.2.0
4. d	4.5.0
5. a	4.5.0
6. d	5.0.0; Glossary
7. d	5.0.0
8. b	5.0.0
9. b	5.0.0; Glossary
10. a	6.1.0
11. a	6.1.0; Glossary
12. a	6.1.0
13. b	6.1.0
14. c	6.2.0
15. c	6.2.0
16. a	6.2.0
17. a	6.3.1; Figure 21
18. b	6.3.1
19. a	6.5.0
20. b	6.6.0
21. b	6.6.1
22. c	6.6.1
23. b	6.6.2
24. c	6.6.2
25. b*	7.0.0
26. b*	7.0.0
27. c*	7.0.0
28. a	7.0.0
29. d*	7.0.0
30. b	8.0.0

*Please see the following page for math calculations.

MODULE 21301-05 — REVIEW QUESTION CALCULATIONS

25. 17,950 lbs (gross capacity)
– 604 lbs (23 ft – 38 ft. telescoping jib stowed)
– 300 lbs (15-ton, 2-sheave block)
– 150 lbs (5-ton headache ball)
– 100 lbs (auxiliary boom head)
– 65 lbs (rigging)
16,731 lbs (net capacity)

26. 4,330 lbs (gross capacity)
– 381 lbs (23 ft A-frame jib stowed)
– 285 lbs (12-ton, 1-sheave (12⅛" OD) block)
– 70 lbs (rigging)
3,594 lbs (net capacity)

27. 29,850 lbs (gross capacity)
– 490 lbs (22-ton, 3-sheave block)
– 45 lbs (rigging)
29,315 lbs (net capacity)

29. 24,500 lbs (gross capacity)
– 1,500 lbs (2-sheave boom point)
– 865 lbs (15-ton hook block)
– 2,065 lbs (100-ton block)
– 345 lbs (rigging)
– 1,433 lbs (rope)
18,292 lbs (net capacity)

CONTREN® LEARNING SERIES — USER UPDATE

The NCCER makes every effort to keep these textbooks up-to-date and free of technical errors. We appreciate your help in this process. If you have an idea for improving this textbook, or if you find an error, a typographical mistake, or an inaccuracy in NCCER's Contren® textbooks, please write us, using this form or a photocopy. Be sure to include the exact module number, page number, a detailed description, and the correction, if applicable. Your input will be brought to the attention of the Technical Review Committee. Thank you for your assistance.

Instructors – If you found that additional materials were necessary in order to teach this module effectively, please let us know so that we may include them in the Equipment/Materials list in the Annotated Instructor's Guide.

Write: Product Development and Revision
National Center for Construction Education and Research
P.O. Box 141104, Gainesville, FL 32614-1104

Fax: 352-334-0932

E-mail: curriculum@nccer.org

Craft ______________________ Module Name ______________________

Copyright Date ____________ Module Number ____________ Page Number(s) ____________

Description

____

____

____

____

(Optional) Correction

____

____

____

(Optional) Your Name and Address

____

____

____

Module 21302-05

Telescopic Boom Attachment Assembly and Disassembly

NCCER STANDARDIZED CRAFT TRAINING PROGRAM

The National Center for Construction Education and Research (NCCER) provides a standardized national program of accredited craft training. Key features of the program include instructor certification, competency-based training, and performance testing. The program provides trainees, instructors, and companies with a standard form of recognition through a National Craft Training Registry. The program is described in full in the *Guidelines for Accreditation*, published by the NCCER. For more information on standardized craft training, contact the NCCER by writing us at P.O. Box 141104, Gainesville, FL 32614-1104; calling 352-334-0911; or e-mailing info@nccer.org. More information may be found at our Web site, www.nccer.org.

HOW TO USE THIS ANNOTATED INSTRUCTOR'S GUIDE

Each page presents two sections of information. The larger section displays each page exactly as it appears in the Trainee Module. The narrow column ties suggested trainee and instructor actions to each page and provides icons (detailed below) to call your attention to material, safety, audiovisual, or testing requirements. The bottom of each page includes space for your notes.

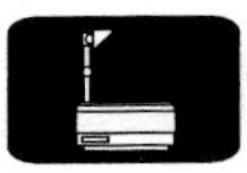

The **Audiovisual** icon indicates an appropriate time to show a transparency or other audiovisual aid.

The **Classroom** icon prompts you to define a term, stress a point, ask trainees to explain a concept, or give examples.

The **Demonstration** icon directs you to show trainees how to perform tasks.

The **Examination** icon tells you to administer the written module examination.

The **Homework** icon is placed where you may wish to assign reading for the next class, to assign a project, or to advise trainees to prepare for an examination.

The **Laboratory** icon is used when trainees are to practice performing tasks.

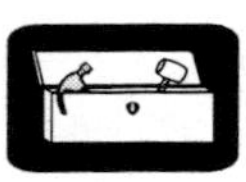

The **Materials** icon is a reminder for you to gather materials needed for classes, labs, and testing.

The **Performance Testing** icon tells you to administer a performance test or a portion thereof.

The **Safety** icon is used to emphasize safety issues. It is often keyed to *Caution* and *Warning* statements in the Trainee Module.

The **Teaching Tip** icon indicates additional guidance is available, such as how to conduct an exercise, get the most educational value from a field trip, or encourage class participation. Teaching Tips may expand on a feature (*Think About It*, *Did You Know?*) or provide Quick Quizzes or similar exercises. You will be referred to the Teaching Tips section at the back of the module if there is additional material.

The **Combination** icon indicates that the laboratory listed corresponds with a performance task. If desired, you can note the proficiency of the trainees during the laboratory and use it to satisfy performance testing requirements.

PREPARATION

Before teaching this module, you should review the Objectives, Performance Tasks, Materials and Equipment List, and the Module Outline. Be sure to allow ample time to prepare your own training or lesson plan and gather all required materials and equipment.

Telescopic Boom Attachment Assembly and Disassembly Annotated Instructor's Guide

Module 21302-05

MODULE OVERVIEW

This module discusses the stowing and erection of the swing-away extension, A-frame jib, and auxiliary single-sheave boom head, as well as the assembly and removal of intermediate boom sections.

PREREQUISITES

Prior to training with this module, it is recommended that the trainee shall have successfully completed *Core Curriculum; Mobile Crane Operations Level One; Mobile Crane Operations Level Two;* and *Mobile Crane Operations Level Three,* Module 21301-05.

OBJECTIVES

Upon completion of this module, the trainee will be able to do the following:

1. Determine if there is adequate space and resources for crane assembly and disassembly.
2. Define the relationship of the counterweight to the assembly and disassembly of the boom.
3. Assemble and disassemble various crane attachments and components.

PERFORMANCE TASKS

Under the supervision of the instructor, the trainee should be able to do the following:

1. Analyze the site for assembly/disassembly room.
2. Analyze and determine counterweight requirements needed for boom assembly.
3. Determine the ground stability and possible ground support needed for assembly and disassembly.
4. Assemble and disassemble crane attachments and components.
5. Install and remove an auxiliary single-sheave boom head.
6. Properly install wire rope and all lifting attachments.

MATERIALS AND EQUIPMENT LIST

Overhead projector and screen

Transparencies

Blank acetate sheets

Transparency pens

Whiteboard/chalkboard

Markers/chalk

Pencils and scratch paper

Copies of site safety manual or procedures

Copies of mobile crane operator's manuals

Mobile crane

Swing-away lattice extension

A-frame jib

Intermediate boom sections

Power-pinned fly

Manufacturer's assembly and disassembly instructions for crane and components

Auxiliary single-sheave boom head

Wire rope

Blocking

Tools and accessories to attach and stow crane components

Copies of the Quick Quiz*

Module Examinations**

Performance Profile Sheets**

*Located in the back of this module.

**Located in the Test Booklet.

SAFETY CONSIDERATIONS

Ensure that the trainees are equipped with appropriate personal protective equipment and know how to use it properly. This module requires trainees to assemble and disassemble cranes. Review site safety procedures and site evacuation procedures. Ensure that all trainees are familiar with hand signals and other site communication procedures. Brief trainees on pinching and crushing hazards attendant with assembling and disassembling cranes.

TEACHING TIME FOR THIS MODULE

An outline for use in developing your lesson plan is presented below. Note that each Roman numeral in the outline equates to one session of instruction. Each session has a suggested time period of 2½ hours. This includes 10 minutes at the beginning of each session for administrative tasks and one 10-minute break during the session. Approximately 20 hours are suggested to cover *Telescopic Boom Attachment Assembly and Disassembly*. You will need to adjust the time required for hands-on activity and testing based on your class size and resources. Because laboratories often correspond to Performance Tasks, the proficiency of the trainees may be noted during these exercises for Performance Testing purposes.

Topic	Planned Time
Sessions I and II. Introduction, Pre- and Post-Assembly Considerations	
A. Introduction	____________
B. Pre- and Post-Assembly Considerations	____________
C. Counterweight Considerations	____________
D. Laboratory – Trainees practice analyzing the site for assembly/disassembly room, and determining the ground stability and possible ground support needed for assembly and disassembly. Trainees also practice analyzing and determining counterweight requirements needed for boom assembly. This laboratory corresponds to Performance Tasks 1, 2, and 3.	____________
Session III. Swing-Away Lattice Extension	
A. Installing a Swing-Away Lattice Extension	____________
B. Stowing a Swing-Away Lattice Extension	____________
C. Laboratory – Trainees practice installing and stowing a swing away lattice extension. This laboratory corresponds to Performance Task 4.	____________
Session IV. A-Frame Jib	
A. Installing an A-Frame Jib	____________
B. Stowing an A-Frame Jib	____________
C. Laboratory – Trainees practice installing and stowing an A-frame jib. This laboratory corresponds to Performance Task 4.	____________
Sessions V. Intermediate Boom Sections	
A. Installing Intermediate Boom Sections	____________
B. Removing Intermediate Boom Sections	____________
C. Laboratory – Trainees practice installing and removing intermediate boom sections. This laboratory corresponds to Performance Task 4.	____________
Sessions VI. Power-Pinned Fly	
A. Extending a Power-Pinned Fly	____________
B. Retracting a Power-Pinned Fly	____________
C. Laboratory – Trainees practice extending and retracting a power-pinned fly. This laboratory corresponds to Performance Task 4.	____________

Sessions VII. Auxiliary Single-Sheave Boom Head and Wire Rope

A. Auxiliary Single-Sheave Boom Head ____________

B. Laboratory – Trainees practice installing and removing an auxiliary single-sheave boom head. This laboratory corresponds to Performance Task 5. ____________

C. Wire Rope ____________

D. Laboratory – Trainees practice properly installing wire rope and all lifting attachments. This laboratory corresponds to Performance Task 6. ____________

Session VIII. Review, Module Examination, and Performance Testing

A. Review ____________

B. Module Examination ____________

1. Trainees must score 70 percent or higher to receive recognition from NCCER.
2. Record the testing results on Craft Training Report Form 200, and submit the results to the Training Program Sponsor.

C. Performance Testing ____________

1. Trainees must perform each task to the satisfaction of the instructor to receive recognition from NCCER. If applicable, proficiency noted during laboratory exercises can be used to satisfy the Performance Testing requirements.
2. Record the testing results on Craft Training Report Form 200, and submit the results to the Training Program Sponsor.

Module 21302-05

Telescopic Boom Attachment Assembly and Disassembly

Instructor's Notes:

COURSE MAP

This course map shows all of the modules in the third level of the *Mobile Crane Operations* curriculum. The suggested training order begins at the bottom and proceeds up. Skill levels increase as you advance on the course map. The local Training Program Sponsor may adjust the training order.

Assign reading of Module 21302-05.

MOBILE CRANE OPERATIONS LEVEL THREE

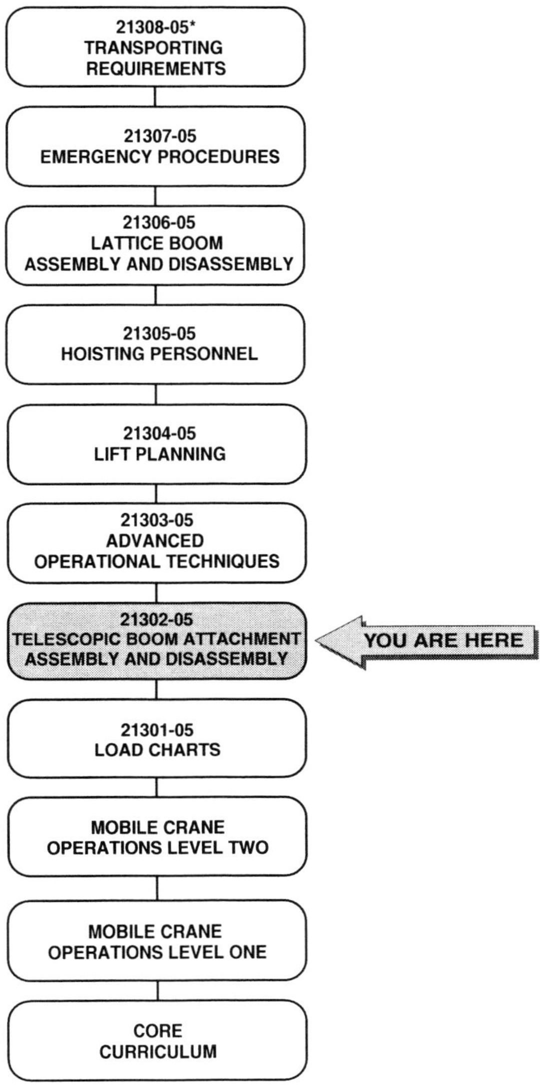

*Please note that 21308-05 is an elective module. It is not necessary to complete this module in order to advance in your training.

302CMAP.EPS

Copyright © 2005 National Center for Construction Education and Research, Gainesville, FL 32614-1104. All rights reserved. No part of this work may be reproduced in any form or by any means, including photocopying, without written permission of the publisher.

TELESCOPIC BOOM ATTACHMENT ASSEMBLY AND DISASSEMBLY 2.iii

Instructor's Notes:

MODULE 21302-05 CONTENTS

TELESCOPIC BOOM ATTACHMENT ASSEMBLY AND DISASSEMBLY 2.v

Figures

Instructor's Notes:

MODULE 21302-05

Telescopic Boom Attachment Assembly and Disassembly

Objectives

When you have completed this module, you will be able to do the following:

1. Determine if there is adequate space and resources for crane assembly and disassembly.
2. Define the relationship of the counterweight to the assembly and disassembly of the boom.
3. Assemble and disassemble various crane attachments and components.

Prerequisites

Before you begin this module, it is recommended that you successfully complete *Core Curriculum; Mobile Crane Operations Level One; Mobile Crane Operations Level Two*; and *Mobile Crane Operations Level Three*, Module 21301-05.

Required Trainee Materials

1. Pencil and paper
2. Appropriate personal protective equipment

1.0.0 ◆ INTRODUCTION

Over half of all mobile crane accidents result from human errors made during crane assembly. For this reason, there is a need for the operator and the site engineer or management to make sure the boom and its attachments are assembled and disassembled properly. Specific guidelines for installing various boom attachments are contained in the operator's safety manual for each crane. The purpose of this module is to give the trainee an overview of the assembly and disassembly procedures for telescoping boom crane attachments.

2.0.0 ◆ PRE- AND POST-ASSEMBLY CONSIDERATIONS

An erection area should be roped off and large enough to allow crane components to be stacked and handled. It should also allow a crane to be assembled without interfering with or posing a hazard to other site personnel. In addition, as the crane swings, there needs to be at least two feet of clearance between the counterweight and the nearest obstacle.

Before assembling the crane, the operator and the person responsible for the lift must consider the following:

- Is there enough room for the crane to maneuver into position on the site?
- Is there a large enough area in which to erect or extend the boom?
- Can the trucks hauling boom sections get into place and be unloaded safely?
- Is there enough timber blocking to support the boom when it is being assembled and disassembled?

3.0.0 ◆ COUNTERWEIGHT CONSIDERATIONS

It is imperative that the operator and any personnel involved in the assembly and disassembly of hydraulic boom cranes be aware of counterweight requirements. When attaching boom extensions, the telescope status and superstructure position on some cranes depend on the mounted counterweight. The crane may overturn if the specified maximum telescope status is exceeded or if the position of the superstructure is not observed.

Ensure that you have everything required to teach the course. Check the Materials and Equipment list at the front of this module.

Explain that terms shown in bold are defined in the Glossary at the back of this module.

Discuss the pre- and post-assembly survey. Discuss counterweight requirements.

See the general Teaching Tip at the end of this module.

Show Transparency 1, Objectives, and Transparency 2, Performance Tasks. Review the goals of the module, and explain what will be expected of trainees.

Demonstrate how to analyze a site for room and ground support and determine counterweight requirements.

Have trainees practice analyzing a site for room and ground support and determining counterweight requirements. Note the proficiency of each trainee. This laboratory corresponds to Performance Tasks 1, 2, and 3.

Have trainees review Sections 4.0.0–4.2.0.

Discuss pinching and crushing hazards inherent in crane operations. Review personal protective equipment, communication signals, and site safety procedures.

Describe the swing-away lattice extension. Explain that all parts must have the same serial number. Explain that the crane and load moment indicator will not function properly if other extensions are used.

Have trainees describe some of the variations for the swing-away extension on different cranes.

Discuss the importance of following the manufacturer's instructions and procedures for the crane being used.

Discuss the importance of the operator and other personnel observing the proper communication and safety procedures.

Provide a copy of the crane operator's manual and site safety manual for the trainees to examine.

See the Teaching Tip for Section 4.0.0 at the end of this module.

> **WARNING!**
> Assembly and disassembly personnel must always wear the appropriate personal protective equipment and observe safe working practices when working with boom extensions.

4.0.0 ◆ SWING-AWAY LATTICE EXTENSION

A swing-away lattice extension, sometimes called a fly, is designed for the crane with which it is delivered. These extension parts are serial-number specific. The operator should only operate the crane with those parts of the extension which have the same serial number or other identifier as the crane. If the crane is equipped with a load moment indicator (LMI), the indicator is only set for this particular extension. If any other extension is used, the crane will not function properly and damage may occur. *Figures 1* and *2* show a two-piece adjustable fly. This type of swing-away extension can be assembled straight on the boom or at various offsets to the boom. The crane can be used with the extension stowed on the side of the boom if the extension is not required.

4.1.0 Installing a Swing-Away Lattice Extension

The length of non-adjustable swing-away extensions can vary depending on the type of crane and the manufacturer. *Figure 3* shows a rough-terrain crane with only a swing-away extension. Some cranes are equipped with both a swing-away and an A-frame jib on the side of their main booms. A jib attachment adds steps to the swing-away attachment procedure.

302F01.EPS

Figure 1 ◆ Stowed two-piece adjustable fly.

Always follow the manufacturer's instructions and procedures for the crane being used. The specific crane may require different processes than those discussed in this module. The following is a general procedure for installing a non-adjustable swing-away extension. Note that this reflects the product of a particular manufacturer.

> **WARNING!**
> The operator and any other personnel involved in boom assembly and installation should observe the proper warning and safety information provided in the crane manual or site safety manual.

302F02.EPS

Figure 2 ◆ Two-piece adjustable fly assembled at an offset.

302F03.EPS

Figure 3 ◆ Non-adjustable swing-away extension.

Instructor's Notes:

Step 1 Extend and set the outriggers or crawlers according to the manufacturer's recommendations. Install proper counterweights as necessary for the required load and boom length.

Step 2 Fully retract all boom sections, and lower the boom to minimum elevation to permit installation of **pins** and access to the boom nose.

Step 3 Rig either the main hoist or optional **auxiliary** hoist cable for a single part of line. Lay the cable off to the left side of the boom while erecting the swing-away.

Step 4 To deploy the stop rail (see *Figure 4*), remove the retaining clip from the stop-rail pin, and pull out the stop-rail pin. Hold the open end of the stop rail with your hand to prevent the stop rail from swinging open.

Step 5 Rotate the stop rail until it is in a horizontal position, and insert the stop-rail pin into the outer bore hole (*Figure 5*). Secure the stop-rail pin in place with its retaining pin. Check that the section is locked with the stop-rail pin and a pivot pin in the bearing point on the stop rail.

For the following steps, refer to *Figure 6*:

Step 6 Ensure the latch lock pins (1 and 2) and latch plates (3 and 4) that secure the A-frame (5 ref) to the front and rear support brackets (6 and 7) are properly engaged.

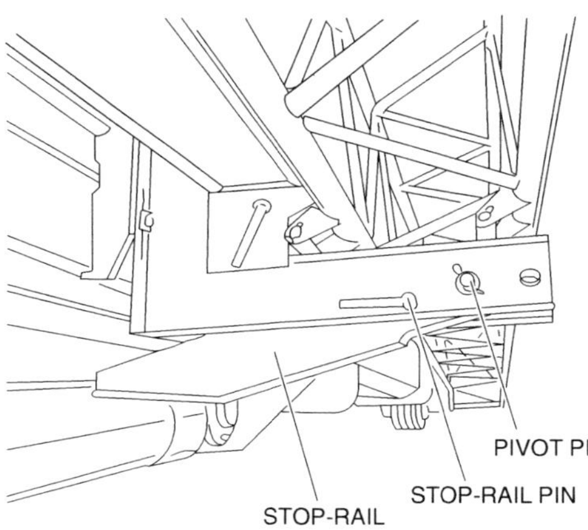

302F04.EPS

Figure 4 ◈ Stop rail viewed from under stowed extension.

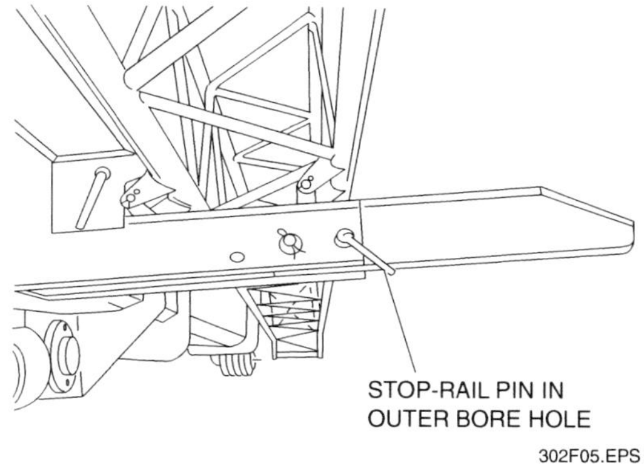

302F05.EPS

Figure 5 ◈ Stop rail extended under stowed extension.

Step 7 At the base end of the A-frame (5 ref), remove the pin (8) securing the A-frame to the swing-away (9 ref). Stow the pin in the A-frame attaching link (13).

Step 8 At the sheave end of the A-frame (5), remove the pin (10) securing the A-frame to the swing-away. Position the attaching link (11) in the stowed position (12) on the A-frame, and secure it with the pin (10).

Step 9 On the swing-away rear support bracket (7), pull the latch lock pin (15), and swing the latch plate (16) out of the way.

Step 10 With the front support bracket (6) acting as a pivot, pull outward on the swing-away tip to position the swing-away anchor fittings (18) onto the attach fittings (17) on the right side of the boom nose.

Step 11 Remove the attach pins (14) and their retainer clips stowed in the base of the swing-away, and insert them though the attach and anchor fittings on the right side of the boom nose. Install the retainer clips in the pins.

Step 12 Attach a guide rope to the swing-away. This ensures that the swing-away extension swings around safely.

WARNING!

The boom must be raised to horizontal to prevent the swing-away from swinging around after removing the attach pins from the front support bracket. When erecting the swing-away, make sure all personnel and equipment are kept clear of the swing path.

Show Transparencies 3 through 6 (Figures 4 through 7). Describe the steps for installing a swing-away lattice extension.

Discuss the dangers when the boom is being raised and the precautions that must be taken to keep all personnel clear of the swing path.

Provide a copy of the crane manufacturer's swing-away assembly instructions for the trainees to examine.

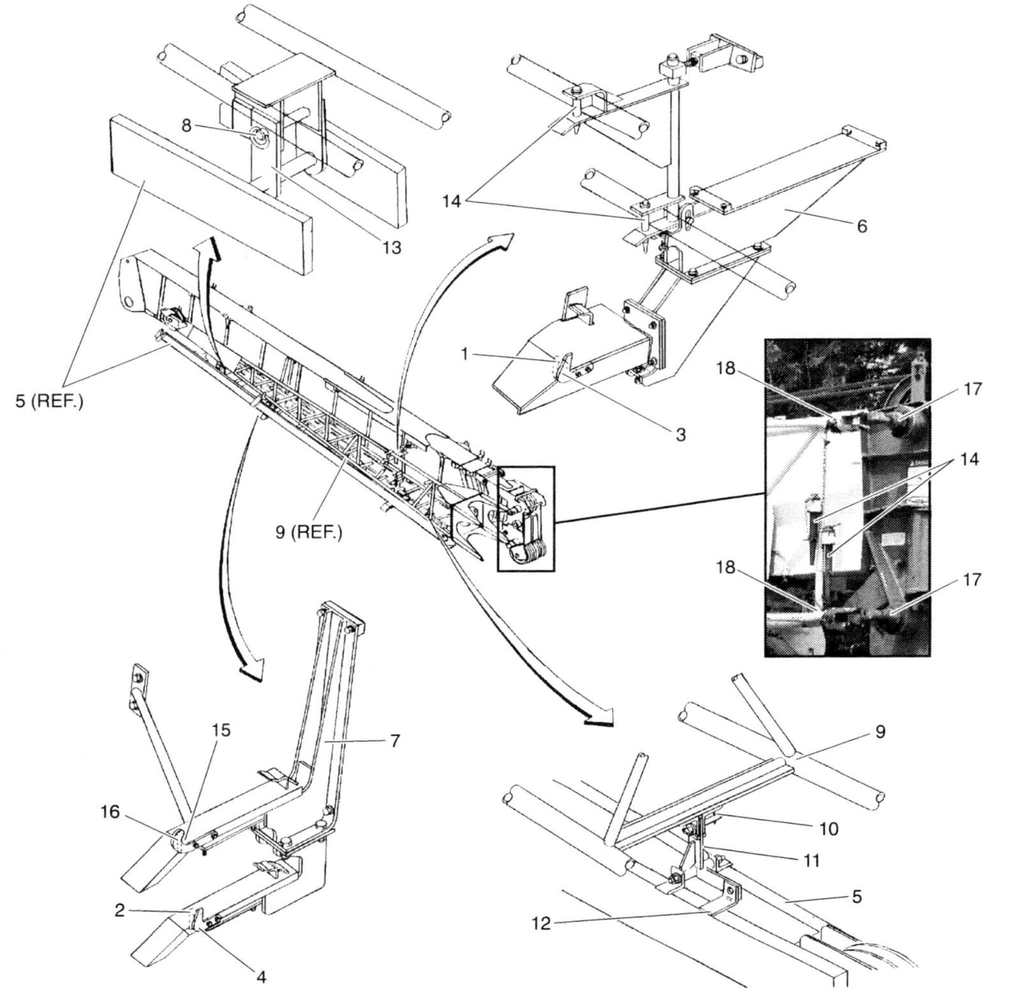

1. LATCH LOCK PIN
2. LATCH LOCK PIN
3. LATCH PLATE
4. LATCH PLATE
5. A-FRAME
6. FRONT SUPPORT BRACKET
7. REAR SUPPORT BRACKET
8. PIN
9. SWING-AWAY
10. PIN
11. ATTACHING LINK
12. ATTACHING LINK STOWAGE POSITION
13. ATTACHING LINK
14. ATTACH PIN
15. LATCH LOCK PIN
16. LATCH PLATE
17. ATTACH FITTINGS
18. ANCHOR FITTINGS

302F06.EPS

Figure 6 ◈ Swing-away assembly, part one.

Instructor's Notes:

Step 13 Raise the boom to a level, horizontal position. Then, remove the attach pins (14) and their retainer clips from the front support bracket (6).

For the following steps, refer to *Figure 7*:

Step 14 Swing the swing-away into place in front of the boom nose, engaging the anchor fittings (18) with the attach fittings (17) on the left side of the boom nose.

NOTE

To swing the extension around, use the outriggers to move the crane into a diagonal position and incline the boom slightly. If there is difficulty in aligning the swing-away extension's attach points to the boom nose, assembly personnel should place blocking under the tip of the swing-away to relieve the weight.

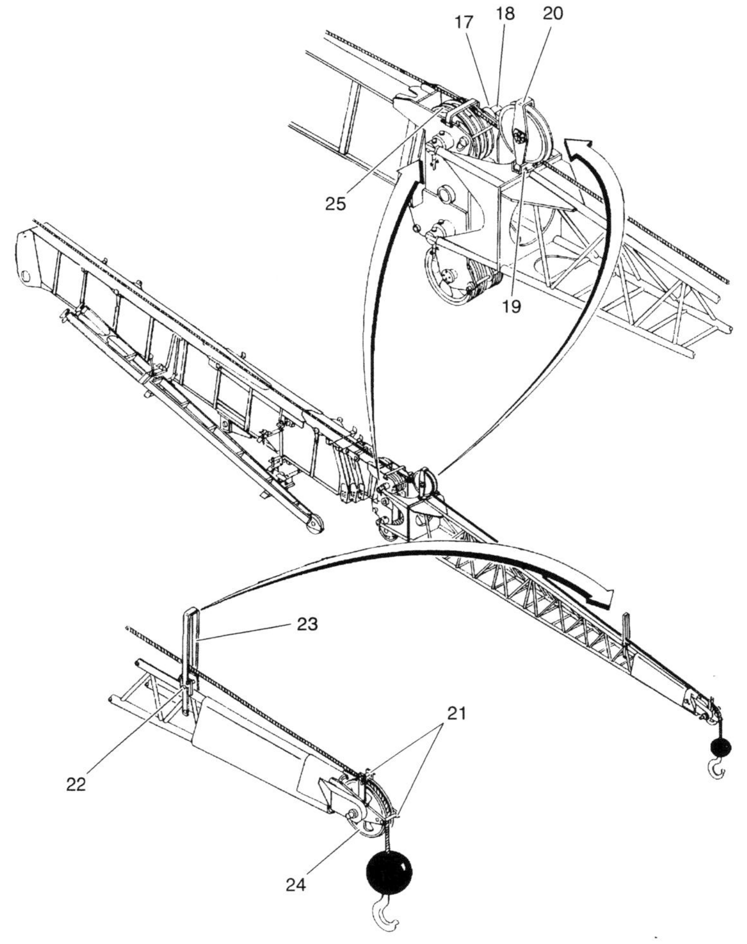

17. ATTACH FITTINGS
18. ANCHOR FITTINGS
19. PIN
20. CABLE-ROLLER SHEAVE
21. CABLE-RETAINING PINS
22. PIN
23. CABLE GUIDE
24. EXTENSION SHEAVE
25. BOOM-HEAD SHEAVE

302F07.EPS

Figure 7 ◆ Swing-away assembly, part two.

TELESCOPIC BOOM ATTACHMENT ASSEMBLY AND DISASSEMBLY 2.5

Explain the steps for stowing a swing-away lattice extension.

Reiterate the importance of following the specific manufacturer's instructions for stowing a swing-away lattice extension.

Demonstrate how to install and stow a swing-away lattice extension.

Have trainees practice installing and stowing a swing-away lattice extension. Note the proficiency of each trainee. This laboratory corresponds to Performance Task 4.

Have trainees review Sections 5.0.0–5.2.0.

Ensure that you have everything required for teaching this session.

Describe an A-frame jib.

Step 15 Lock the swing-away to the main boom by inserting the attach pins (pins removed from the front support bracket) into the anchor (18) and attach fittings (17) on the left side of the boom. Install the retainer clips in the pins.

Step 16 Remove a pin (19) and retainer clip from the cable-roller sheave (20). Move the cable-roller sheave (20) into working position in accordance with the prescribed inclination. Lock it into place with a pin (19), and secure the pin with its retainer clip.

Step 17 Reel off the hoist rope from the main or auxiliary hoist.

Step 18 Lay the hoist cable over the boom-head sheave (25), under the cable-roller sheave (20), and over the extension sheave (24).

Step 19 Remove the pin (22) and retainer clip securing the cable guide (23) in the stowed position. Position the cable guide in the vertical position, with the cable passing through the guide. Reinstall the pin through the second hole in the guide, and install the clip.

Step 20 Make sure that cable-retaining pins (21) are installed.

Step 21 Reeve the hoist cable on the headache ball.

Step 22 Establish electrical connections to the main boom.

Step 23 Insert the anemometer (if equipped) at the head of the swing-away, and establish an electrical connection.

Step 24 Attach the lifting limit switch weight. Feed the hoist rope through the weight.

Step 25 Move the stop rail into transport position.

4.2.0 Stowing a Swing-Away Lattice Extension

Always follow the manufacturer's instructions for stowing a swing-away lattice extension. For general guidelines, you can refer to the following abbreviated procedure:

Step 1 Fully retract the telescopic sections of the boom.

Step 2 Turn the superstructure to the back or the side.

Step 3 Lower the main boom until the load moment indicator (LMI) switches off.

Step 4 Override the LMI.

Step 5 Lower the main boom until the head sheave is approximately five feet above the ground.

Step 6 Unreeve the hoist rope from the load block, and remove it from the head sheave.

Step 7 Remove the lifting limit switch weight.

Step 8 Remove the anemometer (if equipped).

Step 9 Move the main boom into 0-degree position, if necessary.

Step 10 Disconnect the electrical connections to the main boom.

Step 11 Move the cable-roller sheave into the transport position.

Step 12 Wind in the hoist rope.

Step 13 Move the stop rail into assembly position.

Step 14 Attach the guide rope to the swing-away.

Step 15 Remove the connecting pins between the swing-away and main boom on the left-hand side, insert in the holder, and lock.

Step 16 Swing the swing-away on the main boom until the pivot points on the pivot bearing align.

Step 17 Insert pins in the pivot bearing.

Step 18 Remove the connecting pins between the swing-away and main boom on the right-hand side, insert in the holder, and lock.

Step 19 Move the stop rail into transport position.

5.0.0 ◈ A-FRAME JIB

The A-frame jib consists of two side rails and cross members made of rectangular tubing and welded together to form an A-frame. On some models of cranes, it is the only boom extension attachment. On others, it is attached to the swing-away and will go out with the swing-away. This section provides a procedure for one particular crane. Be sure to follow the procedure recommended by the manufacturer of your crane.

Instructor's Notes:

5.1.0 Installing an A-Frame Jib

When erecting or stowing the jib, the operator or personnel installing the jib should follow the manufacturer's instructions and the warning decals installed on the swing-away and/or A-frame. A general procedure to install an A-frame jib is as follows:

Step 1 Extend and set the outriggers or crawlers according to the manufacturer's recommendations.

Step 2 Swing the boom to the proper operational quadrant.

> **WARNING!**
> The boom must be completely retracted before engaging the swing-away boom extension with the boom nose.

Step 3 Fully retract all boom sections, and lower the boom to minimum elevation.

Step 4 Rig either the main hoist or optional auxiliary hoist cable for single part of line with nothing but the **becket** on the end of the cable.

For the following steps, refer to *Figure 8*:

> **WARNING!**
> Do not release the A-frame latch plates unless the pins (2 and 6) attaching the A-frame to the swing-away are in place.

Step 5 Pull the latch lock pins on the front and rear support brackets, and swing the A-frame latch plates out of the way.

Step 6 Assemble the swing-away extension to the boom as described in a previous section, but do not release the A-frame jib from the swing-away extension.

Step 7 Telescope the boom, and elevate it as necessary to prevent damage to the swing-away nose.

Step 8 Reeve the hoist cable over the boom nose center **idler** sheave, down between the boom nose and the base of the swing-away, and connect the becket to the cable-retaining pin (1) on the A-frame nose.

Step 9 Slowly reel in the hoist cable to relieve pressure on the pin (2) securing the A-frame nose to the base of the swing-away.

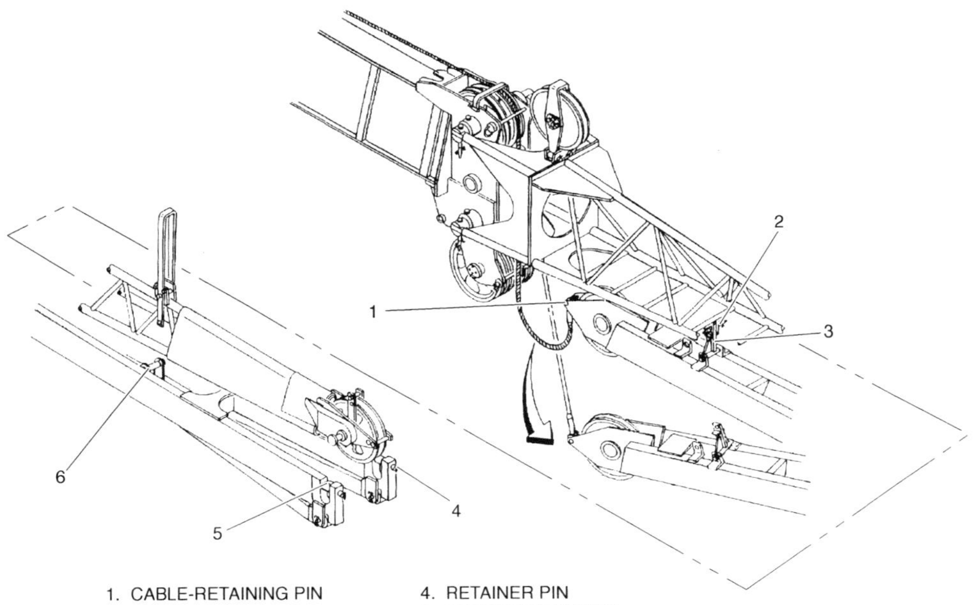

1. CABLE-RETAINING PIN
2. PIN
3. ATTACHING LINK
4. RETAINER PIN
5. RETAINER BUSHING
6. PIN

302F08.EPS

Figure 8 ◈ A-frame jib assembly, part one.

See the Teaching Tip for Section 5.0.0 at the end of this module.

Explain that the boom must be completely retracted before engaging the swing-away boom extension with the boom nose.

Show Transparencies 7 through 9 (Figures 8 through 10). Describe the steps for installing an A-frame jib.

Describe the danger in releasing A-frame latch plates before the pins attaching the A-frame to the swing-away are in place.

Provide a copy of the manufacturer's A-frame assembly and disassembly instructions for the trainees to examine.

TELESCOPIC BOOM ATTACHMENT ASSEMBLY AND DISASSEMBLY 2.7

Step 10 Remove the pin (2) and its retainer clip. Position the attaching link (3) in the stowed position on the A-frame, and secure it with the pin. Install the pin's retainer clip.

Step 11 Remove the retainer pins (4), their clips, and the retainer **bushings** (5) from the attach fittings on the base of the A-frame.

Step 12 Slowly reel out the hoist cable, allowing the A-frame to pivot on the pin (6) in the front attach link until the A-frame attach fittings engage with the shaft on the swing-away nose. Install the retainer bushings (5) and retainer pins (4). Install the pin clips.

Step 13 Lift the A-frame jib nose up slightly, and remove the pin (6) and its retainer clip, securing the base of the A-frame to the swing-away. Stow the pin in the swing-away, and secure the pin with its retainer clip.

Step 14 Slowly reel out the hoist cable, and lower the A-frame nose to the ground. Disconnect the hoist cable becket from the cable retainer pin (1) in the A-frame nose.

Step 15 Remove the hoist cable from between the boom nose and the swing-away.

Step 16 Elevate the boom until the A-frame is perpendicular to the ground.

Step 17 With the A-frame touching the ground, retract and lower the boom so that the A-frame extends out from the end of the swing-away. It will be necessary to reel in the hoist cable while lowering and retracting the boom.

For the following steps, refer to *Figure 9*:

Step 18 Install offset links (8), using pins (7), in the correct fittings on each side of the boom nose applicable to different degrees of offset. Secure pins (7) with their retainer clips.

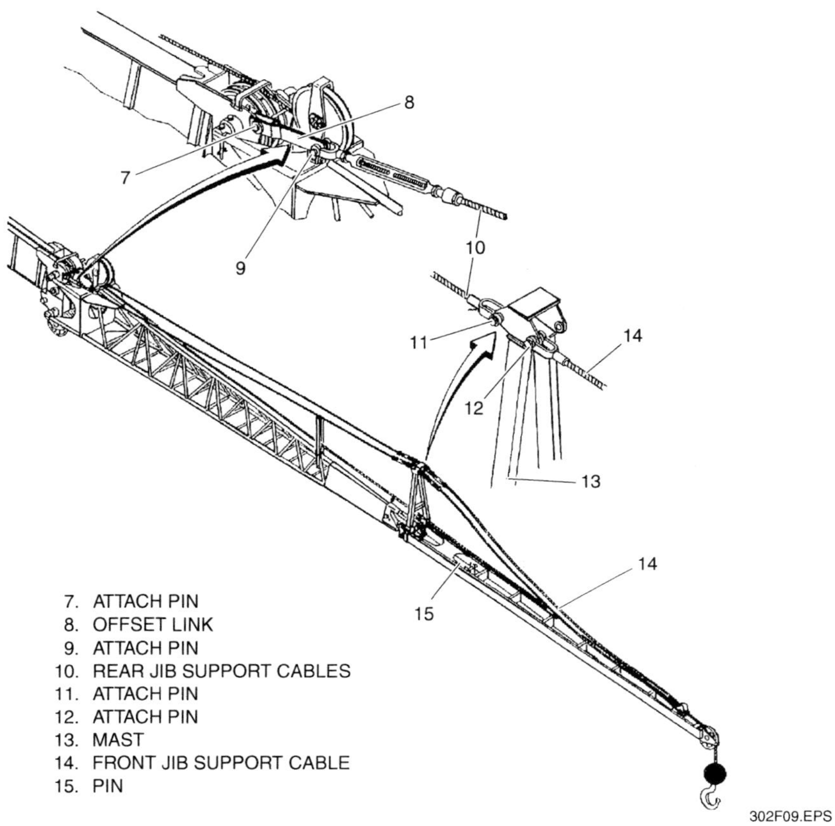

Figure 9 ◈ A-frame jib assembly, part two.

Instructor's Notes:

Step 19 Unhook the front jib support cables (14) from their stowed position.

Step 20 Pull the pin (15) and its retainer clip, latching the **mast** (13) to the A-frame.

Step 21 Reel out the hoist cable, and attach the becket to the **lug** on the mast (13).

Step 22 Reel in the hoist cable, raising the mast. Connect the rear jib support cables (10) to the mast using attach pins (11) and to the offset links (8) using attach pins (9). Secure pins with their retaining clips.

Step 23 Reel out the hoist cable, and disconnect the hoist cable becket from the lug on the mast (13).

Step 24 Set the required jib offset by adjusting the rear jib support cable (10) **turnbuckles**.

For the following steps, refer to *Figure 10*:

Step 25 Reeve the hoist cable over the boom nose center idler sheave, over the extension sheave, and over the A-frame sheave (24).

Step 26 Remove the pin (19) and its retainer clip from the stowed position on the swing-away. Raise the cable roller (20) to an upright position so the hoist cable is under the sheave. Reinstall the pin (19) through the second hole in the guide, and secure the pin with its retainer clip.

Step 27 Remove the pin (22) and its retainer clip, securing the cable guide (23) in the stowed position. Position the cable guide in the vertical position, with the cable passing through the guide. Reinstall the pin through the second hole in the guide, and secure the pin with its retainer clip.

Step 28 Make sure the cable-retaining pin (21) is installed.

Step 29 Rig the hoist cable.

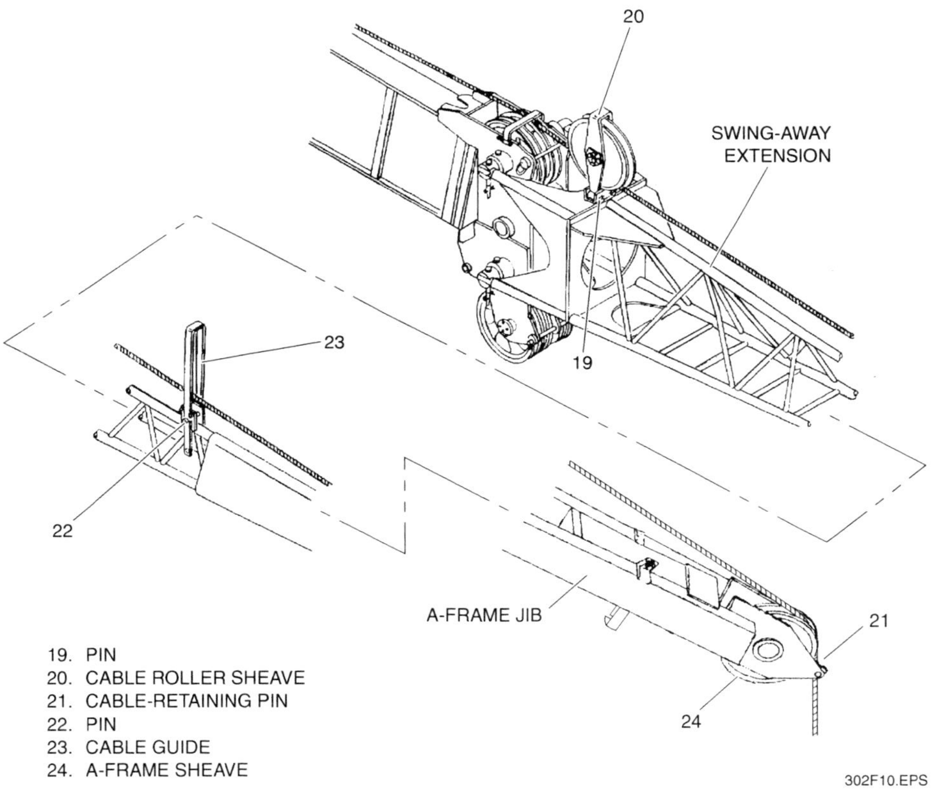

Figure 10 ◈ A-frame jib assembly, part three.

TELESCOPIC BOOM ATTACHMENT ASSEMBLY AND DISASSEMBLY 2.9

Show Transparencies 7 through 9 (Figures 8 through 10) again. Describe the steps for stowing an A-frame jib.

5.2.0 Stowing an A-Frame Jib

Once operations using the jib are completed, it should be stowed. The following are abbreviated steps for stowing an A-frame jib (refer to *Figure 10*), but be sure to follow the manufacturer's procedures for a specific crane:

Step 1 Fully retract the boom. Swing the boom to the appropriate quadrant with the outriggers extended and set and with the crane level.

Step 2 Lower the boom to minimum elevation, and remove the load block or ball.

Step 3 Remove the pin (22) and its retainer clip, securing the cable guide (23). Position the cable guide in the stowed position, and reinsert the pin. Secure the pin with the retainer clip.

Step 4 Remove the pin (19) and its retainer clip, securing the cable roller sheave (20). Lay the cable roller over, and reinsert the pin in the bracket. Secure the pin with its retainer clip.

Step 5 Remove the cable-retaining pin (21) from the A-frame sheave (24). Reel in the hoist cable, removing it from the extension sheave and boom nose idler sheave. Lay the cable off to the side of the boom.

For the following steps, refer to *Figure 9*:

Step 6 Reel out the hoist cable, and connect the hoist cable becket to the lug on the mast (13).

Step 7 Reel in the hoist cable until pressure is relieved from the rear jib support cables (10).

Step 8 Remove the rear jib support cables (10) from the mast (13).

Step 9 Reel out the hoist cable until the mast (13) is laying on the A-frame.

Step 10 Remove the hoist cable becket from the mast lug, and reel in the cable.

Step 11 Secure the mast to the A-frame with the pin (15). Install the pin retainer clip.

Step 12 Stow the front jib support cable (14) in the stowage brackets.

Step 13 Remove the rear jib support cables (10). Also remove the offset links (8) from the fittings on the boom nose if they have been used.

Step 14 Reeve the hoist cable over the boom nose center idler sheave and down between the boom nose and base of the swing-away.

Step 15 Elevate and extend the boom until the A-frame is perpendicular to the ground and the A-frame nose is touching the ground.

For the following steps, refer to *Figure 8*:

Step 16 Reinstall the cable-retaining pin (1) in the A-frame nose. Reel out the hoist cable, and connect the hoist cable becket to the cable retainer pin (1).

Step 17 Reel in the hoist cable, and lower and retract the boom until the A-frame is pulled up under the swing-away.

Step 18 Remove the attach pin (6) and its retainer clip from the base of the A-frame, and attach the base of the A-frame to the nose of the swing-away using the pin. Secure the pin with its retainer clip. Then, unreel the hoist cable slightly.

Step 19 Remove the retainer pins (4), their retainer clips, and the retainer bushings (5) from the fittings on the base of the A-frame.

CAUTION

Make sure the pins and retainer bushings are removed from the base of the A-frame before reeling in the hoist cable. Damage to the A-frame could occur if they are not.

Step 20 Reel in the hoist cable until the attaching link (3) on the A-frame nose can be attached to the swing-away. Remove the pin (2) and its retainer clip, and swing the attaching link (3) up. Install the pin, and secure it with its retainer clip.

Step 21 Reinstall the retainer bushings (5) and retainer pins (4) in the A-frame base fittings. Install the pin retainer clips.

Step 22 Remove the hoist cable becket from the A-frame nose, and reinstall the cable retainer pin (1).

Step 23 Fully retract the boom.

Step 24 Stow the swing-away.

Instructor's Notes:

Step 25 Swing the A-frame latch plates on the front and rear support brackets up, and secure them with the latch lock pins. This secures the A-frame to the brackets.

Step 26 Rig the boom nose and hoist cable as desired, and operate the crane using normal operating procedures.

6.0.0 ◈ INTERMEDIATE BOOM SECTIONS

The boom extension of a hydraulic crane may also include **intermediate** boom sections (see *Figure 11*). This portion of the module will cover intermediate boom sections and the procedures for installing them on one particular model crane.

Before installing the boom extensions, you must establish the rigging mode of the crane. As with the swing-away and A-frame jib, the outriggers must be extended and set, the appropriate amount of counterweight used, and the main boom completely retracted and lowered into the horizontal position.

> **WARNING!**
> The boom extension may only be mounted with the span and counterweight indicated in the lifting capacity table. Otherwise, the assembly personnel risk overturning the crane.
> Assembly and disassembly personnel must follow the manufacturer's instructions and only use authorized rigging with sufficient loadbearing capacity when assembling boom extensions.

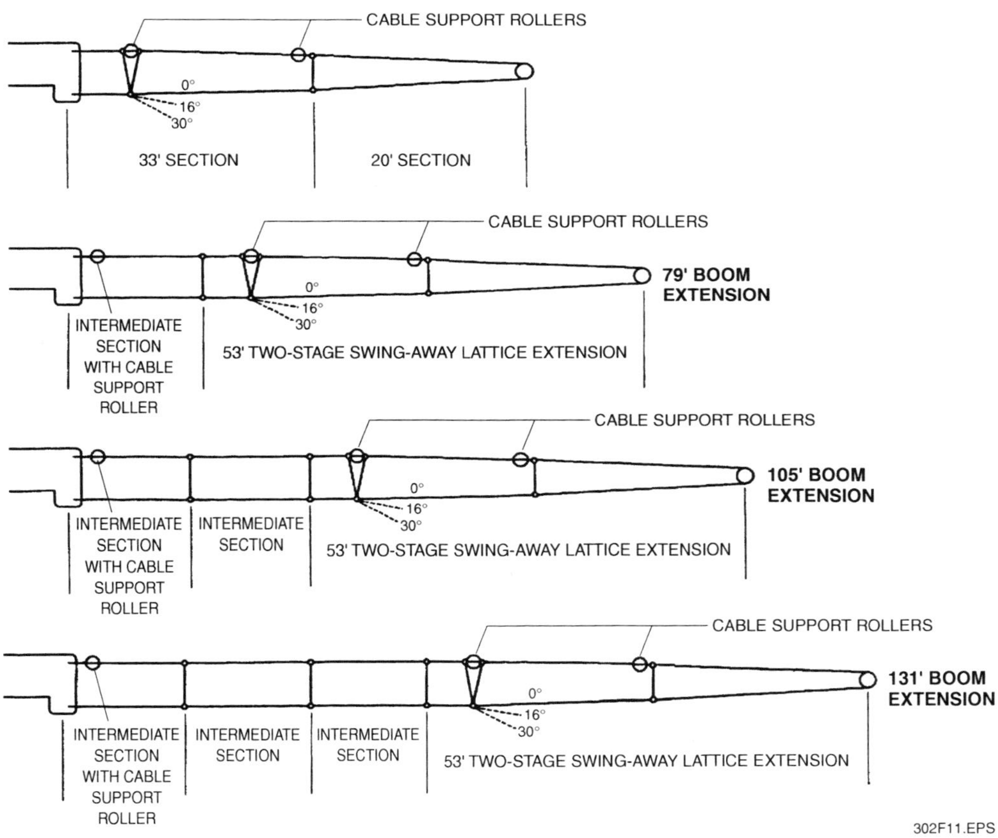

Figure 11 ◈ Intermediate boom extensions.

Demonstrate how to install and stow an A-frame jib.

Have trainees practice installing and stowing an A-frame jib. Note the proficiency of each trainee. This laboratory corresponds to Performance Task 4.

Have trainees review Sections 6.0.0–6.2.0.

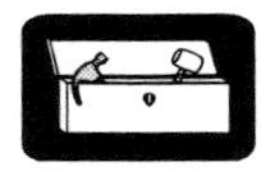

Ensure that you have everything required for teaching this session.

Show Transparency 10 (Figure 11). Describe an intermediate boom section.

Explain how to establish the rigging mode for the crane.

Discuss the importance of using the lifting capacity table to avoid tipping. Reiterate the importance of following the manufacturer's instructions.

TELESCOPIC BOOM ATTACHMENT ASSEMBLY AND DISASSEMBLY 2.11

Provide a copy of the manufacturer's instructions for installing intermediate boom sections for the trainees to examine.

Explain the steps for installing intermediate boom sections.

Explain the steps for removing the intermediate boom sections for offset and non-offset boom extensions.

6.1.0 Installing Intermediate Boom Sections

Use the following steps and an auxiliary crane when attaching intermediate boom sections to the main boom. This abbreviated procedure is for a particular manufacturer's crane. Always consult the manufacturer's guidelines for the crane being used.

Step 1 Before mounting the intermediate boom sections, detach the swing-away and A-frame jib from the main boom, and place them on the ground using an auxiliary crane.

Step 2 Attach the rigging on the intermediate section with the cable support roller. Attachment devices are provided on the intermediate sections for this purpose.

Step 3 Use the auxiliary crane to raise the intermediate section with the cable support roller in front of the boom head until the bores on the intermediate section align with the bearing points on the boom head.

> **CAUTION**
> Always assemble the intermediate section with the cable support roller directly in front of the main boom head. The hoist rope may be damaged on sections of the main boom head if the intermediate section is assembled without the cable support roller.

Step 4 Remove the pins from the holders on the intermediate section, and insert them at the top and bottom in the bearing points between the boom head and the intermediate section. Fasten the pins using retaining clips.

Step 5 Use an auxiliary crane to raise the swing-away in front of the intermediate section until the bores on the foot section of the swing-away align with the bearing points on the intermediate section.

Step 6 Remove the pins from the holders on the swing-away, and insert them at the top and bottom in the bearing points between the intermediate section and the swing-away. Fasten the pins using retaining clips.

Step 7 If an offset is required, use the auxiliary crane to support the end of the assembled extension. Then remove the offset adjustment pins, and lower the end of the extension until the pins can be reinserted for the proper offset. Secure the pins with their retaining clips.

Step 8 Complete all electrical connections.

Step 9 Place the hoist cable through the anti-two-blocking switch weight, and reeve the hook ball or load block.

6.2.0 Removing Intermediate Boom Sections

When operations are completed and the intermediate boom sections are no longer needed, follow the abbreviated procedure in this section to remove the extensions. An auxiliary crane must be used to disassemble and incline the intermediate boom sections.

- Retract all telescopic sections completely with the boom raised.
- Turn the superstructure to the appropriate quadrant, depending on the space available.
- Lower the main boom until the LMI triggers a shutdown.
- Override the LMI, and lower the main boom into the horizontal position.

If the boom extension is offset, proceed as follows:

Step 1 Lower the main boom until the head sheave on the offset boom extension is just resting on a board on the ground.

Step 2 Remove the pins for offset adjustment from the bearing points.

Step 3 Raise the end of the boom extension using the auxiliary crane.

Step 4 Move the boom extension into the 0-degree offset position with the auxiliary crane, and pin the offset adjusting system in this position. Then perform the following steps for a non-offset boom extension.

If the boom extension is not offset, proceed as follows:

Step 1 Remove the anti-two-blocking switch weight from the hoist cable on the boom extension, and unreeve the ball or load-block hook.

Instructor's Notes:

Step 2 Remove the hoist cable from all sheaves on the boom extension.

Step 3 Remove the retaining pin locking the anti-two-block switch from the swing-away.

Step 4 Release the anti-two-block switch lock on the main boom.

Step 5 Disconnect the electrical connection at all connecting points between the distribution boxes at the main boom and the swing-away.

Step 6 Spool the cable onto the appropriate devices.

Step 7 Screw the plugs of the connecting cable onto the dummy sockets of the individual parts of the boom extension.

Step 8 Screw the short-circuit plugs onto the distribution boxes on the main boom and on the swing-away.

Step 9 Screw the dummy plugs onto the dummy socket on the intermediate boom sections.

Step 10 Attach the appropriate lifting gear to the swing-away, and hold it with the auxiliary crane.

Step 11 Remove the pins between the swing-away and the intermediate boom section, and insert the pins in the bracket on the swing-away. Secure the pins with their retaining clips.

Step 12 Place the swing-away on the ground.

Step 13 Attach the lifting tackle to the intermediate boom section, and hold the section with the auxiliary crane.

Step 14 Remove the pins securing the intermediate boom section, and insert the pins into the bracket on the intermediate boom section. Secure the pins with their retaining clips.

Step 15 Place the intermediate boom section on a separate vehicle.

7.0.0 ◈ POWER-PINNED FLY

The power-pinned fly is another type of boom extension. It is used on hydraulic telescoping-boom cranes. The following sections cover steps for extending and retracting a particular manufacturer's power-pinned fly.

7.1.0 Extending a Power-Pinned Fly

This is a procedure for a particular manufacturer's crane. Always follow the manufacturer's instructions for the crane being used. Follow these steps for extending a power-pinned fly:

Step 1 Position the machine to allow the boom to be extended with the boom over the front, side, or rear of the carrier.

Step 2 Extend the outriggers, and level the machine.

Step 3 Fully retract the boom, and make sure that the manual load pin is engaged in the appropriate section (*Figure 12*).

Step 4 Move the manual extension switch to the ON position. The switch is located in the upper cab. This will energize the solenoid on the flow divider, which blocks flow to the lower telescope cylinder and locks the lower section.

Step 5 Push the toe of the boom extend pedal until the upper telescope cylinder is fully extended. The rear clearance hole in section two and the manual extension pin hole in section three should now be aligned with the access holes in the base section and section one. The pins on the end of the manual extension gear reducer must now be visible when viewed through the access holes in the boom sections (*Figure 13*).

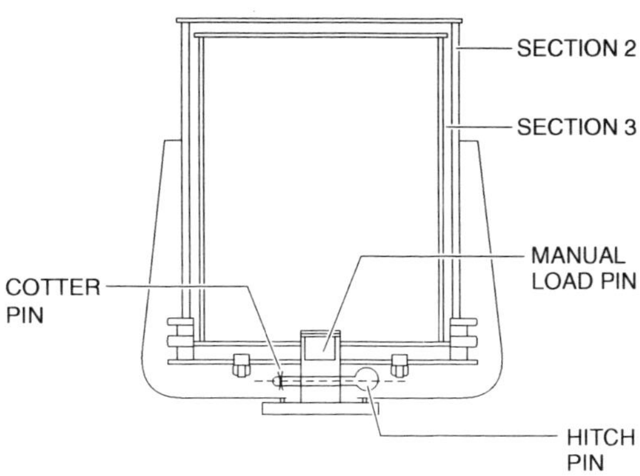

©PPM Cranes, Inc. 302F12.EPS

Figure 12 ◈ Power-pinned fly extension – manual load pin.

Demonstrate how to install and remove intermediate boom sections.

Have trainees practice installing and removing intermediate boom sections. Note the proficiency of each trainee. This laboratory corresponds to Performance Task 4.

Have trainees review Sections 7.0.0–7.2.0.

Ensure that you have everything required for teaching this session.

Describe a power-pinned fly.

Show Transparencies 11 through 14 (Figures 12 through 15). Describe the steps for extending a power-pinned fly.

See the Teaching Tip for Section 7.0.0 at the end of this module.

TELESCOPIC BOOM ATTACHMENT ASSEMBLY AND DISASSEMBLY 2.13

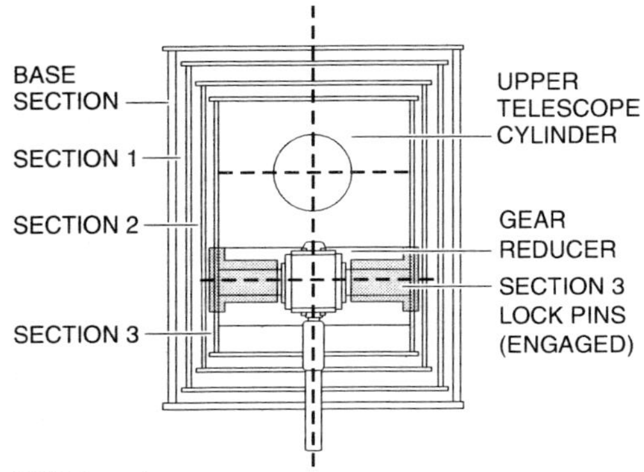

©PPM Cranes, Inc. 302F13.EPS

Figure 13 ◆ Power-pinned fly extension – pin and gear reducer.

Step 6 Insert the hand crank through the opening in the bottom of the boom, and turn the input shaft of the gear reducer clockwise to extend the gear reducer lock pins into section three (*Figure 14*). This will pin section three to the lower telescope cylinder. Visually inspect through the access holes to ensure that the lock pins are fully extended into section three. Remove the crank when the pins are fully extended.

Step 7 Disengage the manual load pin by pushing up on the bottom of the pin and removing the cotter pin and the hitch pin (refer to *Figure 12*).

CAUTION

Do not retract the upper telescope cylinder until the manual load pin is fully disengaged from section three. Boom components could be damaged if the load pin is not fully disengaged.

Step 8 Fully retract the upper telescope cylinder by depressing the heel of the boom telescope pedal. Section two will fully retract, and section three will stay extended since it is pinned to the end of the lower telescope cylinder. When fully retracted, the manual load pin in section two will align with the hole located at the base of section three.

Step 9 Engage the manual load pin by pushing up on the bottom of the pin and inserting the hitch pin and cotter pin (refer to *Figure 12*).

Step 10 Insert the hand crank through the opening in the bottom of the boom, and turn the input shaft of the gear reducer counterclockwise to retract the lock pins from section three (*Figure 15*). This will unpin section three from the lower telescope cylinder. Make a visual inspection through the access holes to ensure that the lock pins are fully retracted from section three. Remove the crank when the pins are fully retracted.

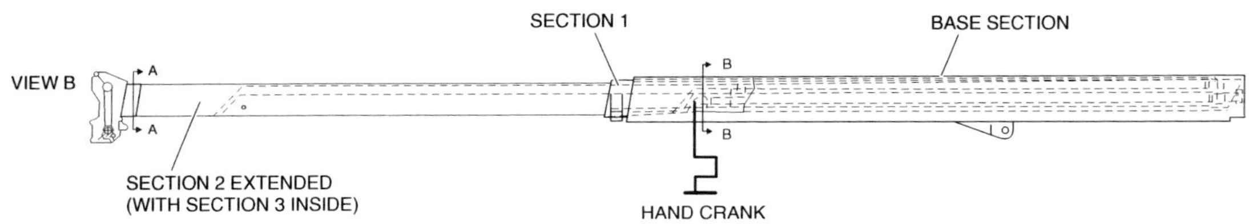

©PPM Cranes, Inc. 302F14.EPS

Figure 14 ◆ Section two extended with section three inside.

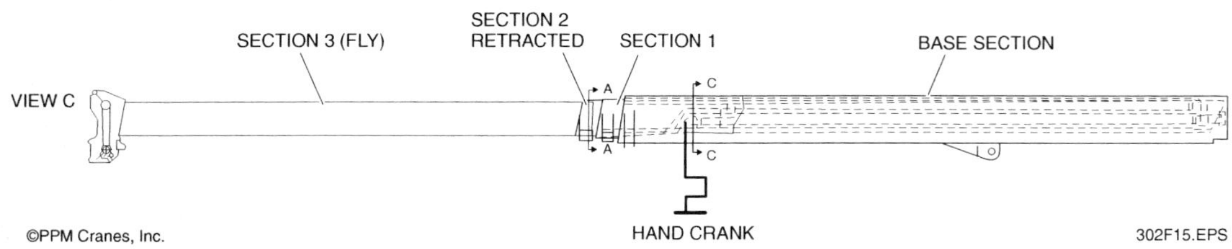

©PPM Cranes, Inc. HAND CRANK 302F15.EPS

Figure 15 ◆ Section three extended with section two retracted.

Instructor's Notes:

Step 11 Move the manual extension switch to the OFF position. This will disengage the solenoid on the flow divider to return equal flow to the telescope cylinders for synchronous operation.

Step 12 The manual section three is now extended and pinned to section two. Powered sections one and two will operate normally.

7.2.0 Retracting a Power-Pinned Fly

Once operations involving the power-pinned fly are completed, it should be retracted and secured. To retract the power-pinned fly, follow these steps:

Step 1 Extend the outriggers, and level the machine.

Step 2 Fully retract the boom to align the pin holes in section three with the lock pins on the gear reducer.

Step 3 Insert the hand crank through the opening in the bottom of the boom, and turn the input shaft of the gear reducer clockwise to extend the gear reducer lock pins into section three (refer to *Figure 15*). This will pin section three to the lower telescope cylinder. Visually inspect through the access holes to ensure that the lock pins are fully extended into section three. Remove the crank when the pins are fully extended.

Step 4 Disengage the manual load pin by pushing up on the bottom of the pin (refer to *Figure 12*). Remove the cotter pin and hitch pin.

Step 5 Move the manual boom extension switch to the ON position. The switch is located in the upper cab. This will energize the solenoid on the flow divider, which blocks flow to the lower telescope cylinder and locks the lower section.

Step 6 Depress the toe of the boom telescope pedal until the upper telescope cylinder is fully extended.

Step 7 Engage the manual load pin by pushing up on the bottom of the pin and inserting the hitch pin and cotter pin (refer to *Figure 12*).

Step 8 Insert the hand crank through the opening in the bottom of the boom, and turn the input shaft of the gear reducer counterclockwise to retract the lock pins from section three (refer to *Figure 14*). This will unpin section three from the lower telescope cylinder. Make a visual inspection through the access holes to ensure that the lock pins are fully retracted from section three. Remove the crank when the pins are fully retracted.

CAUTION

Do not retract the upper telescope cylinder until the lock pins are fully disengaged from section three. Boom components could be damaged if the lock pins are not fully disengaged.

Step 9 Fully retract the upper telescope cylinder by depressing the heel of the boom telescope pedal. The boom is now in the fully retracted position.

Step 10 Move the manual extension switch to the OFF position. This will disengage the solenoid on the flow divider to return equal flow to the telescope cylinders for synchronous operation.

Step 11 The manual section is retracted and pinned to section two. The powered sections operate normally.

8.0.0 ◆ AUXILIARY SINGLE-SHEAVE BOOM HEAD

The auxiliary single-sheave boom head (*Figure 16*) is an additional piece of equipment that can be installed on the boom. In order to attach it to the boom, the crane must be on outriggers or crawlers, the superstructure must be turned to a suitable position, and the main boom must be retracted and set horizontal. To install an auxiliary single-sheave boom head, use the steps that follow.

302F16.EPS

Figure 16 ◆ Auxiliary boom head.

Describe the steps for retracting a power-pinned fly.

Demonstrate how to extend and retract a power-pinned fly.

Have trainees practice extending and retracting a power-pinned fly. Note the proficiency of each trainee. This laboratory corresponds to Performance Task 4.

Have trainees review Sections 8.0.0–9.0.0.

Ensure that you have everything required for teaching this session.

Describe an auxiliary single-sheave boom head and how the crane must be positioned prior to installation.

Provide a copy of the manufacturer's instructions for installing an auxiliary single-sheave boom head for the trainees to examine.

TELESCOPIC BOOM ATTACHMENT ASSEMBLY AND DISASSEMBLY 2.15

Describe the steps for installing an auxiliary single-sheave boom head.

Demonstrate how to install and remove an auxiliary single-sheave boom head.

Have trainees practice installing and removing a single-sheave boom head. Note the proficiency of each trainee. This laboratory corresponds to Performance Task 5.

Show Transparency 15 (Figure 17). Explain how to properly install wire rope.

Demonstrate how to install wire rope on all crane attachments.

Have trainees practice installing wire rope on all crane attachments. Note the proficiency of each trainee. This laboratory corresponds to Performance Task 6.

Summarize the major concepts presented in the module.

Step 1 Swing the boom top in front of the end of the last boom section. Insert a securing pin through the upper and lower bearing points, and secure it with a retaining clip.

Step 2 Screw the short-circuit plug into the dummy socket on the boom head.

Step 3 Remove the rope guide bars from the head of the main boom and the boom head.

Step 4 Lay the hoist rope over the boom head and over the appropriate sheaves.

Step 5 Reinsert the rope guide bars, and secure them with retaining pins.

Step 6 Hang the hook tackle on the auxiliary hoist rope.

Step 7 Suspend a lifting limit switch weight both on the lifting limit switch on the auxiliary single-sheave boom head and on the lifting unit switch on the main boom.

Step 8 Guide the hoist rope through the lifting limit switch weight.

9.0.0 ◆ ROPE INSTALLATION

The primary concern when installing wire rope is getting the rope on the equipment without trapping any twist that may have been induced during handling or installation. The preferred technique for installing rope onto a crane is as follows:

Step 1 Unload the rope properly, and relieve any twists.

Step 2 Attach the end of the rope to the drum. Pull the rope over the point sheave and attach the end to the drum.

Step 3 Spool the rope onto the underside of the drum slowly and carefully (*Figure 17*).

Step 4 Spool the first layer tightly. The first layer forms the foundation, or grooving, for the remaining upper layers. If it is not tight, wraps in upper layers will pull down between wraps already on the drum. This can lead to rope damage and reduced rope strength and service life.

Step 5 Spool multiple layers with sufficient tension. The tension load should range from one to two percent of the rope's nominal strength.

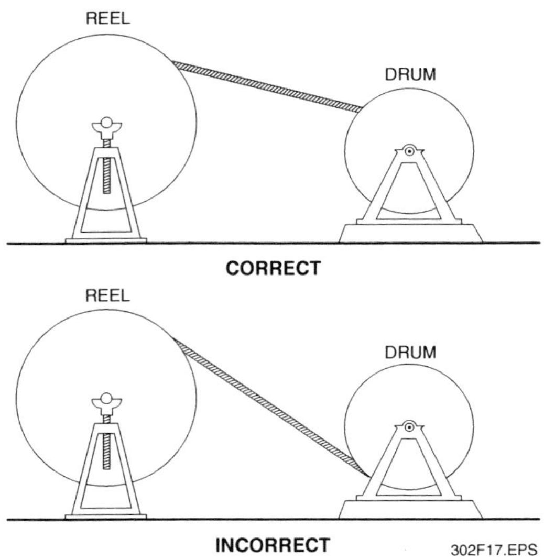

Figure 17 ◆ Storage reel to drum.

Step 6 If the ropes are in a multi-part system, reeve the traveling block and boom tip sheaves so that the rope spacing is maximized. This is done so the traveling block hangs straight and level, and to assure block stability.

Step 7 Check the rope for twists.

Step 8 Cycle the new rope several times, allowing it to form itself to the drum before placing it into service.

Summary

Lifting operations depend on proper assembly of the boom of the crane. On telescopic cranes, this can involve the proper extension of a power-pinned fly or the proper attachment of the swing-away, A-frame jib, and any intermediate boom extensions to the main boom. Assembly and disassembly personnel must be aware of clearance requirements on the job site and follow the appropriate safety requirements associated with this work.

The procedures given in this module are general steps only; always adhere to the manufacturer's instructions when performing attachment assembly.

Instructor's Notes:

Review Questions

1. Work site personnel can find information for installing various boom attachments on a particular crane _____.
   a. on the boom parts
   b. on the crane manufacturer's website
   c. on the load charts
   d. in the operator's safety manual

2. Assembly errors cause _____ of all crane accidents.
   a. half
   b. more than half
   c. less than half

3. Proper counterweights are important in boom assembly because _____.
   a. the crane could overturn during assembly
   b. extra boom attachments, not specific to the crane, can then be attached
   c. they eliminate the need for proper ground support
   d. they allow room for error in assembly

4. When the swing-away extension covered in this module is assembled to the boom, the _____ is used as the first pivot point.
   a. stop rail
   b. attach fitting
   c. front support bracket
   d. A-frame bracket

5. During attachment of the swing-away extension covered in this module, the electrical connections should be established to the main boom immediately after _____.
   a. reeling off the hoist rope from the main hoist
   b. laying the hoist rope over both head sheaves
   c. reeving the hoist rope on the headache ball
   d. attaching the lifting limit switch weight

Have trainees complete the Review Questions, and go over the answers prior to administering the Module Examination.

Administer the Module Examination. Record the results on Craft Training Report Form 200, and submit the results to the Training Program Sponsor.

Administer the Performance Test, and fill out Performance Profile Sheets for each trainee. If desired, trainee proficiency noted during laboratory sessions may be used to complete the Performance Test. Record the results on Craft Training Report Form 200, and submit the results to the Training Program Sponsor.

TELESCOPIC BOOM ATTACHMENT ASSEMBLY AND DISASSEMBLY 2.17

Using *Figure 1*, match the callout numbers on the figure to the correct descriptions.

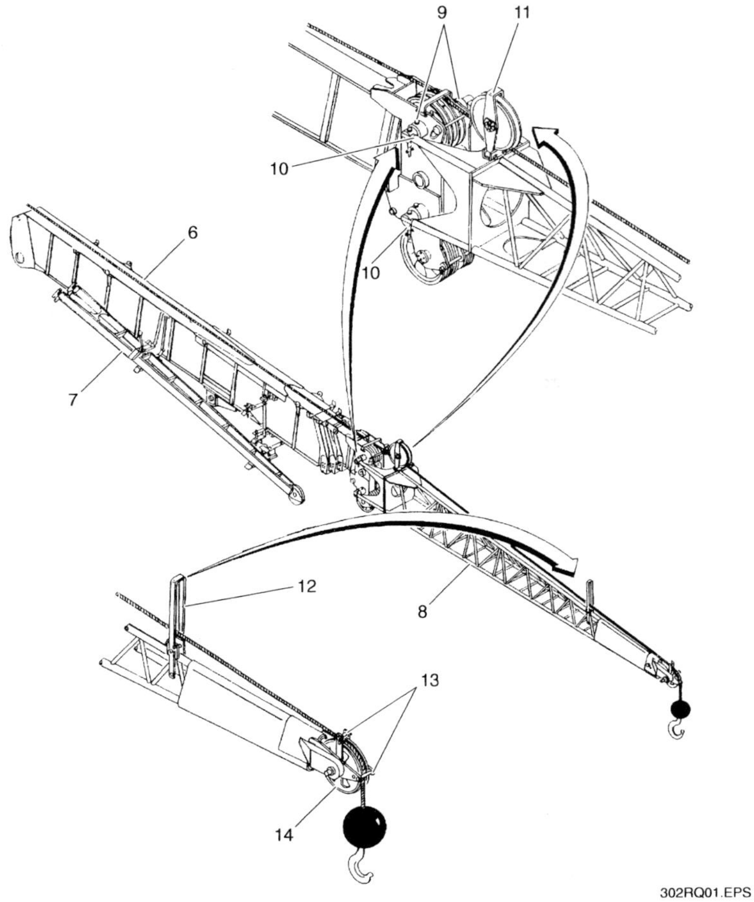

Figure 1

6. _____
7. _____
8. _____
9. _____
10. _____
11. _____
12. _____
13. _____
14. _____

a. Retainer bushing and pin
b. Swing-away extension
c. Cable-retaining pins
d. Attach fitting
e. Cable-roller sheave
f. A-frame jib
g. Telescoping boom
h. Anchor fitting
i. Attaching link
j. Cable guide
k. Extension sheave

Instructor's Notes:

15. The first step in stowing the swing-away extension is to _____.

a. lower the main boom until the LMI switches off
b. override the LMI
c. fully retract the telescopic sections of the boom
d. turn the superstructure to the back or to the side

16. The A-frame jib covered in this module is assembled to the nose of the swing-away extension using _____.

a. offset links
b. turnbuckles
c. retainer bushings and pins
d. an adjustable jib mast

17. The final offset for the A-frame jib covered in this module is made using _____.

a. offset links
b. turnbuckles
c. retainer bushings and pins
d. an adjustable jib mast

18. A replaceable steel tube mounted in a case or housing as a bearing is called a(n) _____.

a. becket
b. bushing
c. idler
d. lug

19. A(n) _____ sheave is used to equalize tension in opposite parts of a rope.

a. idler
b. turnbuckle
c. pin
d. lug

20. When stowing the A-frame jib covered in this module, tension on the rear jib support cables is relieved by _____.

a. reeling in a hoist cable attached to the jib mast
b. releasing the forward jib support cables
c. loosening the turnbuckles
d. using supports under the A-frame jib

21. Assembly and disassembly personnel must only use ______ with sufficient load-bearing capacity when assembling boom extensions.

a. protective equipment
b. authorized rigging
c. non-stick lubrication
d. operator aids

22. For the intermediate boom extensions covered in this manual, any intermediate boom extension can be assembled to the boom head.

a. True
b. False

23. When removing intermediate boom sections, if the boom extension is offset you must _____.

a. lower the main boom until the head sheave on the boom extension is touching the board on the ground
b. remove the hoist rope from all sheaves
c. release the lifting limit switch lock on the main boom
d. spool the cable

24. In the power-pinned fly covered in this module, the section of the boom that is powered to extend or retract the fly is the _____.

a. first section
b. second section
c. third section
d. base section

25. When attaching an auxiliary single-sheave boom head to the main boom, the _____ need to be removed before laying the hoist rope for the auxiliary single-sheave boom head.

a. mast attachments
b. sheave retaining pins
c. sheave attachment pins
d. rope guide bars

TELESCOPIC BOOM ATTACHMENT ASSEMBLY AND DISASSEMBLY 2.19

Instructor's Notes:

GLOSSARY

Trade Terms Introduced in This Module

Auxiliary: As used in the text, a secondary hoist rope system used either in conjunction with, or independently of, the main hoist system.

Becket: A wedge socket that is used as an end preparation for wire rope.

Bushing: A replaceable hardened steel tube mounted in a case or housing as a bearing.

Idler: As used in the text, a sheave used to equalize tension in opposite parts of a rope.

Intermediate: Being situated between two points.

Lug: A projecting piece by which something is held or supported.

Mast: A frame hinged at or near the boom point for use in supporting a jib.

Pin: A piece of metal used to fasten, support, or attach components or members.

Turnbuckle: A link or sleeve used as a means of coupling and tightening two parts.

TELESCOPIC BOOM ATTACHMENT ASSEMBLY AND DISASSEMBLY 2.21

Instructor's Notes:

Figure Credits

Link-Belt Construction Equipment Company	302F01, 302F02
Topaz Publications, Inc.	302F03
Grove Crane	302F04–302F11, 302RQ01
PPM Terex	302F12–302F15
Tony Guidry	302F16
Wire Rope Technical Board	302F17

TELESCOPIC BOOM ATTACHMENT ASSEMBLY AND DISASSEMBLY 2.23

MODULE 21302-05 — TEACHING TIPS

The following are suggested activities or instructional methods to help you teach the material in this module.

General

When you call on someone to answer a question, the rest of the class relaxes or even tunes out because they expect that the question and answer will take place only between you and the trainee you called on. Instead, use this technique to involve more trainees in answering questions and to keep them on their toes.

1. Ask trainees to define a term or explain a concept.
2. After one trainee has answered, ask a trainee seated nearby if the answer is right. Then ask whether a trainee in the back of the room agrees.
3. Ask trainees to explain why they think an answer is right or wrong.
4. Use the session to clear up incorrect ideas and encourage trainees to learn from their mistakes.

Section 4.0.0 *Swing-Away Lattice Extensions*

Swing-away lattice extensions from various models or manufacturers are different. Trainees will need appropriate personal protective equipment, pencils, and paper. Arrange for an opportunity to observe several different types of swing-away lattice extensions. Alternatively, have a manufacturer's representative give a presentation on the various types of swing away-lattice extensions for their cranes. Allow 1 to 2 hours for this exercise.

1. Before the tour discuss at least one method for attaching a swing-away extension with trainees.
2. Ask the site safety technician to explain how to attach the swing-away extension. Have trainees take notes and write down questions during the tour.
3. Have the trainees identify differences in how the various extensions are attached.
4. After the tour answer any questions the trainees may have.

Section 5.0.0 *A-Frame Jibs*

This Quick Quiz will familiarize trainees with the terms and definitions that are commonly used in working with A-frame jibs. You will need photocopies of the quiz provided on the following page. Trainees will need pencils. If you allow trainees to use the Trainee Module, decrease the amount of time you give them to complete the quiz.

1. Make a photocopy of the quiz for each trainee.
2. Give trainees between 5 and 10 minutes to complete the quiz.
3. Go over the answers to the quiz.
4. Ask trainees if they have questions.

Answers to Quick Quiz

1. d
2. c
3. e
4. f
5. a
6. b

Quick Quiz *A-Frame Jibs*

For each description listed, identify the term that the text best describes. Write the corresponding letter in the blank provided.

_____ **1.** A sheave used to equalize tension in opposite parts of a rope

_____ **2.** A replaceable hardened steel tube mounted on a case or housing as a bearing

_____ **3.** A frame hinged at or near the boom point for use in supporting a jib

_____ **4.** A projecting piece by which something is held or supported

_____ **5.** A link or sleeve used as a means of coupling and tightening two parts

_____ **6.** A wedge socket that is used as an end preparation for wire rope

a. Turnbuckle
b. Becket
c. Bushing
d. Idler
e. Mast
f. Lug

Copyright © 2005 National Center for Construction Education and Research. Permission is granted to reproduce this page provided that copies are for local use only and that each copy contains this notice.

Section 7.0.0 *Power-Pinned Fly*

Different models of power-pinned flies or flies from different crane manufacturers will vary. Trainees will need appropriate personal protective equipment, pencils, and paper. Arrange for an opportunity to observe several different types of power-pinned flies. Alternatively, have a manufacturer's representative give a presentation on the various types of power-pinned flies for their cranes. Allow 1 to 2 hours for this exercise.

1. Before the tour discuss at least one method for attaching a power-pinned fly with trainees.
2. Ask the site safety technician to explain how to attach the power-pinned fly. Have trainees take notes and write down questions during the tour.
3. Have the trainees identify differences in how the various flies are attached.
4. After the tour answer any questions the trainees may have.

MODULE 21302-05 — ANSWERS TO REVIEW QUESTIONS

Answer	Section
1. d	1.0.0
2. b	1.0.0
3. a	3.0.0
4. c	4.1.0
5. c	4.1.0
6. g	4.1.0
7. f	4.1.0
8. b	4.1.0
9. d	4.1.0
10. h	4.1.0
11. e	4.1.0
12. j	4.1.0
13. c	4.1.0
14. k	4.1.0
15. c	4.2.0
16. c	5.1.0
17. b	5.1.0
18. b	5.1.0; Glossary
19. a	5.1.0; Glossary
20. a	5.2.0
21. b	6.0.0
22. b	6.1.0
23. a	6.2.0
24. b	7.1.0; 7.2.0
25. d	8.0.0

CONTREN® LEARNING SERIES — USER UPDATE

The NCCER makes every effort to keep these textbooks up-to-date and free of technical errors. We appreciate your help in this process. If you have an idea for improving this textbook, or if you find an error, a typographical mistake, or an inaccuracy in NCCER's Contren® textbooks, please write us, using this form or a photocopy. Be sure to include the exact module number, page number, a detailed description, and the correction, if applicable. Your input will be brought to the attention of the Technical Review Committee. Thank you for your assistance.

Instructors – If you found that additional materials were necessary in order to teach this module effectively, please let us know so that we may include them in the Equipment/Materials list in the Annotated Instructor's Guide.

Write: Product Development and Revision
National Center for Construction Education and Research
P.O. Box 141104, Gainesville, FL 32614-1104

Fax: 352-334-0932

E-mail: curriculum@nccer.org

Craft ______________________ Module Name ______________________

Copyright Date __________ Module Number __________ Page Number(s) __________

Description

____

____

____

(Optional) Correction

____

____

(Optional) Your Name and Address

____

____

Module 21303-05

Advanced Operational Techniques

NCCER STANDARDIZED CRAFT TRAINING PROGRAM

The National Center for Construction Education and Research (NCCER) provides a standardized national program of accredited craft training. Key features of the program include instructor certification, competency-based training, and performance testing. The program provides trainees, instructors, and companies with a standard form of recognition through a National Craft Training Registry. The program is described in full in the *Guidelines for Accreditation*, published by the NCCER. For more information on standardized craft training, contact the NCCER by writing us at P.O. Box 141104, Gainesville, FL 32614-1104; calling 352-334-0911; or e-mailing info@nccer.org. More information may be found at our Web site, www.nccer.org.

HOW TO USE THIS ANNOTATED INSTRUCTOR'S GUIDE

Each page presents two sections of information. The larger section displays each page exactly as it appears in the Trainee Module. The narrow column ties suggested trainee and instructor actions to each page and provides icons (detailed below) to call your attention to material, safety, audiovisual, or testing requirements. The bottom of each page includes space for your notes.

The **Audiovisual** icon indicates an appropriate time to show a transparency or other audiovisual aid.

The **Classroom** icon prompts you to define a term, stress a point, ask trainees to explain a concept, or give examples.

The **Demonstration** icon directs you to show trainees how to perform tasks.

The **Examination** icon tells you to administer the written module examination.

The **Homework** icon is placed where you may wish to assign reading for the next class, to assign a project, or to advise trainees to prepare for an examination.

The **Laboratory** icon is used when trainees are to practice performing tasks.

The **Materials** icon is a reminder for you to gather materials needed for classes, labs, and testing.

The **Performance Testing** icon tells you to administer a performance test or a portion thereof.

The **Safety** icon is used to emphasize safety issues. It is often keyed to *Caution* and *Warning* statements in the Trainee Module.

The **Teaching Tip** icon indicates additional guidance is available, such as how to conduct an exercise, get the most educational value from a field trip, or encourage class participation. Teaching Tips may expand on a feature (*Think About It, Did You Know?*) or provide Quick Quizzes or similar exercises. You will be referred to the Teaching Tips section at the back of the module if there is additional material.

The **Combination** icon indicates that the laboratory listed corresponds with a performance task. If desired, you can note the proficiency of the trainees during the laboratory and use it to satisfy performance testing requirements.

PREPARATION

Before teaching this module, you should review the Objectives, Performance Tasks, Materials and Equipment List, and the Module Outline. Be sure to allow ample time to prepare your own training or lesson plan and gather all required materials and equipment.

MODULE OVERVIEW

This module addresses operational techniques beyond simple lifts. These include more advanced lifting operations, such as multiple-crane lifts, critical lifts, blind lifts, and demolition. This module also includes sections on how to use magnet and vacuum lifting devices, and how to operate a mobile crane in extreme weather conditions.

PREREQUISITES

Prior to training with this module, it is recommended that the trainee shall have successfully completed *Core Curriculum*; *Mobile Crane Operations Level One*; *Mobile Crane Operations Level Two*; and *Mobile Crane Operations Level Three*, Module 21301-05 and 21302-05.

OBJECTIVES

Upon completion of this module, the trainee will be able to do the following:

1. Recognize the crane, communication, and personnel requirements for multiple-crane lifts, and identify requirements for personnel lifts.
2. Perform a blind lift under the direction of one or more signal people.
3. Identify the crane requirements and environmental guidelines for critical lifts.
4. Identify safety practices when operating in the path of transmitter energy.
5. Define the regulations for demolition procedures.
6. Describe the guidelines for safely operating a vacuum lifting device for lifting purposes.
7. Define the requirements for operating a magnet for lifting purposes.
8. Define the guidelines for operating a crane in extreme weather conditions.

PERFORMANCE TASKS

Under the supervision of the instructor, the trainee should be able to do the following:

1. Perform a blind lift.
2. Identify the components of a crane that should be observed closely before cold weather operation.

MATERIALS AND EQUIPMENT LIST

Overhead projector and screen

Transparencies

Blank acetate sheets

Transparency pens

Whiteboard/chalkboard

Markers/chalk

Pencils and scratch paper

OSHA 1926.550(g)

Copies of a critical lift plan

Copies of site safety procedures

Mobile crane with rigging and lifting equipment

Crane manufacturer's operator's manual

Signal flags or communications equipment

ANSI A10.6-1983 Safety Requirements for Demolition Operations

Demolition videocassette or DVD

TV and VCR/DVD

Copies of the Quick Quiz*

Module Examinations**

Performance Profile Sheets**

*Located in the back of this module.

**Located in the Test Booklet.

SAFETY CONSIDERATIONS

Ensure that the trainees are equipped with appropriate personal protective equipment and know how to use it properly. This module requires trainees to operate a crane. Review site safety procedures and site evacuation procedures. Ensure that all trainees are familiar with hand signals and other site communication procedures. Brief trainees on pinching and crushing hazards attendant with crane operations. This module may require trainees to visit construction sites. Brief trainees on site safety procedures.

TEACHING TIME FOR THIS MODULE

An outline for use in developing your lesson plan is presented below. Note that each Roman numeral in the outline equates to one session of instruction. Each session has a suggested time period of 2½ hours. This includes 10 minutes at the beginning of each session for administrative tasks and one 10-minute break during the session. Approximately 20 hours are suggested to cover *Advanced Operational Techniques*. You will need to adjust the time required for hands-on activity and testing based on your class size and resources. Because laboratories often correspond to Performance Tasks, the proficiency of the trainees may be noted during these exercises for Performance Testing purposes.

Topic	Planned Time
Session I. Introduction, Personnel Lifts, and Multiple-Crane Lifts	
A. Introduction	____________
B. Personnel Lifts	____________
C. Multiple-Crane Lifts	____________
Sessions II and III. Blind Lifts	
A. Blind Lifts	____________
B. Laboratory – Trainees practice making blind lifts. This laboratory corresponds to Performance Task 1.	____________
Session VI. Critical Lifts and Working Near Transmitters	
A. Critical Lifts	____________
B. Working Near Transmitters	____________
Sessions V and VI. Demolition, Vacuum Lifting, and Electromagnet Lifting	
A. Demolition	____________
B. Vacuum Lifting	____________
C. Electromagnet Lifting	____________
Sessions VII. Cold Weather Operation	
A. Cold Weather Operation	____________
B. Laboratory – Trainees practice identifying the components of a crane that should be observed closely before cold weather operation. This laboratory corresponds to Performance Task 2.	____________
Session VIII. Review, Module Examination, and Performance Testing	
A. Review	____________
B. Module Examination	____________
1. Trainees must score 70 percent or higher to receive recognition from NCCER.	
2. Record the testing results on Craft Training Report Form 200, and submit the results to the Training Program Sponsor.	
C. Performance Testing	____________
1. Trainees must perform each task to the satisfaction of the instructor to receive recognition from NCCER. If applicable, proficiency noted during laboratory exercises can be used to satisfy the Performance Testing requirements.	
2. Record the testing results on Craft Training Report Form 200, and submit the results to the Training Program Sponsor.	

Module 21303-05

Advanced Operational Techniques

Instructor's Notes:

COURSE MAP

This course map shows all of the modules in the third level of the *Mobile Crane Operations* curriculum. The suggested training order begins at the bottom and proceeds up. Skill levels increase as you advance on the course map. The local Training Program Sponsor may adjust the training order.

Assign reading of Module 21303-05.

MOBILE CRANE OPERATIONS LEVEL THREE

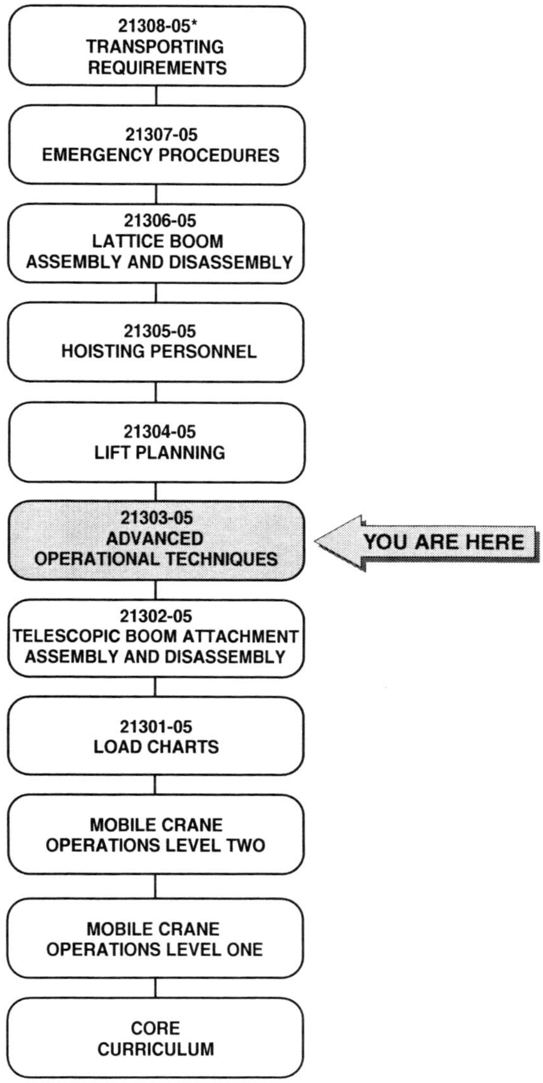

*Please note that 21308-05 is an elective module. It is not necessary to complete this module in order to advance in your training.

303CMAP.EPS

Copyright © 2005 National Center for Construction Education and Research, Gainesville, FL 32614-1104. All rights reserved. No part of this work may be reproduced in any form or by any means, including photocopying, without written permission of the publisher.

ADVANCED OPERATIONAL TECHNIQUES 3.iii

Instructor's Notes:

MODULE 21303-05 CONTENTS

ADVANCED OPERATIONAL TECHNIQUES 3.v

Figures

Instructor's Notes:

MODULE 21303-05

Advanced Operational Techniques

Objectives

When you have completed this module, you will be able to do the following:

1. Recognize the crane, communication, and personnel requirements for multiple-crane lifts, and identify the requirements for personnel lifts.
2. Perform a blind lift under the direction of one or more signal people.
3. Identify the crane requirements and environmental guidelines for critical lifts.
4. Identify safety practices when operating in the path of transmitter energy.
5. Define the regulations for demolition procedures.
6. Describe the guidelines for safely operating a vacuum lifting device for lifting purposes.
7. Define the requirements for operating a magnet for lifting purposes.
8. Define the guidelines for operating a crane in extreme weather conditions.

Prerequisites

Before you begin this module, it is recommended that you successfully complete *Core Curriculum; Mobile Crane Operations Level One; Mobile Crane Operations Level Two*; and *Mobile Crane Operations Level Three*, Modules 21301-05 and 21302-05.

Required Trainee Materials

1. Pencil and paper
2. Appropriate personal protective equipment

1.0.0 ◆ INTRODUCTION

This module addresses operational techniques beyond simple lifts. These include more advanced lifting operations, such as multiple-crane lifts, and the use of different lifting devices, such as personnel lifts. Demolition is discussed, as well as cold weather operation.

2.0.0 ◆ PERSONNEL LIFTS

According to *OSHA 1926.550(g)*, the use of a crane or derrick to hoist employees on a suspended personnel platform is prohibited. However, exception is granted when other means would be more hazardous or not possible. Personnel lifts are always considered **critical lifts**.

Personnel must only be carried in properly designed platforms or cradles (*Figure 1*), sometimes called manbaskets or personnel lifts. The lifts must be provided with suitable means to prevent persons or tools from falling out. Take precautions to prevent the lifts from spinning or tipping.

The lift should be clearly and permanently marked with the maximum load that it can safely carry. The lift should be inspected before use to ensure that it is in a safe condition to carry personnel.

No other crane function can be performed while a person is on a man-lift platform or other personnel-lifting device attached to the crane's load line or auxiliary line.

A more in-depth discussion of suspended personnel platforms and hoisting personnel, including OSHA and ASME standards, will be presented later in the *Hoisting Personnel* module.

Ensure that you have everything required to teach the course. Check the Materials and Equipment list at the front of this module.

Explain that terms shown in bold are defined in the Glossary at the back of this module.

See the general Teaching Tip at the end of this module.

Show Transparency 1, Objectives, and Transparency 2, Performance Tasks. Review the goals of the module, and explain what will be expected of trainees.

Discuss personnel lifts. Explain the importance of using properly designed manbaskets or personnel lifts. Have a trainee define critical lifts.

Provide a copy of *OSHA 1926.550(g)* for the trainees to examine.

Explain that multiple-crane lifts are hazardous and must be carefully planned and closely coordinated. Emphasize that all personnel must be familiar with the lift plan and emergency procedures.

Explain the importance of planning multiple-crane lifts. Discuss important factors to be considered when planning and executing multiple-crane lifts.

See the Teaching Tip for Section 3.0.0 at the end of this module.

Have trainees review Section 4.0.0.

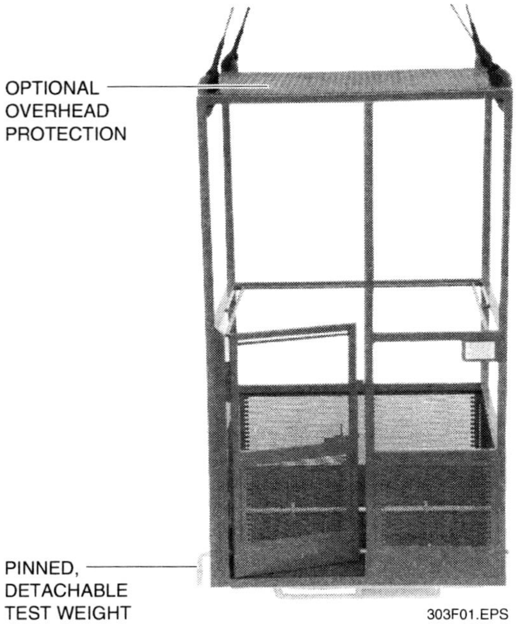

Figure 1 ◆ Personnel lift with overhead protection.

3.0.0 ◆ MULTIPLE-CRANE LIFTS

Multiple-crane lifts are always considered critical lifts, requiring detailed plans and procedures. The effects of the relative motion between cranes may induce additional weight on the cranes, the load, and the lifting gear. Multiple-crane lifting must be used when the physical dimensions, characteristics, mass, or required movement of the load prevent the operation from being carried out by a single crane. Procedures must be in the written form of a lift plan and signed by qualified personnel. These factors should be considered when planning multiple-crane lifts:

- The lift must be planned and carried out under qualified supervision.
- Ground conditions must be stable and level.
- Consideration must be given to the environment and the location of the lift.
- All cranes must rest on a solid, firm, level surface.
- All cranes must be level.
- The load weight must be known.
- The longest load radius for each crane of the operation must be measured.
- The maximum allowable boom length operating radius and minimum boom angle must be known.
- The rated capacity of each crane must be known.
- No crane should be loaded to more than 75 percent of its net capacity.
- Exactly how much of the load to be carried by each crane must be known. The rigging must be arranged to divide the load as planned.
- The line, swing, and boom speeds of the cranes must be matched.
- Swing and booming methods should be kept to a minimum.
- Whenever possible, the cranes should not travel with the load.
- A lift plan must be developed.
- Signal persons, riggers, and operators must know exactly what to do before the lift begins.
- If possible, a dry run should be conducted without a load.
- Communications by radio are preferred when hand signals are not effective.
- One person is required to direct and control the operation.
- Hoist lines must be kept vertical at all times. When the lines are not vertical, the cranes are transferring load to each other, and one crane may sideload the other.

Figure 2 is an example of a non-typical, extremely critical lift that required large cranes and extensive planning to execute.

> **WARNING!**
> Multiple-crane lifts are particularly hazardous. Because the movements of multiple cranes must be closely coordinated, all crane movements must be carefully planned and slowly executed to prevent lift failures. All personnel involved must be thoroughly familiar with the lift plan and any applicable emergency procedures.

Instructor's Notes:

303F02.EPS

Figure 2 ◆ Critical, large-scale multiple-crane lift.

4.0.0 ◆ BLIND LIFTS

A **blind lift** occurs when a crane operator picks up, lifts, or sets a load down at a destination where the operator's view is obstructed (see *Figure* 3). In some situations, a relay person may be required to communicate information from the signal person to the operator. In these cases, the communication between the signal person or relay person and the crane operator should be made through telephone, hard-line devices, or two-way radios.

Ensure that you have everything required for teaching this session.

Describe a blind lift. Discuss the importance of communications and signal people.

Review site safety and emergency procedures with trainees.

Demonstrate how to perform a blind lift.

Have trainees practice performing a blind lift. Note the proficiency of each trainee. This laboratory corresponds to Performance Task 1.

Have trainees review Sections 5.0.0 – 6.0.0.

ADVANCED OPERATIONAL TECHNIQUES 3.3

Ensure that you have everything required for teaching this session.

Describe a critical lift. Discuss necessary precautions for safely performing a critical lift. Explain how to measure the load radius to determine capacity.

Provide a copy of a critical lift plan to the trainees for their examination.

Show Transparency 3 (Figure 4). Discuss the dangers of working near transmitters.

Explain the precautions necessary to work safely around transmitters.

See the Teaching Tip for Section 6.0.0 at the end of this module.

Discuss the potential for fire, explosion, and burn hazards from RF voltages.

Have trainees review Sections 7.0.0–9.4.0.

303F03.EPS

Figure 3 ◆ Blind lift in progress.

5.0.0 ◆ CRITICAL LIFTS

Critical lifts are those where the load requires exceptional care in handling because of the size, weight, close-tolerance installation, high susceptibility to damage, or other unusual factors, including multiple-crane lifts. When lifting loads in these conditions, the following precautions are necessary:

- *Supporting surface* – The ground must be firm and level.
- *Blocking* – Unless the crane sits on a sufficient concrete foundation, outrigger blocking must be used. Crawlers should be on pads.
- *Level* – The crane must be level within 1 percent.
- *Load* – The exact weight must be determined.
- *Center of gravity* – The location of the load's center of gravity must be known. The crane hook must be positioned above it.
- *Load radius, boom length, and boom angle* – These items must be measured exactly.
- *Wind* – Wind effects must be considered and the lift delayed if the wind loads are significant. The operator must consult the manufacturer's recommendations when lifting in windy conditions.
- *Reeving* – The reeving must be balanced.
- *Load rigging* – The exact weight of the rigging must be known.
- *Operation* – All movements shall be made as slowly and smoothly as possible.
- *Lift plan* – Plan the lift to utilize as few movements as possible.

Although the actual load weight may be small compared to the maximum rated capacity of the crane, it may still be considered a critical lift. A 1-ton load on a 50-ton capacity crane may seem insignificant, but if the crane's rated capacity at the load radius is only 2,400 pounds, the lift becomes critical.

In Level One, boom angle indicators were referred to as a necessary operational aid for cranes having booms capable of moving in the vertical plane. On lifts from the main boom, the angle given on the load chart is the loaded boom angle, which indicates what the boom angle should be with the crane loaded to the weight given at any listed load radius on the chart for the applicable boom length. It is for reference only. On lifts from the main boom, the load radius determines the required capacity. It must be measured exactly from the center of crane rotation to the center of gravity of the load. When using load charts, boom length, and load radius to determine crane capacity, always use the next longest load radius listing in the chart if the measured load radius is between two load radius listings for a given boom length. On lifts from a boom extension, such as a jib or fly, it is the angle of the main boom or extension that is usually used in the load charts to determine capacity. An extension loses capacity as its angle to the horizontal decreases.

6.0.0 ◆ WORKING NEAR TRANSMITTERS

Prior to crane operations near **transmitter** towers (*Figure 4*), where a radio frequency (RF) electrical voltage can be conducted in the equipment or materials being handled, the transmitter must be de-energized, or tests must be made to determine if an electrical voltage is induced on the crane. The following precautions must be taken when necessary to dissipate induced voltages:

- The equipment must be provided with an electrical ground attached directly to the upper rotating structure supporting the boom.
- Ground jumper cables must be attached to materials being handled by boom equipment when an electrical charge is induced while working near energized transmitters. Crews must be provided with non-conductive poles having large alligator clips or other similar protection to attach the ground cable to the load.
- Combustible and flammable materials must be removed from the immediate area prior to operation.

Instructor's Notes:

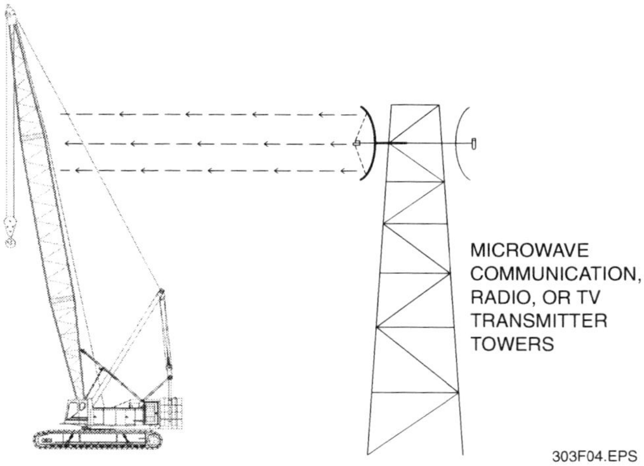

Figure 4 ◆ Working near transmitters.

> **WARNING!**
> While induced radio frequency (RF) voltages from transmitters are not an electrocution hazard like power line voltages, they can result in sparks to any object or person. The sparks can cause fires or explosions of combustible or flammable materials. They can also result in painful burns and shocks to personnel that come in close contact to the crane or the load being lifted by the crane.

7.0.0 ◆ DEMOLITION

In **demolition** ball operations, a round or pear-shaped weight, known as a wrecking ball (*Figure 5*), is suspended from the hoist rope of a crane and used to strike a building, structure, or other object so that the impact causes collapse or breakage. A square-shaped wrecking block can also be used.

During wrecking ball operations, **dynamic loading** is imposed on the boom structure and other parts of the crane by the movement and impact of the wrecking ball. The magnitude of dynamic loading varies widely depending on wrecking ball usage, the skill of the operator, and the impact resistance of the object being broken. Because of these factors, the manufacturer's recommendations and working loads should be regarded only as a guide.

Usually, you must use several heavy impacts to cause multiple cracking. Follow those with a number of lighter impacts until part of the object is broken. Repeat the process until the job is done.

Figure 5 ◆ Pear-shaped wrecking ball.

> **WARNING!**
> Position any crane used for high-structure demolition as far as practical from the structure during demolition in case of unexpected structure collapse. In some cases, a portable concrete barrier between the structure and crane may be advisable to help prevent falling rubble from contacting the crane. Ensure that the area around the demolition site and crane is roped off to prevent injury or death to personnel due to flying debris or unexpected structure collapse. Make sure proper environmental protection procedures, such as removal of toxic materials and/or water suppression of dust, are implemented before demolition activities.

Ensure that you have everything required for teaching this session.

See the Teaching Tip for Section 7.0.0 at the end of this module.

Describe demolition ball operations. Discuss dynamic loading. Explain that multiple impacts may be needed to break an object.

Discuss the importance of protecting the crane and operator from falling rubble. Have trainees explain the importance of proper environmental procedures, such as removal of toxins and dust suppression.

Provide a copy of *ANSI A10.6-1983 Safety Requirements for Demolition Operations* for the trainees to examine.

Emphasize that all cranes cannot be used for demolition. It is important to check with the manufacturer before using a crane for demolition.

Describe the steps for vertical-drop balling demolition operations. Discuss the importance of maintaining a minimum drop distance. Explain safety precautions in case the ball misses the intended impact site.

Discuss the importance of not applying the crane brake until the ball has struck and stopped. Emphasize the importance of ensuring that there is enough rope on the hoist drum for unexpected ball travel.

CAUTION

Certain manufacturers do not recommend use of their cranes for demolition operations. Always consult the crane manufacturer when planning demolition work.

7.1.0 Vertical-Drop Balling

In this type of operation, demolition is accomplished by raising the ball vertically above the object and then allowing it to drop into the object by force of gravity. The ball may be attached to the hoist rope of a crane with a free-fall capability (*Figure 6*). It should be raised a short distance, held by the hoist brake, and then permitted to fall until the object is struck. Do not apply the brake until the ball has stopped moving; structural damage or overturning of the crane could result. At the lower right of *Figure 6*, note the water spray used for dust suppression.

Where the point of impact is above ground level or above an underground void, there is the possibility of the ball missing the intended point of impact. It may pass through the object or bounce off. In this situation, the drop distance should be kept to a minimum. At the intended point of impact, there should be sufficient length of rope left on the crane hoist drum to allow the ball to be brought to rest by the brake to avoid endangering the crane.

Some highly specialized cranes are available that are designed for the dynamics caused by heavy vertical impact loading. The crane shown in *Figure 7* is

WARNING!

To prevent crane damage or tipping during vertical-drop balling, do not apply the crane brake until the ball has struck an intended area and stopped. If the ball is still moving due to a glancing blow or if it has passed through an object, it may be possible to apply the crane brake slowly to bring the ball to a stop. Make sure that sufficient rope is on the hoist drum to allow for any excessive and unexpected ball travel.

Figure 6 ◆ Vertical-drop balling.

Instructor's Notes:

a high-cycle dynamic compactor used for earth compaction at depths up to 40 feet. It can also be used for certain types of vertical-drop demolition. It is constructed entirely of high-strength, low-alloy steels. The upperworks is hard-mounted to the carbody and cannot be rotated. This eliminates damage to any swing-circle bearings caused by heavy loading and unloading forces. The boom is secured with pendants fore and aft to prevent boom whip due to the sudden loading and unloading forces. A high-speed hoist uses a large-diameter hoist drum and disc-style braking for the drum to eliminate overspooling problems and increase cycle speeds. These cranes are capable of using drop weights up to 20 tons from heights exceeding 100 feet.

7.2.0 Swinging the Ball In-Line with the Boom

For this demolition procedure, an additional rope attached to the ball from a second drum on the crane is used to pull the ball in towards the crane boom. When the pulling rope is released, the ball swings outward, in the plane of the boom, to strike the object (*Figure 8*). The pulling rope serves a second purpose in that it also controls and limits the outward swing of the ball.

This method is the most advisable for demolishing low-level objects because it produces the least strain on the crane. However, precautions should always be taken. The capacity of the crane should not be exceeded by allowing the ball to swing outwards beyond the maximum operating radius.

7.3.0 Swinging the Ball by Slewing

In this operation, the ball is suspended about ten feet or more below the boom head. The slew motion is engaged, causing the ball to swing in an arc and strike the object. The slew motion should be disengaged and the slew brake applied to stop the boom at a point in line with the point of impact. As with the boom method, a second rope should be used to prevent the ball from swinging out of the safe radius. This process is not recommended because it results in large torsion loadings applied to the boom and other elements of the crane. However, they can be reduced by good operator control.

303F07.EPS

Figure 7 ◈ Dynamic earth compactor.

303F08.EPS

Figure 8 ◈ Release of a 3-ton ball during in-line demolition.

Describe a high-cycle dynamic compactor. Discuss the capabilities and limitations.

Describe the demolition procedure of swinging the ball in-line with the boom. Explain when this procedure is advisable and the necessary precautions required.

Describe the demolition procedure of swinging the ball by slewing. Discuss why this procedure is not usually recommended.

ADVANCED OPERATIONAL TECHNIQUES 3.7

Describe a vacuum lifter. Discuss the capabilities and limitations. Explain the importance of performing a load test.

Describe normal lifting operations and safety precautions.

Explain the hazards of standing under a load.

Ask a trainee to describe an electromagnet, and explain the types of loads normally lifted with one.

Emphasize that electromagnets pose an electrocution hazard. Discuss precautions.

See the Teaching Tip for Section 9.0.0 at the end of this module.

8.0.0 ◆ VACUUM LIFTING

A **vacuum lifter** (*Figure 9*) is commonly used to lift loads that have a surface suitable for vacuum lifting pads. This means the load surface must be clean, flat, dry, and nonporous. The vacuum lifter is constructed so that the following conditions are met:

- The rated load must not be greater than 50 percent of the ultimate pad capacity (UPC) for horizontal surface lifts.
- The rated load must not be greater than 25 percent of the ultimate pad capacity (UPC) for vertical surface lifts.

If the vacuum lifter is being used for the first time or is being placed back in service after repairs have been made to it, a load test must be conducted. The pad contact surface must be clean and free of any loose particles. The load rating should not be more than 80 percent of the maximum load sustained during the test. The test load must not exceed 125 percent of the rated load of the system unless the manufacturer or a qualified individual recommends otherwise.

For normal operations, the operator must make sure the pad surface is clean, the vacuum level indicator has reached the required level, and the vacuum lines are not kinked, twisted, or wrapped around portions of the lifter that will move during the lift. Care should be taken to make certain the load is correctly distributed for the lifter being used. These checks are done before the actual lift is made. The operator is required to lift the load a few inches to make sure the vacuum lifting device has been correctly applied and that a stable vacuum level exists.

Every vacuum lifter is fitted with a device that gives a visual indication to the crane operator of the state of the vacuum at any time. If, during lifting operations, the vacuum-inducing device fails, the operator must immediately warn all persons in the vicinity of the lifter and land the load if at all possible to do so.

It is imperative for the operator to be careful with the load. There is to be no sudden acceleration or deceleration of the load, the load and the lifter are not to come into contact with any obstruction, and carrying the load over people is prohibited.

> **WARNING!**
> Due to possibility of a sudden vacuum failure, personnel should never position themselves under or near a vacuum lifter load.

303F09.EPS

Figure 9 ◆ Vacuum lifter.

9.0.0 ◆ ELECTROMAGNET LIFTING

Electromagnets (*Figure 10*) are used for lifting **ferrous metals**. Some of these magnets are large and powerful enough to lift thick, solid, flat, or block-shaped ferrous loads of nearly 60 tons. However, load characteristics, including the type and shape of ferrous metal, greatly affect the amount of load that a magnet will lift.

> **WARNING!**
> Electromagnet lifting poses an electrocution hazard. Before connecting or disconnecting a crane electromagnet, check the controls to ensure that the electromagnet circuit is not energized.

303F10.EPS

Figure 10 ◆ Electromagnet.

Instructor's Notes:

9.1.0 Load Thickness

The greater the number of lines of magnetic force flowing from a magnet into a ferrous load, the greater the effectiveness of the magnet. The thicker the ferrous load, the more lines of magnetic force are able to flow. After a certain thickness of load, no additional lines of force will flow because the magnet has reached its full capacity.

Thin material means there is less ferrous metal, so fewer lines of magnetic force flow from the magnet into the load. The lifting capacity of the magnet is reduced with thinner loads.

9.2.0 Load Alloy

Low-carbon steels, such as SAE 1020 steel, are nearly as good conductors of magnetic lines of force as pure iron. However, many other **alloys** contain non-magnetic materials that reduce the ability of magnetic lines of force to flow into the load. Non-magnetic metals include copper, aluminum, brass/bronze, stainless steel, lead, zinc, manganese, tin, and nickel.

9.3.0 Other Load Characteristics

In addition to load thickness and load alloy, other load characteristics affect magnetic lifting capacity. These characteristics include the following:

- *Load surface conditions* – Anything that creates an air gap between the magnet and the load limits the flow of magnetic force and thus reduces the lifting capacity of a magnet. Paper, dirt, rust, paint, and **scale** act the same as air. A rough surface finish is the same as dirt because it creates an air gap between the magnet and the load.
- *Load length or width* – As the length or width of the load increases, it ceases to remain flat when lifted and the edges droop. This drooping or sagging of the load, called peel, can create an air gap between the load and the magnet, reducing the lifting capacity. Multiple magnets attached to one or more spreader beams are available for lifting long or wide loads.
- *Attitude of load* – As the **attitude** of the lifting surface changes from horizontal to vertical, the lifting capacity of the magnet reaches a minimum and becomes dependent upon the **coefficient** of friction for the lifting surface (10 percent steel to steel).
- *Load temperature* – The temperature of the load can cause damage to the magnet and, if high enough, can even change the magnetic characteristics of the load.

9.4.0 Lifting Guidelines

The lifting guidelines charts supplied by the magnet manufacturer show the effects of air gap, load thickness, load length, and load width on lifting capacity. As the thickness of the load decreases, so does the rated lifting capacity of the magnet. Guidelines show the maximum weight or load size that can be lifted for each thickness under varying air gap conditions. Do not exceed either the maximum weight or size for each thickness.

When first lifting materials with a magnet, lift the load a couple of inches to establish that the load is securely attached to the magnet. Also, any adjustable input control should be switched to the full power or full ON position before the load is raised more than a couple of inches. After making these checks, raise the load to the desired height, and follow through on the lift. The adjustable input control should remain in the full power or full ON position until the load is removed from the magnet. All movements are to be smooth, free of sudden acceleration or deceleration, and clear of any obstructions. The operator must not carry the load over people.

When using a magnet to pick up scrap metal, work from deep piles. Allow the magnet to settle on the pile before energizing it. If this is not done, two situations may occur: first, smaller pieces will jump on the magnet and block off the rest of the metal so you will not be able to get a full load; second, unwanted materials could be attracted to the magnet. After you turn the power on, let the magnet sit for five to ten seconds to build up full power.

> **WARNING!**
> Due to possibility of a sudden electrical failure, personnel should never position themselves under or near an electromagnet load.

Do not exceed the **duty cycle** rating of the magnet. Exceeding the rating will reduce the lifting capability because of the excessive heat that will be built up by the magnet. The duty cycle rating is defined as (time on × 100) ÷ (time off + time on), with time being measured in minutes. It is expressed as a percent:

3 minutes on, 1 minute off

$$(3 \times 100) \div (1 + 3) = 75 \text{ percent}$$

5 minutes on, 5 minutes off

$$(5 \times 100) \div (5 + 5) = 50 \text{ percent}$$

Explain how load thickness affects the lifting capacity of the magnet.

Describe how different alloys react to magnets.

Discuss how load surface conditions, load length or width, attitude of load, and load temperature can affect the lift.

Discuss lifting guidelines for electromagnetic lifts. Explain the importance of lifting the load a few inches to establish that the load is secure. Discuss the importance of letting the magnet settle on a scrap pile before energizing.

Warn trainees against standing under or near an electromagnetic load due to the possibility of a sudden electrical failure.

Describe the duty cycle rating of a magnet. Explain how this affects lifting capacity.

ADVANCED OPERATIONAL TECHNIQUES 3.9

Have trainees review Section 10.0.0.

Ensure that you have everything required for teaching this session.

Explain the effects of cold weather on crane lubricants and crane operations. Discuss specific operating procedures for working in cold weather.

Review the process for warming up a crane and beginning crane operations.

Demonstrate how to identify the components of a crane that should be observed closely before cold weather operation.

Have trainees practice identifying the components of a crane that should be observed closely before cold weather operation. Note the proficiency of each trainee. This laboratory corresponds to Performance Task 2.

The maximum limit for time on to avoid overheating the magnet is 10 minutes. To prolong the life of the magnet, keep it dry. Never cool it with water. Water is an electrical conductor and could electrically short out the magnet. Also, never disconnect the magnet from its power source while it is energized. Electrical **arcing** may occur.

10.0.0 ◆ COLD WEATHER OPERATION

Cold weather places an extra stress on crane operations. Lubricants within the crane lose their **viscosity**, and the engine takes longer to warm up. Therefore, it is important that the operator consult the manufacturer's manual and read the hydraulic oil temperature gauge prior to operating when weather conditions warrant it. Mobile hydraulic crane operators should follow these guidelines when working in cold temperatures:

- Do not start or operate a crane when the air temperature is below the minimum specified by the manufacturer.
- Do not touch metal surfaces that could freeze to your skin.
- Clean the crane, especially the boom, of all ice and snow.
- Allow ample time for hydraulic oil to warm up.
- Keep the engine at average speed until the hydraulic oil has reached 50°F.
- Carry out all crane functions at least twice once the oil is heated in order to displace the cold oil in all parts of the hydraulic system.
- During freezing weather, frequently check all air tanks for water.
- Always handle propane tanks according to the supplier's instructions.
- Never store flammable materials on the crane.
- Before lifting, ensure that the load is not frozen to the ground.
- Be extremely careful when using aerosol spray or starting fluids containing **ether** or other volatile components that can cause explosions or fire.

In addition, truck crane operations should follow these guidelines:

- Start the truck and run at the manufacturer's recommended idle speed for proper warmup.
- After a sufficient warmup time, depress the clutch pedal and engage the power take-off (PTO).
- At engine idle, release the clutch pedal slowly until full engagement of the PTO is realized.
- With the PTO fully engaged and the truck engine running at idle speed, let the hydraulic system oil circulate.

During the warmup, do not race the truck engine and over-speed the hydraulic pumps. Pump **cavitation** and permanent damage may result. If, at any time during oil circulation and especially during the initial warmup, any hydraulic pump noise, such as metal grinding or popping, is heard, shut down the unit immediately. Check that the hydraulic oil line leading to the suction port on the pump is not clogged and that the hydraulic oil is not in a jelling condition.

After the warmup period, begin crane operation as follows:

Step 1 Slowly extend horizontal outrigger cylinders approximately 6 inches and retract; extend again approximately halfway and retract; then extend full stroke.

Step 2 Follow the procedure on crane deployment, and extend cylinders.

Step 3 When completed, begin crane swing operation by slowly rotating approximately ⅛ revolution one way, return to the previous position, and rotate in the opposite direction. Do this several times, and then rotate 90 degrees and return.

Step 4 If equipped with a winch, begin the winch operation by slowly opening the control valve, allowing several revolutions in one direction. Then, reverse the control valve, and operate several revolutions in the opposite direction. Repeat this several times, allowing longer operation intervals.

CAUTION

Valves should be engaged slowly during the warmup cycle to prevent sudden hydraulic spikes, which may damage hydraulic components.

Instructor's Notes:

Summary

This module discusses operating and safety guidelines for more advanced crane operations that the crane operator may be required to perform on a job site. Requirements for critical lifts, including personnel and multiple-crane lifts, are described. The module also discusses the safety requirements for operating near transmitter towers. Electromagnetic and vacuum lifting techniques are advanced crane operations that a crane operator may be required to learn. Finally, guidelines for operating a crane in cold weather are defined.

Review Questions

1. According to _____, the use of a crane or derrick to hoist employees on a suspended personnel platform is prohibited with certain exceptions.

a. *OSHA 1910.179*
b. *OSHA 1910.180*
c. *OSHA 1926.550 (g)*
d. *OSHA 1926.552*

2. During a multiple-crane lift, the longest load radius for each crane of the operation must be measured.

a. True
b. False

3. Multiple-crane lifts are *not* classified as critical lifts.

a. True
b. False

4. In a multiple-crane lift, no crane can be loaded to more than 75 percent of its capacity.

a. True
b. False

5. When a load that is being picked up or set down is not visible to the operator, the lift is classified as a critical lift.

a. True
b. False

6. In a critical lift, the crane hook must be positioned above the _____.

a. middle of the load
b. center line of rotation
c. load's center of gravity
d. crane's center of gravity

7. In demolition operations, it is advisable to use several _____ in order to break the object.

a. heavy impacts followed by a number of lighter impacts
b. heavy impacts
c. light impacts
d. light impacts to break the initial surface followed by a number of heavy impacts

8. Dynamic compactors use _____ -style brakes for the hoist drum.

a. electric
b. disc
c. friction band
d. hydraulic

9. Dynamic compactor boom whip is eliminated by _____.

a. front and rear pendants
b. hard mounting of the upperworks
c. a special drum brake
d. rapid cycling

10. During vertical-drop balling, the hoist brake should normally *not* be applied until _____.

a. the ball has finished moving
b. the ball has made contact with the structure
c. the ball has swung back towards the boom structure
d. just before the ball makes contact with the structure

11. The demolition method that is the most advisable for demolishing low-level objects is _____ demolition.

a. vertical-drop
b. in-line
c. slew
d. radial

12. A rope used to control a demolition ball during in-line or slewing demolition is called a _____ rope.

a. boom
b. slewing
c. ball
d. pulling

13. A vacuum lifter is constructed so that the rated load is no greater than _____ percent of the ultimate pad capacity for vertical surface lifts.

a. 25
b. 30
c. 35
d. 50

Summarize the major concepts presented in the module.

Have trainees complete the Review Questions, and go over the answers prior to administering the Module Examination.

Administer the Module Examination. Record the results on Craft Training Report Form 200, and submit the results to the Training Program Sponsor.

Administer the Performance Test, and fill out Performance Profile Sheets for each trainee. If desired, trainee proficiency noted during laboratory sessions may be used to complete the Performance Test. Record the results on Craft Training Report Form 200, and submit the results to the Training Program Sponsor.

ADVANCED OPERATIONAL TECHNIQUES 3.11

14. When performing a load test on a vacuum lifter, the test load must not exceed _____ percent of the rated maximum load of the system unless the manufacturer or a qualified person recommends otherwise.
    a. 90
    b. 100
    c. 110
    d. 125

15. The greater the number of lines of magnetic force flowing from a magnet into the load, the _____ of the magnet.
    a. greater the effectiveness
    b. greater the reach
    c. the more room for operator error with regard to the operation
    d. lower the lifting capacity

16. A metal composed of two or more metals is called a(n) _____.
    a. alloy
    b. compound
    c. mixture
    d. amalgam

17. A metal that can be lifted with an electromagnet is _____.
    a. stainless steel
    b. nickel
    c. low-carbon steel
    d. aluminum

18. The _____ of an object is its position relative to the horizontal.
    a. viscosity
    b. attitude
    c. rotation
    d. slew

19. A rough surface finish on ferrous material will _____ the lifting capacity of a magnet.
    a. increase
    b. not affect
    c. decrease

20. The condition that occurs when there is a drooping or sagging of the load when lifting with a magnet is called _____.
    a. peel
    b. friction
    c. coagulation
    d. reeming

21. The bright light that is formed when a strong current jumps a gap in a circuit or between electrodes is called _____.
    a. arcing
    b. cavitation
    c. laser light
    d. photoelectric light

22. When lifting materials with a magnet, the operator should _____.
    a. first try lifting with the least magnet power
    b. first try lifting with full magnet power
    c. decrease magnet power as the lift progresses
    d. increase magnet power as the lift progresses

23. The duty cycle rating of a magnet is defined as _____.
    a. (time on × 100) ÷ (time off × 100)
    b. (time on + 100) ÷ (time off + 100)
    c. (time on × 100) ÷ (time off + time on)
    d. (time on + 100) ÷ (time off – time on)

24. During freezing weather, frequently check the _____ for water accumulation.
    a. gears
    b. propane tanks
    c. clutches
    d. air tanks

25. The property of a fluid that resists the force tending to cause the fluid to flow is called _____.
    a. scale
    b. ether
    c. attitude
    d. viscosity

Instructor's Notes:

GLOSSARY

Trade Terms Introduced in This Module

Alloy: A substance composed of two or more metals.

Arcing: A bright light that is formed when a strong current jumps a gap in a circuit or between two electrodes.

Attitude: Position of the three axes of an object relative to the horizontal.

Blind lift: A lift where the crane operator cannot see where to pick up, lift, or set down the load.

Cavitation: The rapid forming and collapse of vapor pockets in liquid in low-pressure areas that can cause damage to a motor.

Coefficient: A number that is constant for a given substance, body, or process under certain specified conditions, serving as a measure of one of its properties.

Critical lift: A lift where the load requires exceptional care in handling because of the size, weight, close-tolerance installation, high susceptibility to damage, or other unusual factors.

Demolition: The act or instance of demolishing or destroying.

Duty cycle: An 8-, 16-, or 24-hour rating expressed as a percentage.

Dynamic loading: Loads introduced into the crane or its components due to accelerating or decelerating forces.

Electromagnet: A laminated soft iron core surrounded by a coil of wire. The device temporarily becomes a magnet to attract ferrous metals when an electric current flows through the wire of the coil.

Ether: A colorless, highly volatile, flammable liquid.

Ferrous metals: Metals that are iron or steel.

Scale: An oxide occurring in scaly form on the surface of metal brought to a high temperature.

Transmitter: That part of a telephonic or telegraphic apparatus converting sound waves or mechanical movements into corresponding electric waves or impulses.

Vacuum lifter: A below-the-hook lifting device for lifting and transporting loads in a fixed attitude using a holding force by means of a vacuum. The vacuum operates by pressure less than ambient atmospheric pressure.

Viscosity: The property of a fluid that resists the forces tending to cause the fluid to flow.

ADVANCED OPERATIONAL TECHNIQUES 3.13

Instructor's Notes:

Figure Credits

Lifting Technologies, Inc.	303F01
Lampson International LLC	303F02, 303F07
Link-Belt Construction Equipment Company	303F03
E & K Equipment, Inc.	303F05
Frattalone Companies of Saint Paul, MN	303F06, 303F08
©Anver Corp. www.anver.com	303F09
Reprinted with permission from Eriez Magnetics, Inc.	303F10

ADVANCED OPERATIONAL TECHNIQUES 3.15

MODULE 21303-05 — TEACHING TIPS

The following are suggested activities or instructional methods to help you teach the material in this module.

General

When you call on someone to answer a question, the rest of the class relaxes or even tunes out because they expect that the question and answer will take place only between you and the trainee you called on. Instead, use this technique to involve more trainees in answering questions and to keep them on their toes.

1. Ask trainees to define a term or explain a concept.
2. After one trainee has answered, ask a trainee seated nearby if the answer is right. Then ask whether a trainee in the back of the room agrees.
3. Ask trainees to explain why they think an answer is right or wrong.
4. Use the session to clear up incorrect ideas and encourage trainees to learn from their mistakes.

Section 3.0.0 *Multiple-Crane Lifts*

Multiple-crane lifts must be carefully planned and executed. Trainees will need appropriate personal protective equipment, pencils, and paper. Arrange for an opportunity to observe a multiple-crane lift. Alternatively, have a manufacturer's representative give a presentation. Allow 60 to 90 minutes for this exercise.

1. Before the presentation discuss important lift considerations with trainees.
2. If possible meet with the site safety supervisor, and review the lift plan.
3. Observe the lift. Point out safety considerations, signal people, and other factors. Have trainees take notes and write down questions during the tour.
4. After the lift answer any questions the trainees may have.

Section 6.0.0 *Working Near Transmitters*

Transmitter towers present specific hazards for crane operations. Trainees will need pencils and paper. Invite a safety technician from the utility company to give a presentation to the class on working safely around transmitters. Allow 60 to 90 minutes for this exercise.

1. Before the presentation discuss safety precautions with trainees.
2. Have the utility representative give a presentation on working safely around transmitters, covering grounding, flammable materials handling, and grounding equipment.
3. After the presentation, have the presenter answer any questions the trainees may have.

Section 7.0.0 *Demolition*

Demolition can be one of the most dangerous tasks for a crane operator. Trainees will need appropriate personal protection, pencils, and paper. Obtain a demolition safety video and/or arrange to observe a wrecking ball demolition of a building. Allow 45 to 120 minutes for this exercise. Demolition videos are available from the following:

Crane Safety, NADC, 1992, videocassette, 36 minutes. National Association of Demolition Contractors, www.demolitionassociation.com, $100.00.

The Wrecking Ball, A&E Home Video, 100 minutes. www.historychannel.com, $24.95.

Modern Marvels: Demolition, The History Channel, 50 minutes. www.historychannel.com, $24.95.

1. Before the presentation discuss crane safety during demolition operations with trainees.
2. Show the video and/or observe the demolition.
3. After the presentation answer any questions the trainees may have.

Section 9.0.0 *Electromagnetic Lifting*

This Quick Quiz will familiarize trainees with the terms and definitions that are commonly used in electromagnetic lifting. You will need photocopies of the quiz provided on the following page. Trainees will need pencils. If you allow trainees to use the Trainee Module, decrease the amount of time you give them to complete the quiz.

1. Make a photocopy of the quiz for each trainee.
2. Give trainees between 5 and 10 minutes to complete the quiz.
3. Go over the answers to the quiz.
4. Ask trainees if they have questions.

Answers to Quick Quiz

1. c
2. e
3. d
4. a
5. f
6. b

Quick Quiz *Electromagnetic Lifting*

For each description listed, identify the term that the text best describes. Write the corresponding letter in the blank provided.

_____ **1.** An oxide occurring on the surface of metal brought to a high temperature

_____ **2.** A bright light that is formed when a strong current jumps a gap in a circuit or between two electrodes

_____ **3.** Position of the three axes on an object relative to the horizontal

_____ **4.** Metals that are iron or steel

_____ **5.** An 8-, 16-, or 24-hour rating expressed as a percentage

_____ **6.** A substance composed of two or more metals

a. Ferrous metals
b. Alloy
c. Scale
d. Attitude
e. Arcing
f. Duty cycle

Copyright © 2005 National Center for Construction Education and Research. Permission is granted to reproduce this page provided that copies are for local use only and that each copy contains this notice.

MODULE 21303-05 — ANSWERS TO REVIEW QUESTIONS

Answer	Section
1. c	2.0.0
2. a	3.0.0
3. b	3.0.0
4. a	3.0.0
5. b	4.0.0
6. c	5.0.0
7. a	7.0.0
8. b	7.1.0
9. a	7.1.0
10. a	7.1.0
11. b	7.2.0
12. d	7.2.0; 7.3.0
13. a	8.0.0
14. d	8.0.0
15. a	9.1.0
16. a	9.2.0; Glossary
17. c	9.2.0
18. b	9.3.0; Glossary
19. c	9.3.0
20. a	9.3.0
21. a	9.4.0; Glossary
22. b	9.4.0
23. c	9.4.0
24. d	10.0.0
25. d	10.0.0; Glossary

CONTREN® LEARNING SERIES — USER UPDATE

The NCCER makes every effort to keep these textbooks up-to-date and free of technical errors. We appreciate your help in this process. If you have an idea for improving this textbook, or if you find an error, a typographical mistake, or an inaccuracy in NCCER's Contren® textbooks, please write us, using this form or a photocopy. Be sure to include the exact module number, page number, a detailed description, and the correction, if applicable. Your input will be brought to the attention of the Technical Review Committee. Thank you for your assistance.

Instructors – If you found that additional materials were necessary in order to teach this module effectively, please let us know so that we may include them in the Equipment/Materials list in the Annotated Instructor's Guide.

Write: Product Development and Revision
National Center for Construction Education and Research
P.O. Box 141104, Gainesville, FL 32614-1104

Fax: 352-334-0932

E-mail: curriculum@nccer.org

Craft ______________________ Module Name ______________________

Copyright Date ____________ Module Number ____________ Page Number(s) ____________

Description

____

____

____

(Optional) Correction

____

____

(Optional) Your Name and Address

____

____

Module 21304-05

Lift Planning

NCCER STANDARDIZED CRAFT TRAINING PROGRAM

The National Center for Construction Education and Research (NCCER) provides a standardized national program of accredited craft training. Key features of the program include instructor certification, competency-based training, and performance testing. The program provides trainees, instructors, and companies with a standard form of recognition through a National Craft Training Registry. The program is described in full in the *Guidelines for Accreditation*, published by the NCCER. For more information on standardized craft training, contact the NCCER by writing us at P.O. Box 141104, Gainesville, FL 32614-1104; calling 352-334-0911; or e-mailing info@nccer.org. More information may be found at our Web site, www.nccer.org.

HOW TO USE THIS ANNOTATED INSTRUCTOR'S GUIDE

Each page presents two sections of information. The larger section displays each page exactly as it appears in the Trainee Module. The narrow column ties suggested trainee and instructor actions to each page and provides icons (detailed below) to call your attention to material, safety, audiovisual, or testing requirements. The bottom of each page includes space for your notes.

The **Audiovisual** icon indicates an appropriate time to show a transparency or other audiovisual aid.

The **Classroom** icon prompts you to define a term, stress a point, ask trainees to explain a concept, or give examples.

The **Demonstration** icon directs you to show trainees how to perform tasks.

The **Examination** icon tells you to administer the written module examination.

The **Homework** icon is placed where you may wish to assign reading for the next class, to assign a project, or to advise trainees to prepare for an examination.

The **Laboratory** icon is used when trainees are to practice performing tasks.

The **Materials** icon is a reminder for you to gather materials needed for classes, labs, and testing.

The **Performance Testing** icon tells you to administer a performance test or a portion thereof.

The **Safety** icon is used to emphasize safety issues. It is often keyed to *Caution* and *Warning* statements in the Trainee Module.

The **Teaching Tip** icon indicates additional guidance is available, such as how to conduct an exercise, get the most educational value from a field trip, or encourage class participation. Teaching Tips may expand on a feature (*Think About It, Did You Know?*) or provide Quick Quizzes or similar exercises. You will be referred to the Teaching Tips section at the back of the module if there is additional material.

The **Combination** icon indicates that the laboratory listed corresponds with a performance task. If desired, you can note the proficiency of the trainees during the laboratory and use it to satisfy performance testing requirements.

PREPARATION

Before teaching this module, you should review the Objectives, Performance Tasks, Materials and Equipment List, and the Module Outline. Be sure to allow ample time to prepare your own training or lesson plan and gather all required materials and equipment.

Lift Planning
Annotated Instructor's Guide

Module 21304-05

MODULE OVERVIEW

This module provides an in-depth discussion of lift plan implementation. The topics covered include reference information, calculations, single- and multiple-crane lifting, critical lifts, and engineering.

PREREQUISITES

Prior to training with this module, it is recommended that the trainee shall have successfully completed *Core Curriculum; Mobile Crane Operations Level One; Mobile Crane Operations Level Two;* and *Mobile Crane Operations Level Three,* Modules 21301-05 through 21303-05.

OBJECTIVES

Upon completion of this module, the trainee will be able to do the following:

1. Provide the necessary information requested on a lift plan.
2. Reference available material that will assist in a safe lifting operation.
3. Calculate additions and deductions involved in lifting operations.
4. Identify existing operations that need special approval.
5. Identify engineering considerations in a lift plan.
6. Identify the various types of lift plans and their differences.
7. Identify the importance of lift plan implementation.
8. Describe the importance of following and adhering to a lift plan.

PERFORMANCE TASKS

Under the supervision of the instructor, the trainee should be able to do the following:

1. Fill out a lift plan based on a given scenario.
2. Perform single-crane lifting calculations.
3. Perform multiple-crane lifting calculations.

MATERIALS AND EQUIPMENT LIST

Overhead projector and screen

Transparencies

Blank acetate sheets

Transparency pens

Whiteboard/chalkboard

Markers/chalk

Pencils and scratch paper

OSHA 29 CFR 1910.180

OSHA 29 CFR 1926.550

ASME B30.5

Crane(s) and operator's manuals available for performing lifts

Copies of site emergency procedures

Copies of sample pre-lift worksheets and lift plans

Module Examinations*

Performance Profile Sheets*

Performance Profile Examination Worksheets*

*Located in the Test Booklet.

SAFETY CONSIDERATIONS

Ensure that the trainees are equipped with appropriate personal protective equipment and know how to use it properly. This module may require trainees to visit construction sites. Brief trainees on site safety procedures.

TEACHING TIME FOR THIS MODULE

An outline for use in developing your lesson plan is presented below. Note that each Roman numeral in the outline equates to one session of instruction. Each session has a suggested time period of 2½ hours. This includes 10 minutes at the beginning of each session for administrative tasks and one 10-minute break during the session. Approximately 20 hours are suggested to cover *Lift Planning*. You will need to adjust the time required for hands-on activity and testing based on your class size and resources. Because laboratories often correspond to Performance Tasks, the proficiency of the trainees may be noted during these exercises for Performance Testing purposes.

Topic	Planned Time
Sessions I through III. Introduction and Lift Plan	
A. Introduction	____________
B. Lift Plan	____________
C. Laboratory – Trainees practice filling out a lift plan based on a given scenario. This laboratory corresponds to Performance Task 1.	____________
Sessions II and III. Calculations I	
A. Calculations for Single-Crane Lifts	____________
B. Laboratory – Trainees practice performing the calculations for single-crane lifts. This laboratory corresponds to Performance Task 2.	____________
Sessions IV through VI. Calculations II	
A. Calculations for Multiple-Crane Lifts	____________
B. Laboratory – Trainees practice performing the calculations for multiple-crane lifts. This laboratory corresponds to Performance Task 3.	____________
Session VII. Critical Lifts, Engineering Considerations, and Lift Plan Implementation	
A. Critical Lifts	____________
B. Engineering Considerations	____________
C. Lift Plan Implementation	____________
Session VIII. Review, Module Examination, and Performance Testing	
A. Review	____________
B. Module Examination	____________
1. Trainees must score 70 percent or higher to receive recognition from NCCER.	
2. Record the testing results on Craft Training Report Form 200, and submit the results to the Training Program Sponsor.	
C. Performance Testing	____________
1. Trainees must perform each task to the satisfaction of the instructor to receive recognition from NCCER. If applicable, proficiency noted during laboratory exercises can be used to satisfy the Performance Testing requirements.	
2. Record the testing results on Craft Training Report Form 200, and submit the results to the Training Program Sponsor.	

Module 21304-05

Lift Planning

Instructor's Notes:

COURSE MAP

This course map shows all of the modules in the third level of the *Mobile Crane Operations* curriculum. The suggested training order begins at the bottom and proceeds up. Skill levels increase as you advance on the course map. The local Training Program Sponsor may adjust the training order.

Assign reading of Module 21304-05.

MOBILE CRANE OPERATIONS LEVEL THREE

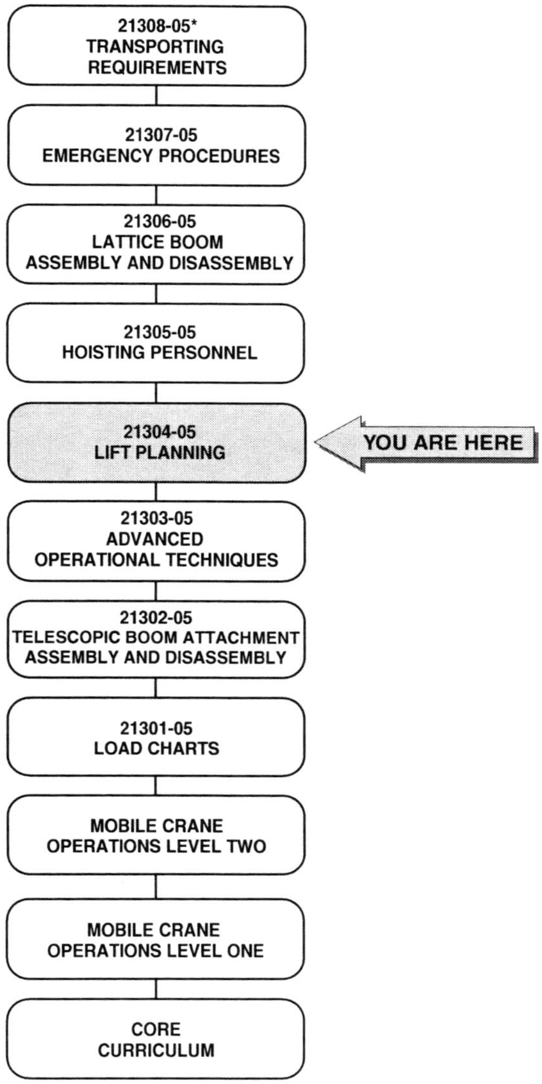

*Please note that 21308-05 is an elective module. It is not necessary to complete this module in order to advance in your training.

304CMAP.EPS

Copyright © 2005 National Center for Construction Education and Research, Gainesville, FL 32614-1104. All rights reserved. No part of this work may be reproduced in any form or by any means, including photocopying, without written permission of the publisher.

LIFT PLANNING 4.iii

Instructor's Notes:

MODULE 21304-05 CONTENTS

Figures

LIFT PLANNING 4.v

Instructor's Notes:

MODULE 21304-05

Lift Planning

Objectives

When you have completed this module, you will be able to do the following:

1. Provide the necessary information requested on a lift plan.
2. Reference available material that will assist in a safe lifting operation.
3. Calculate additions and deductions involved in lifting operations.
4. Identify existing operations that need special approval.
5. Identify engineering considerations in a lift plan.
6. Identify the various types of lift plans and their differences.
7. Identify the importance of lift plan implementation.
8. Describe the importance of following and adhering to a lift plan.

Prerequisites

Before you begin this module, it is recommended that you successfully complete *Core Curriculum; Mobile Crane Operations Level One; Mobile Crane Operations Level Two;* and *Mobile Crane Operations Level Three*, Modules 21301-05 through 21303-05.

Required Trainee Materials

1. Pencil and paper
2. Appropriate personal protective equipment
3. Calculator

1.0.0 ◈ INTRODUCTION

Before conducting a lift, most construction sites or companies require that a **lift plan** be completed and signed by competent personnel. Lift plans are mandatory for steel erection and multiple-crane lifts. A lift plan contains information relative to the crane, load, and rigging, and it lists any special instructions or restrictions for the lift. It is important to remember that any deviation from the original lift plan, no matter how small, requires a new lift plan.

2.0.0 ◈ LIFT PLAN

Critical lifts always require a lift plan. As a matter of policy, all operators should require or prepare a plan for all lifts. A typical lift plan for either a critical or ordinary (minor) lift breaks down the following elements and information (see *Appendix A*):

- Weight
  - Equipment condition
  - Weight empty
  - Weight of overhead ball
  - Weight of block
  - Weight of lifting bar
  - Weight of slings and shackles
  - Weight of jib (if erected or stowed and not used)
  - Weight of overhaul ball on jib
  - Weight of wire rope
  - Allowance for unaccounted material in equipment
- Jib
  - Jib to be used
  - Length of jib
  - Angle of jib
  - Rated capacity of jib
- Crane placement
  - Smooth solid foundation within one percent of grade
  - Electrical hazards in area
  - Obstacles or obstructions to lift or swing
  - Swing direction and degree
  - Area of operation (over-the-front, over-the-rear, over-the-side)

Ensure that you have everything required to teach the course. Check the Materials and Equipment list at the front of this module.

See the general Teaching Tip at the end of this module.

Explain that terms shown in bold are defined in the Glossary at the back of this module.

Show Transparency 1, Objectives, and Transparency 2, Performance Tasks. Review the goals of the module, and explain what will be expected of trainees.

Discuss the importance of using a lift plan. Explain how a lift plan is developed and executed. Refer to *Appendix A* and describe the specific elements of a typical lift plan.

Discuss important reference documents to consider when planning lifts.

Provide copies of *OSHA 29 CFR 1910.180*, *OSHA 29 CFR 1926.550*, and *ASME B30.5*. Have trainees examine them and look up sample specifications.

Demonstrate how to fill out a lift plan based on a scenario.

Have trainees practice filling out a lift plan based on a scenario. Note the proficiency of each trainee. This laboratory corresponds to Performance Task 1.

Have trainees review Sections 3.0.0 and 3.1.0.

Ensure that you have everything required for teaching this session.

Describe the considerations for planning a single-crane lift.

Emphasize the importance of following the manufacturer's load charts and recommendations.

- Rope
  - Number of parts of rope
  - Size of rope
  - Reeving capacity
- Sizing and types of slings
  - Sling selection
  - Shackle selection
  - Sizing and types
- Crane
  - Type of crane
  - Crane capacity
  - Lifting arrangement
- Other
  - Special instructions or restrictions for crane, rigging, or lift
  - Diagrams for crane, load placement, and rigging configuration
  - Signature of job supervisor, lift plan checker, and operator
  - Derigging plan

2.1.0 Reference Information

The operator should consult reference information that will assist in a safe lifting operation. This includes crane specifications, conditions in which the lift is taking place (weather, lift area space), and all applicable compliance and consensus standards. Standards covering mobile cranes include the following:

- *OSHA 29 CFR 1910.180*
- *OSHA 29 CFR 1926.550*
- *ASME B30.5*

3.0.0 ◆ CALCULATIONS

There are numerous calculations involved in lift planning. The operator and site supervisor must take certain considerations into account:

- Available lifting-point height above the lifting location
- Load pick-up height above or below the lifting location
- Load set-down height above or below the lifting location
- Maximum wire rope weight to be experienced during the lift
- Weight of load to be lifted, not including rigging
- Weight of crane components below the lifting point (deductions)
  - Length of line from point to load block
  - Any parts of line that are considered part of the load
  - Load block height
  - Rigging on the block, including lifting beams, slings, and shackles
- Effective weight of crane components past the lifting point
  - Jib
  - Line to jib top
  - Length of line from point to load block
- Weight of any additional crane components and/or boom components

3.1.0 Single-Crane Lifting

The information provided is for instructional purposes only. The load/capacity charts presented here are derived from manufacturer sales information that may not be complete or machine-specific. They are intended to give you practice in lift calculations. On a job site, always follow the manufacturer's load charts and recommendations.

To illustrate lift plan calculations, consider this example. What is the net capacity of the Manitowoc 4100W-2 in the following configuration?

- 22C boom, 190-ft main boom, no offset
- 55 ft of load radius
- 2-sheave boom point attached
- 15-ton hook block 20 ft below boom point
- 100-ton block reeved with 4 parts of line
- Crawlers extended
- Rigging of 365 lbs
- Lifting from the main boom

Refer to the Manitowoc 4100W-2 load/capacity charts shown in *Figures 1* through *3*.

The gross capacity of the crane at 190 ft of boom length and 55 ft of load radius is 88,000 lbs. Deductions from this weight include:

- 2-sheave boom point = 1,500 lbs
- 15-ton hook block or hook and weight ball = 865 lbs
- 100-ton block = 2,065 lbs
- Rigging = 365 lbs
- Rope = 1,827 lbs

The total weight of the deductions is 6,622 lbs. The net capacity of the crane in this configuration is the gross capacity (88,000 lbs) minus the deductions (6,622 lbs), which is 81,378 lbs.

Instructor's Notes:

MANITOWOC 4100 W-2

LIFTCRANE CAPACITIES

BOOM NO. 22C WITH OPEN THROAT TOP
146,400 LB. CRANE COUNTERWEIGHT
60,000 LB. CARBODY COUNTERWEIGHT
26'6" CRAWLERS EXTENDED

WARNING: This chart will apply only when two 12,000 lb. side ctwts. and two 30,000 lb. carbody ctwts. bear MEC registered Serial Numbers.

LIFTING CAPACITIES: Capacities for various boom lengths and operating radii may be based on percent of tipping, strength of structural components, operating speeds and other factors.

Capacities are for freely suspended loads and do not exceed 75% of a static tipping load. Capacities based on structural competence are shown by shaded areas.

Capacities are shown in pounds. Deduct 1200 pounds from capacities listed when single sheave upper boom point is attached and 1500 pounds when two sheave upper boom point is attached. To comply with B30.5 requirements, upper boompoint cannot be used on the 260 ft. boom. Weight of jib, (see chart A), all load blocks, hooks, weight ball, slings, hoist lines beneath boom and jib point sheaves, etc., is considered part of the main boom load. Boom is not to be lowered beyond radii where combined weights are greater than rated capacity. Where no capacity is shown, operation is not intended or approved.

OPERATING CONDITIONS: Machine to operate in a level position on a firm surface with crawlers fully extended and gantry in working position and be rigged in accordance with and under conditions referred to in rigging drawing No. 190693 and load line specification chard No. 6592-A.

Crane operator judgement must be used to allow for dynamic load effects of swinging, hoisting or lowering, travel, as well as adverse operating conditions & physical machine depreciation.

OPERATOR RADIUS: Operating is the horizontal distance from the axis of rotation to the center of vertical hoist line or load block with the load freely suspended. Add 14" to boom point radius for radius of sheave when using single part hoist line.

Boom angle is the angle between horizontal and centerline of boom butt and inserts and is an indication of operating radius. In all cases, operating radius shall govern capacity.

BOOM POINT ELEVATION: Boom point elevation, in feet, is the vertical distance from ground level to centerline of boom point shaft.

MACHINE EQUIPMENT: Machine equipped with 26'6" extendible crawlers, 48" treads, 17' retractable gantry, 12 part boom hoist reeving, four 1 3/8" boom pendants, 1st ctwt. 41,900 lbs., 2nd ctwt. 41,500 lbs., 3rd ctwt. 39,000 lbs., two 12,000 lbs. side ctwt's. and two 30,000 lbs. carbody ctwt's.

LOAD AND WHIP LINE SPECIFICATIONS	
LOAD LINE:	1-1/8" - 6 x 31 Warrington-Seale, Extra Improved Plow Steel, Regular Lay, IWRC. Minimum Breaking Strength 65 Ton. (Approx.Weight Per Ft. in Lbs. 2.34)
WHIP LINE:	1-1/8" - Warrington-Seale, Improved Plow Steel, Regular Lay, IWRC. Minimum Breaking Strength 56.5 Ton. Maximum Load - 28,300 Lbs. Per Line. (Approx. Weight Per Ft. in Lbs. 2.34)

HOIST REEVING FOR MAIN LOAD BLOCK						
No.Parts Of Line	1	2	3	4	5	6
Max Load - Lbs.	32,500	65,000	97,500	130,000	162,500	195,000
No. Parts of Line	7	8	9	10	11	12
Max. Load - Lbs.	227,500	260,000	292,500	325,000	357,500	400,000
No. Parts of Line	13					
Max. Load - Lbs.	430,000					

MAXIMUM BOOM AND JIB LENGTH LIFTED UNASSISTED				DEDUCT FROM CAPACITIES WHEN JIB IS ATTACHED	
OVER FRONT OF BLOCKED CRAWLERS		OVER SIDE OF EXTENDED CRAWLERS			
BOOM LENGTH	JIB NO. 123	BOOM LENGTH	JIB NO. 123	JIB LENGTH	JIB NO. 123
260'	--	260'	--	30'	3,000 lbs.
250'	--	250'	--	40'	3,600 lbs.
240'	40'	240'	40'	50'	4,200 lbs.
230'	60'	230'	60'	60'	4,900 lbs.
Load block, hook and weight ball on ground to start.					

FOR JIB CAPACITIES, CONSULT JIB CHART.

BOOM LGTH FEET	OPER. RAD. FEET	BOOM ANG. DEG.	BOOM POINT ELEV.	CAPACITY: CRAWLERS EXTENDED
70	16.5	79.7	75.9	460,000
	17	79.3	75.8	400,000
	18	78.5	75.6	380,100
	19	77.6	75.4	363,000
	20	76.8	75.1	347,300
	22	75.1	74.6	319,600
	24	73.4	74.1	293,400
	26	71.7	73.5	266,100
	28	69.9	72.8	237,500
	30	68.2	72.0	214,300
	32	66.4	71.2	195,100
	34	64.6	70.2	178,900
	36	62.8	69.3	165,200
	38	60.9	68.2	153,300
	40	59.1	67.0	143,000
	45	54.1	63.7	122,100
	50	48.9	59.8	106,300
	55	43.2	54.9	93,900
	60	36.9	49.0	84,000
	65	29.4	41.3	75,800
	70	19.5	30.3	63,900

BOOM LGTH FEET	OPER. RAD. FEET	BOOM ANG. DEG.	BOOM POINT ELEV.	CAPACITY: CRAWLERS EXTENDED
80	17	80.6	85.9	392,800
	18	79.9	85.8	378,900
	19	79.2	85.6	361,800
	20	78.5	85.4	346,100
	22	77.0	84.9	318,400
	24	75.5	84.5	292,500
	26	74.0	83.9	265,600
	28	72.5	83.3	237,000
	30	71.0	82.7	213,800
	32	69.5	81.9	194,600
	34	68.0	81.2	178,500
	36	66.4	80.3	164,700
	38	64.8	79.4	152,800
	40	63.3	78.4	142,400
	45	59.2	75.7	121,500
	50	54.9	72.5	105,700
	55	50.4	68.6	93,300
	60	45.6	64.1	83,400
	65	40.3	58.8	75,200
	70	34.4	52.2	68,300
	75	27.4	43.9	52,500
	80	18.2	32.0	53,900

BOOM LGTH FEET	OPER. RAD. FEET	BOOM ANG. DEG.	BOOM POINT ELEV.	CAPACITY: CRAWLERS EXTENDED
90	18	81.1	95.9	355,400
	19	80.4	95.7	346,900
	20	79.8	95.6	336,900
	22	78.5	95.2	317,400
	24	77.2	94.7	291,700
	26	75.9	94.3	264,800
	28	74.5	93.7	236,600
	30	73.2	93.2	213,400
	32	71.9	92.5	194,200
	34	70.5	91.9	178,000
	36	69.2	91.1	164,200
	38	67.8	90.3	152,300
	40	66.4	89.5	142,000
	45	62.9	87.1	121,100
	50	59.3	84.4	105,200
	55	55.5	81.2	92,800
	60	51.5	77.5	82,900
	65	47.3	73.2	74,700
	70	42.8	68.2	67,800
	75	37.9	62.3	62,000
	80	32.4	55.2	57,000
	85	25.8	46.2	52,600
	90	17.1	33.5	45,900

BOOM LGTH FEET	OPER. RAD. FEET	BOOM ANG. DEG.	BOOM POINT ELEV.	CAPACITY: CRAWLERS EXTENDED
100	19	81.4	105.9	332,900
	20	80.8	105.7	327,100
	22	79.6	105.4	316,200
	24	78.5	105.0	290,800
	26	77.3	104.5	263,900
	28	76.1	104.1	236,200
	30	74.9	103.6	212,900
	32	73.7	103.0	193,700
	34	72.5	102.4	177,500
	36	71.3	101.7	163,700
	38	70.1	101.0	151,800
	40	68.9	100.3	141,400
	45	65.8	98.2	120,500
	50	62.6	95.8	104,700
	55	59.3	93.0	92,300
	60	55.9	89.8	82,300
	65	52.4	86.2	74,100
	70	48.7	82.1	67,200
	75	44.8	77.4	61,400
	80	40.5	72.0	56,400
	85	35.9	65.6	52,000
	90	30.7	58.0	48,200
	95	24.5	48.5	44,900
	100	16.3	35.0	39,300

CAUTION! CHECK AMOUNT OF COUNTERWEIGHT ON MACHINE BEFORE USE OF THIS CHART

© Manitowoc 1997

304F01.EPS

Figure 1 ◈ Load/capacity chart one.

Show Transparencies 3 through 5 (Figures 1 through 3). Explain how to determine net capacity in the examples provided.

LIFTCRANE CAPACITIES – 4100W SERIES 2 BOOM NO 22C WITH OPEN-THROAT TOP, CONTINUED

BOOM LGTH FEET	OPER. RAD. FEET	BOOM ANG. DEG.	BOOM POINT ELEV.	CAPACITY: CRAWLERS EXTENDED
110	22	82.3	115.7	240,500
	24	81.3	115.4	232,100
	26	80.2	115.0	224,400
	28	79.1	114.6	217,300
	30	78.1	114.1	210,700
	32	77.0	113.6	193,300
	34	75.9	113.1	177,100
	36	74.8	112.5	163,300
	38	73.7	111.8	151,400
	40	72.6	111.2	141,000
	45	69.9	109.3	120,100
	50	67.0	107.2	104,200
	55	64.1	104.7	91,800
	60	61.2	101.9	81,800
	65	58.1	98.8	73,600
	70	54.9	95.3	66,800
	75	51.6	91.3	60,900
	80	48.1	86.8	55,900
	85	44.4	81.7	51,600
	90	40.4	75.9	47,800
	95	36.1	69.2	44,400
	100	31.1	61.1	41,400
	105	25.3	51.1	38,700
	110	17.6	37.2	33,900
120	22	83.0	125.9	228,700
	24	82.0	125.5	220,500
	26	81.0	125.2	213,000
	28	80.0	124.8	206,100
	30	79.1	124.4	199,700
	32	78.1	123.9	192,800
	34	77.1	123.4	176,600
	36	76.1	122.9	162,800
	38	75.1	122.3	150,900
	40	74.1	121.7	140,500
	45	71.6	120.0	119,500
	50	69.1	118.1	103,700
	55	66.5	115.9	91,300
	60	63.8	113.4	81,300
	65	61.1	110.6	73,100
	70	58.3	107.5	66,200
	75	55.4	104.0	60,400
	80	52.4	100.1	55,300
	85	49.2	95.8	51,000
	90	45.9	91.0	47,200
	95	42.4	85.5	43,800
	100	38.6	79.4	40,800
	105	34.4	72.2	38,100
	110	29.7	63.7	35,700
	115	24.1	53.2	33,500
130	24	82.6	135.7	212,700
	26	81.7	135.3	205,300
	28	80.8	135.0	198,600
	30	79.9	134.6	192,300
	32	79.0	134.2	186,500
	34	78.1	133.7	176,300
	36	77.2	133.2	162,500
	38	76.3	132.7	150,600
	40	75.4	132.1	140,200
	45	73.1	130.6	119,300
	50	70.8	128.8	103,400
	55	68.4	126.8	91,000
	60	66.0	124.6	81,000
	65	63.5	122.0	72,800
	70	61.0	119.2	65,900
	75	58.4	116.1	60,000
	80	55.7	112.7	55,000
	85	53.0	108.9	50,700
	90	50.1	104.8	46,800
	95	47.1	100.1	43,500
	100	43.9	95.0	40,500
	105	40.6	89.2	37,800
	110	36.9	82.6	35,300
	115	33.0	75.1	33,100
	120	28.5	66.1	31,100
	125	23.1	55.1	29,300
140	26	82.3	145.5	198,400
	28	81.5	145.1	191,800
	30	80.7	144.8	185,600
	32	79.8	144.4	179,900
	34	79.0	144.0	174,700
	36	78.1	143.5	162,000
	38	77.3	143.0	150,100
	40	76.5	142.5	139,700
	45	74.3	141.1	118,700
	50	72.2	139.4	102,800
	55	70.0	137.6	90,400
	60	67.8	135.5	80,400
	65	65.6	133.2	72,200
	70	63.3	130.7	65,300
	75	60.9	127.9	59,500
	80	58.5	124.8	54,400
	85	56.1	121.4	50,100
	90	53.5	117.7	46,200
	95	50.9	113.7	42,900
	100	48.1	109.2	39,900
140	105	45.3	104.2	37,200
	110	42.2	98.8	34,700
	115	39.0	92.7	32,500
	120	35.5	85.7	30,500
	125	31.7	77.8	28,700
	130	27.4	68.5	27,000
	135	22.2	57.0	25,500
150	28	82.1	155.3	185,600
	30	81.3	154.9	179,600
	32	80.5	154.6	174,000
	34	79.7	154.2	168,800
	36	79.0	153.8	161,600
	38	78.2	153.3	149,600
	40	77.4	152.8	139,200
	45	75.4	151.5	118,200
	50	73.4	150.0	102,400
	55	71.4	148.3	89,900
	60	69.4	146.4	79,900
	65	67.3	144.3	71,700
	70	65.2	141.9	64,800
	75	63.0	139.4	59,000
	80	60.9	136.6	53,900
	85	58.6	133.5	49,600
	90	56.3	130.1	45,700
	95	54.0	126.5	42,400
	100	51.5	122.5	39,400
	105	49.0	118.2	36,700
	110	46.4	113.4	34,200
	115	43.6	108.2	32,000
	120	40.7	102.4	30,000
	125	37.6	96.0	28,200
	130	34.2	88.8	26,500
	135	30.5	80.5	25,000
	140	26.4	70.8	23,500
	145	21.4	58.8	22,200
160	28	82.6	165.4	179,800
	30	81.8	165.1	173,900
	32	81.1	164.7	168,400
	34	80.4	164.4	163,300
	36	79.7	164.0	158,600
	38	78.9	163.6	149,100
	40	78.2	163.1	138,700
	45	76.3	161.9	117,700
	50	74.5	160.5	101,800
	55	72.6	158.9	89,300
	60	70.7	157.1	79,300
	65	68.8	155.1	71,100
	70	66.8	153.0	64,200
	75	64.9	150.6	58,400
	80	62.9	148.0	53,300
	85	60.8	145.2	49,000
	90	58.7	142.2	45,100
	95	56.6	138.8	41,800
	100	54.4	135.2	38,700
	105	52.1	131.4	36,000
	110	49.7	127.1	33,600
	115	47.3	122.5	31,400
	120	44.8	117.5	29,400
	125	42.1	112.0	27,600
	130	39.3	105.9	25,900
	135	36.3	99.2	24,300
	140	33.1	91.7	22,900
	145	29.5	83.0	21,600
	150	25.5	72.9	20,400
	155	20.7	60.5	19,200
170	30	82.3	175.2	169,000
	32	81.6	174.9	163,600
	34	81.0	174.5	158,700
	36	80.3	174.2	154,000
	38	79.6	173.8	148,800
	40	78.9	173.3	138,400
	45	77.2	172.2	117,400
	50	75.4	170.9	101,500
	55	73.7	169.4	89,100
	60	71.9	167.7	79,000
	65	70.1	165.9	70,800
	70	68.3	163.9	63,900
	75	66.4	161.7	58,100
	80	64.6	159.3	53,000
	85	62.7	156.7	48,700
	90	60.8	153.9	44,900
	95	58.8	150.8	41,500
	100	56.8	147.5	38,500
	105	54.7	144.0	35,800
	110	52.6	140.2	33,300
	115	50.4	136.0	31,100
	120	48.1	131.5	29,100
	125	45.8	126.7	27,300
	130	43.3	121.4	25,600
	135	40.8	115.7	24,100
	140	38.1	109.3	22,600
	145	35.2	102.3	21,300
	150	32.0	94.5	20,100
	155	28.6	85.5	18,900
	160	24.7	75.1	17,800
	165	20.0	62.2	16,400
180	32	82.1	185.0	158,700
	34	81.5	184.7	154,000
	36	80.8	184.3	149,500
	38	80.2	184.0	145,200
	40	79.5	183.6	137,900
	45	77.9	182.5	116,900
	50	76.3	181.2	101,000
	55	74.6	179.8	88,500
	60	72.9	178.3	78,500
	65	71.3	176.6	70,300
	70	69.6	174.7	63,400
	75	67.8	172.6	57,500
	80	66.1	170.4	52,500
	85	64.3	168.0	48,100
	90	62.5	165.4	44,300
	95	60.7	162.5	40,900
	100	58.9	159.5	37,900
	105	57.0	156.2	35,200
	110	55.0	152.7	32,700
	115	53.0	149.0	30,500
	120	51.0	144.9	28,500
	125	48.9	140.5	26,700
	130	46.7	135.8	25,000
	135	44.4	130.7	23,500
	140	42.0	125.2	22,000
	145	39.6	119.2	20,700
	150	36.9	112.6	19,500
	155	34.1	105.3	18,300
	160	31.1	97.2	17,200
	165	27.7	87.9	16,200
	170	23.9	77.1	15,300
	175	19.4	63.8	13,500
190	32	82.5	195.1	153,100
	34	81.9	194.8	148,800
	36	81.3	194.5	144,800
	38	80.7	194.1	140,800
	40	80.1	193.8	137,000
	45	78.5	192.7	116,400
	50	77.0	191.5	100,500
	55	75.4	190.2	88,000
	60	73.9	188.8	78,000
	65	72.3	187.1	69,700
	70	70.7	185.4	62,800
	75	69.1	183.4	57,000
	80	67.4	181.4	52,000
	85	65.8	179.1	47,600
	90	64.1	176.6	43,800
	95	62.4	174.0	40,400
	100	60.7	171.2	37,400
	105	58.9	168.2	34,700
	110	57.1	164.9	32,200
	115	55.3	161.4	30,000
	120	53.4	157.7	28,000
	125	51.5	153.7	26,200
	130	49.5	149.5	24,500
	135	47.5	144.9	22,900
	140	45.3	140.0	21,500
	145	43.1	134.6	20,200
	150	40.8	128.9	18,900
	155	38.4	122.6	17,800
	160	35.9	115.8	16,700
	165	33.2	108.3	15,700
	170	30.2	99.8	14,800
	175	27.0	90.3	13,900
	180	23.3	79.1	13,100
	185	18.9	65.4	11,000
200	34	82.3	204.9	143,100
	36	81.7	204.6	139,500
	38	81.2	204.3	136,000
	40	80.6	203.9	132,500
	45	79.1	203.0	115,900
	50	77.7	201.8	100,000
	55	76.2	200.6	87,500
	60	74.7	199.2	77,400
	65	73.2	197.7	69,200
	70	71.7	196.0	62,300
	75	70.2	194.2	56,400
	80	68.6	192.2	51,400
	85	67.1	190.1	47,000
	90	65.5	187.8	43,200
	95	63.9	185.3	39,800
	100	62.3	182.7	36,800
	105	60.6	179.8	34,100
	110	59.0	176.8	31,600
	115	57.3	173.6	29,400
	120	55.5	170.2	27,400
	125	53.0	166.5	25,600
	130	51.9	162.6	23,900
	135	50.1	158.4	22,300
	140	48.1	153.9	20,900
	145	46.2	149.1	19,600
	150	44.1	144.0	18,300
	155	42.0	138.4	17,200
	160	39.7	132.5	16,100
	165	37.4	126.0	15,100
	170	34.9	118.9	14,200
200	175	32.3	111.1	13,300
	180	29.4	102.4	12,400
	185	26.2	92.5	11,700
	190	22.6	81.0	10,900
	195	18.4	67.0	8,500
210	36	82.1	214.7	134,200
	38	81.6	214.4	131,000
	40	81.0	214.1	127,900
	45	79.6	213.2	115,500
	50	78.3	212.1	99,600
	55	76.9	210.9	87,100
	60	75.4	209.6	77,100
	65	74.0	208.1	68,800
	70	72.6	206.6	61,900
	75	71.2	204.8	56,100
	80	69.7	203.0	51,000
	85	68.2	201.0	46,700
	90	66.7	198.8	42,800
	95	65.2	196.5	39,400
	100	63.7	194.0	36,400
	105	62.2	191.3	33,700
	110	60.6	188.5	31,300
	115	59.0	185.5	29,100
	120	57.4	182.3	27,100
	125	55.7	178.9	25,200
	130	54.1	175.2	23,500
	135	52.3	171.4	22,000
	140	50.6	167.3	20,600
	145	48.8	162.9	19,200
	150	46.9	158.2	18,000
	155	45.0	153.2	16,800
	160	43.0	147.9	15,800
	165	40.9	142.1	14,800
	170	38.7	135.9	13,800
	175	36.5	129.2	12,900
	180	34.0	121.9	12,100
	185	31.5	113.9	11,300
	190	28.7	104.9	10,600
	195	25.6	94.7	9,900
	200	22.1	82.9	8,800
	205	17.9	68.5	6,500
220	38	82.0	224.6	125,700
	40	81.4	224.2	122,800
	45	80.1	223.3	115,000
	50	78.8	222.3	99,100
	55	77.5	221.2	86,600
	60	76.1	220.0	76,600
	65	74.8	218.6	68,300
	70	73.4	217.1	61,400
	75	72.0	215.4	55,500
	80	70.7	213.7	50,500
	85	69.3	211.8	46,100
	90	67.9	209.7	42,300
	95	66.4	207.5	38,900
	100	65.0	205.2	35,800
	105	63.6	202.7	33,100
	110	62.1	200.0	30,700
	115	60.6	197.2	28,500
	120	59.1	194.2	26,500
	125	57.5	191.0	24,600
	130	55.9	187.6	23,000
	135	54.3	184.0	21,400
	140	52.7	180.2	20,000
	145	51.0	176.1	18,600
	150	49.3	171.8	17,400
	155	47.6	167.3	16,300
	160	45.7	162.4	15,200
	165	43.9	157.2	14,200
	170	41.9	151.7	13,200
	175	39.9	145.7	12,300
	180	37.8	139.3	11,500
	185	35.6	132.4	10,700
	190	33.2	124.8	10,000
	195	30.7	116.6	9,300
	200	28.0	107.3	8,600
	205	25.0	96.9	8,000
	210	21.5	84.8	6,500

CAPACITIES CONTINUED ON NEXT PAGE

These load charts are intended for instructional purposes only. They were derived from manufacturer sales information which may not be complete or machine specific. Not responsible for typographical errors.

© MANITOWOC 1997

304F02.EPS

Figure 2 ◈ Load/capacity chart two.

Instructor's Notes:

WEIGHTS 4100W Cont'd

DESCRIPTION	APPROX. WEIGHT (IN LBS.)
JIB NO. 123	
Jib Top - 15' (w/Jib point)	695
Jib Butt - 15'	690
Jib Insert - 10'	340
Basic Pendant - 33' 3-3/4" (2 Req'd)	115 each
Pendant - 10' (2 Per insert)	65 each
Jib Backstay Pendant	155 each
Jib Strut - 12'- 6"	365
COMPONENTS	
Hook Rollers (6) - w/shafts	1,020
Light Plant - 6.5KW - w/mounting platform	1,390
Catwalk - left and right side w/rails	1,320
Lagging - 27-5/8" dia. plain	1,410
Boom Hoist Rope - 12 Part - 760' of 7/8" - 6 x 26	1,080
Wire Rope Guide Assembly - Lower	325
Wire Rope Guide Assembly - Upper	510
Rope Guide Roller Assembly	55 each
2-Part Gantry w/Telescopic Backhitch	7,805
Equalizer	2,000
Hoist Line - 1-1/8" - 6 x 31	2.34 lbs./ft.
Whip Line - 1-1/8" - 6 x 31	2.34 lbs./ft.
15-Ton Hook and Weight Ball	865
100-Ton Hook Block Assembly	2,065
200-Ton Hook Block Assembly	4,900
230-Ton Hook Block Assembly	5,375
Boom Stop - Telescopic Air Cushioned	675
Dragline Fairlead - Revolving	1,910
Dragline Fairlead - Hinged	9,330

NOTE: The above weights may fluctuate up or down 5% due to manufacturing tolerances.

© Manitowoc 1997

304F03.EPS

Figure 3 ◆ Load/capacity chart three.

3.2.0 Multiple-Crane Lifting

Multiple cranes are often used to lift very heavy loads. *Figure 4* shows a three-crane lift with two of the cranes using an equalizer beam with a special load attachment. The equalizer beam and attachment split the load between the two cranes. Equalizer beams are covered later in this module.

Whenever two or more cranes are used to lift a load, the location of each crane's hook block above the load has to be chosen carefully so that the load is shared between the cranes. Because each crane used will usually have a different net capacity, the method for properly locating each crane's hook block above the load (*Figure 5*) for a two-crane lift is as follows:

Step 1 Determine the lowest value of rated net capacity that Crane One will have during the whole operation. When lifting from the main boom or an extension, consider all factors, including configuration, quadrant of operation, boom length, boom angle, load radius, weight of rigging, and all capacity deductions.

Step 2 Repeat Step 1 for Crane Two.

Step 3 Ensure that the sum of the lowest rated net capacities of the cranes exceeds the load weight.

Step 4 Locate the position of the load's center of gravity.

Demonstrate how to perform single-crane lifting calculations.

Have trainees practice performing single-crane lifting calculations. See the suggested exercise for Performance Task 2 located in the Test Booklet. Note the proficiency of each trainee. This laboratory corresponds to Performance Task 2.

Have trainees review Section 3.2.0.

Ensure that you have everything required for teaching this session.

Describe multiple-crane lifts. Emphasize the importance of choosing the location of each crane's hook block.

Show Transparency 6 (Figure 5). Explain the steps for properly locating each crane's hook block above the load.

See the Teaching Tip for Section 3.2.0 at the end of this module.

Show Transparencies 7 through 11 (Figures 6 through 10). Review the steps for ensuring that neither crane in a multiple-crane lift is loaded beyond its lowest rated net capacity.

304F04.EPS

Figure 4 ◈ Three-crane lift.

Step 5 Each crane's load blocks must be positioned relative to the load's center of gravity according to the following formula:

(Net capacity of Crane One) × A = (net capacity of Crane Two) × B

Step 6 The crane with the greatest net capacity must always be closest to the load's center of gravity. Measure the distance of this point from the center of gravity. If you have measured distance A, then:

$$B = \frac{\text{net capacity of Crane One}}{\text{net capacity of Crane Two}} \times A$$

If you have measured distance B then:

$$A = \frac{\text{net capacity of Crane Two}}{\text{net capacity of Crane One}} \times B$$

Step 7 Determine the share of the load weight that will be carried by each crane.

Load on Crane One

$$\frac{B}{A + B} \times \text{load weight}$$

Load on Crane Two

$$\frac{A}{A + B} \times \text{load weight}$$

Step 8 Ensure that neither crane is loaded beyond its lowest rated net capacity.

NOTE

Each lifting operation is a unique situation. The numbers provided in this example are randomly selected for instructional purposes.

An example of this procedure is as follows:

Step 1 Lowest rated net capacity of Crane One = 156,000 lbs – 500 lbs (beam clamp) = 155,500 lbs.

Step 2 Lowest rated net capacity of Crane Two = 90,500 lbs – 500 lbs (beam clamp) = 90,000 lbs.

Step 3 Net capacity = 155,500 lbs + 90,000 lbs = 245,500 lbs.

Step 4 Locate the load's center of gravity from the ends of the load, as shown in *Figure 6*.

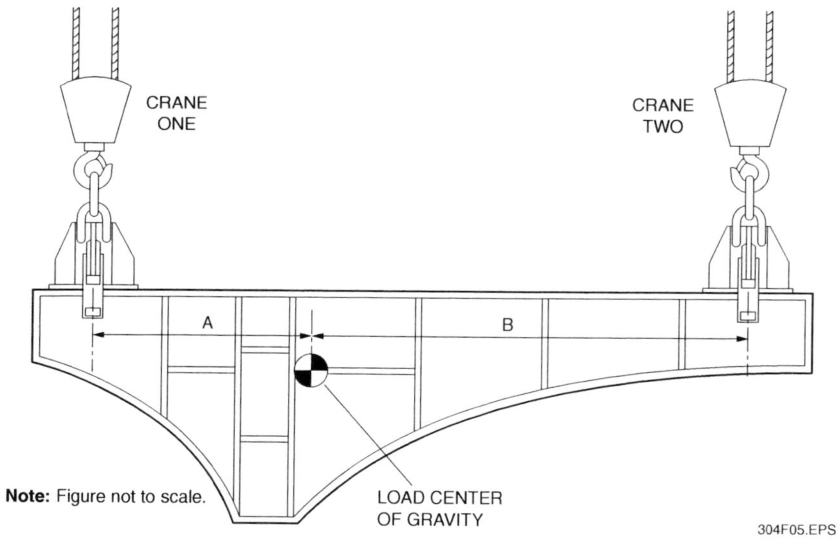

Figure 5 ◈ Multiple-crane lift.

4.6 MOBILE CRANE OPERATIONS LEVEL THREE — TRAINEE MODULE 21304-05

Instructor's Notes:

Step 5 To set the location of the hook blocks, use the formula:

(Net capacity of Crane One) × A = (net capacity of Crane Two) × B

Step 6 Choose a hypothetical length for A that will be within the length of C in *Figure 6*. Start with A = 45 ft (*Figure 7*), for example.

$$\text{Calculate B} = \frac{\text{net capacity of Crane One}}{\text{net capacity of Crane Two}} \times A$$

$$= \frac{155{,}500 \text{ lbs}}{90{,}000 \text{ lbs}} \times 45 \text{ ft}$$

$$= 78 \text{ feet}$$

This configuration will not work because the hook block for B (78 ft) would be beyond the end of the right side of the load (*Figure 8*). Choose another dimension for A: 36 ft (*Figure 9*).

$$\text{Recalculate B} = \frac{\text{net cap. of Crane One}}{\text{net cap. of Crane Two}} \times A$$

$$= \frac{155{,}500 \text{ lbs}}{90{,}000 \text{ lbs}} \times 36 \text{ ft}$$

$$= 62.2 \text{ ft}$$

This dimension fits the load (*Figure 10*).

Step 7 Determine the actual load that each crane will be carrying:

$$\text{Load/Crane One} = \frac{B}{A + B} \times \text{load weight}$$

$$= \frac{62.2}{36 + 62.2} \times 180{,}000 \text{ lbs}$$

$$= 114{,}012.2 \text{ lbs}$$

This load is less than Crane One's rated net capacity of 155,500 lbs, and is under 75 percent (73.3 percent) of the capacity.

$$\text{Load/Crane Two} = \frac{A}{A + B} \times \text{load weight}$$

$$= \frac{36}{36 + 62.2} \times 180{,}000 \text{ lbs}$$

$$= 65{,}988 \text{ lbs}$$

This load is less than Crane Two's rated net capacity of 90,000 lbs and is under 75 percent (73.3 percent) of the capacity.

Step 8 Since neither crane is overloaded nor loaded beyond 75 percent of its capacity, the lift can proceed with A = 36 ft and B = 62.2 ft. Crane One will be 36 ft from the load's center of gravity, and Crane Two will be 62.2 ft from it.

> **NOTE**
> If the load radius of either crane is changed, repeat all of the steps using the new net capacity of the crane(s).

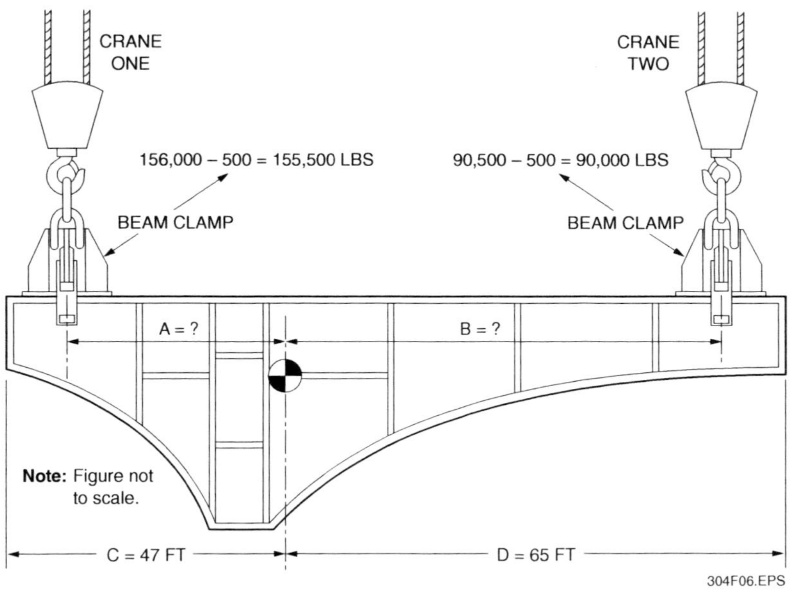

Figure 6 ◈ Multiple-crane lift – center of gravity location.

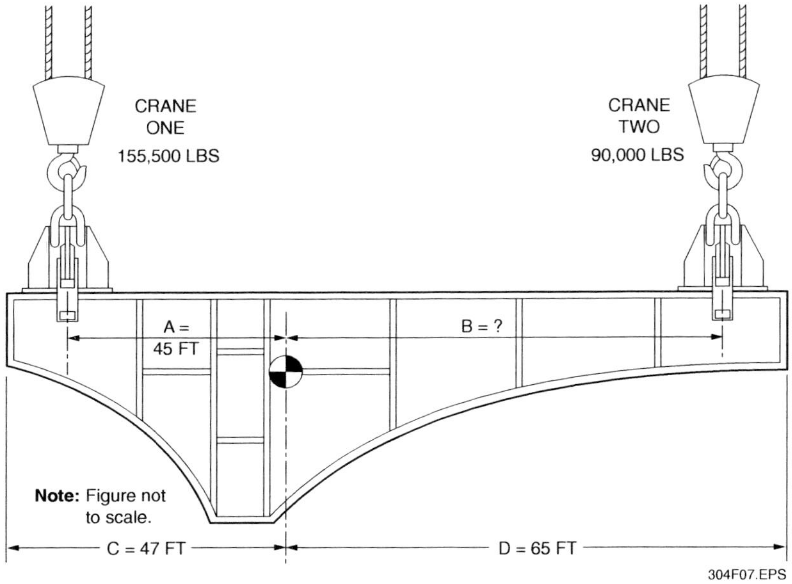

Figure 7 ◈ Multiple-crane lift – distance B, trial 1.

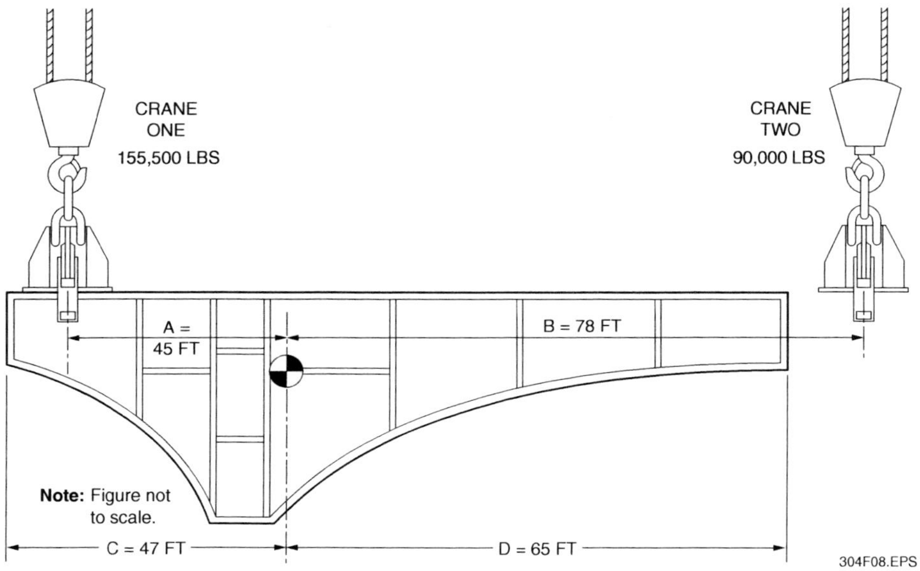

Figure 8 ◈ Multiple-crane lift – distance B, trial 1 results.

Instructor's Notes:

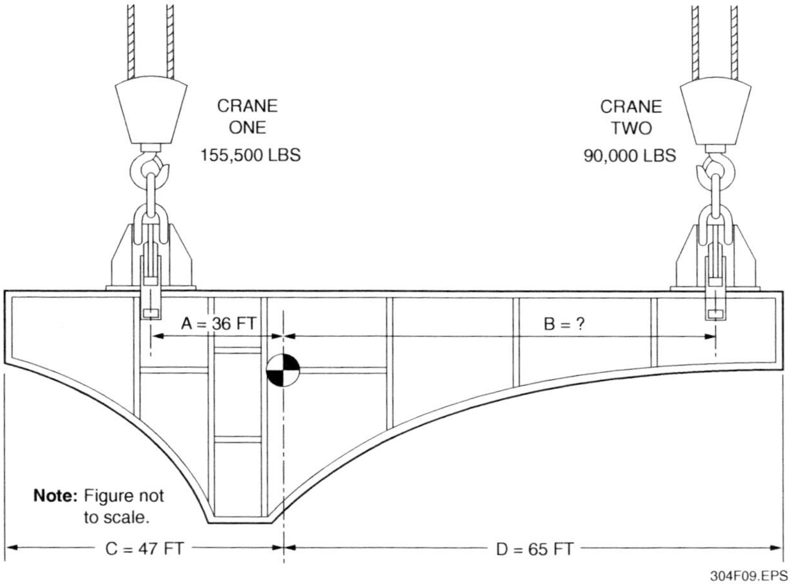

Figure 9 ◈ Multiple-crane lift – distance B, trial 2.

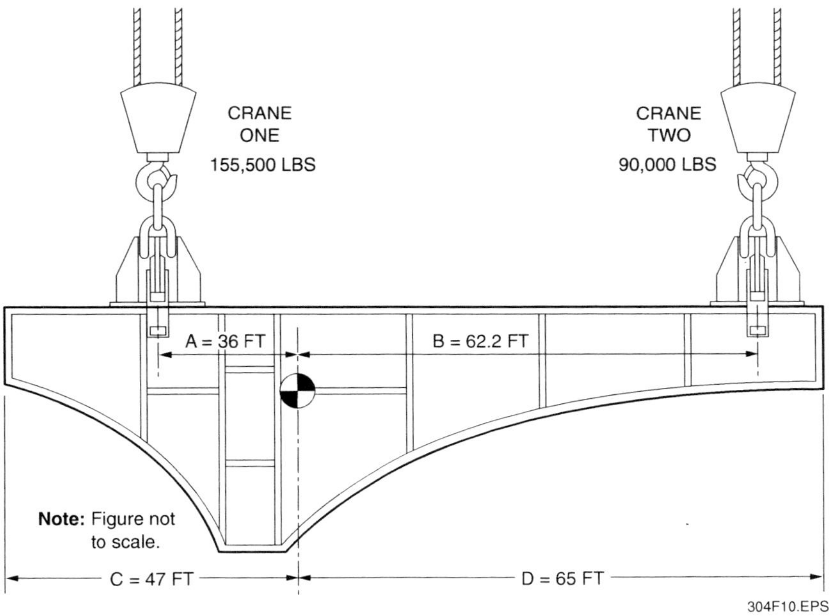

Figure 10 ◈ Multiple-crane lift – distance B, trial 2 results.

LIFT PLANNING 4.9

Describe an equalizer beam and explain how it is used in multiple-crane lifts.

Show Transparency 12 (Figure 11). Work through the procedure for determining the location on the equalizer beam so that each crane shares the load in proportion to its capacity.

Multiple crane lifts can also use **equalizer** beams (*Figure 11*). Equalizer beams distribute the weight of a load between two cranes. By adjusting the crane and/or load attachment points on the beam, it is possible to share the load weight between the cranes in any proportion desired.

The operator can determine the location on the equalizer beam where the load is to be picked up, allowing each crane to share the load proportionately to its capacity, by following this method:

Step 1 Determine the lowest value of rated net capacity that Crane One will have when lifting from the main boom or an extension during the whole operation. Consider all factors, including configuration, operational quadrant, and load radius. Remember the load radius of each crane is measured to its pick-up point on the equalizer, not to the center of gravity of the load.

Step 2 Repeat Step 1 for Crane Two.

Step 3 Check to ensure that the sum of the lowest rated net capacities of the cranes exceeds the load weight plus the equalizer and sling weight. Ideally the weight should not exceed, 75 percent of the sum of their capacities.

Step 4 Measure the distance between the crane pick-up points on the equalizer beam (A).

Step 5 Establish the attachment point of the load on the equalizer beam according to the following formulas:

$$B = \frac{\text{net capacity of Crane Two}}{\text{total net capacity of Cranes One and Two}} \times A$$

$$C = \frac{\text{net capacity of Crane One}}{\text{total net capacity of Cranes One and Two}} \times A$$

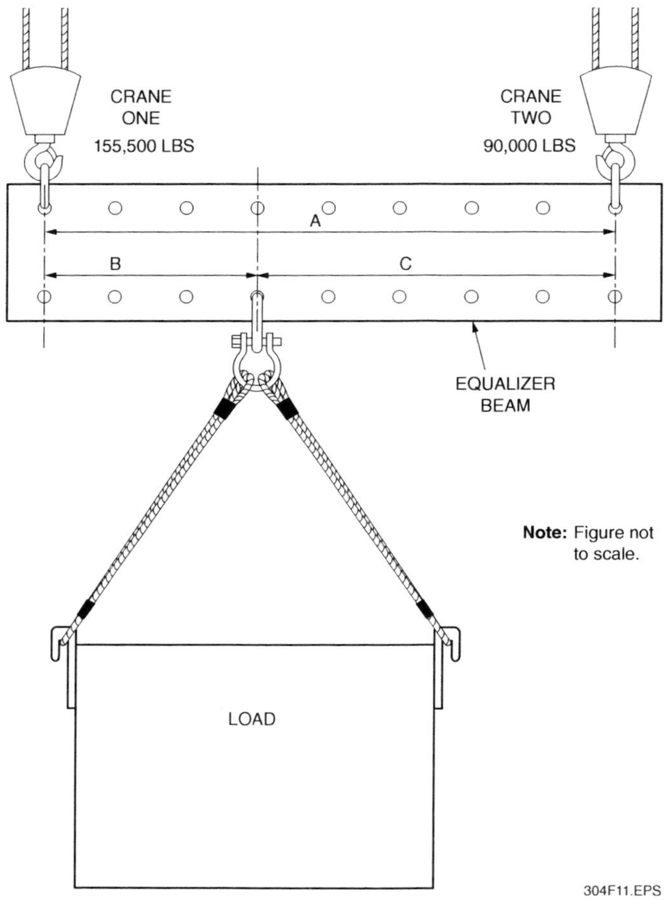

Figure 11 ◆ Multiple-crane lift with equalizer beam.

Instructor's Notes:

Step 6 When the equalizer pick-up points have been established, determine the share of the load weight, including the weight of the equalizer, that will be carried by each crane:

$$\text{Load on Crane One} = \frac{C}{A} \times (\text{load weight})$$

$$\text{Load on Crane Two} = \frac{B}{A} \times (\text{load weight})$$

Check to ensure that neither crane is loaded beyond its lowest rated net capacity. Remember, the best situation is where neither crane is loaded in excess of 75 percent of its lowest rated net capacity. An example of this procedure follows (see *Figure 12*).

Step 1 Check the total capacity of the cranes.

Crane One = 120,000 lbs,
Crane Two = 80,000 lbs

Crane One + Crane Two = total capacity

120,000 lbs + 80,000 lbs = 200,000 lbs total capacity

This is greater than the total load weight of 143,000 lbs (load + equalizer). Also, 75 percent of the capacities' sum exceeds the load weight (75 percent of 200,000 = 150,000 lbs).

Step 2 Establish the load's pick-up point.

$$B = \frac{\text{net capacity of Crane Two}}{\text{total net capacity of Cranes One and Two}} \times A$$

$$B = \frac{80{,}000 \text{ lbs}}{80{,}000 \text{ lbs} + 120{,}000 \text{ lbs}} \times 25 \text{ ft}$$

$$B = \frac{80{,}000 \text{ lbs}}{200{,}000 \text{ lbs}} \times 25 \text{ ft}$$

B = 10 ft ; therefore, C = 15 ft, as shown in *Figure 13*

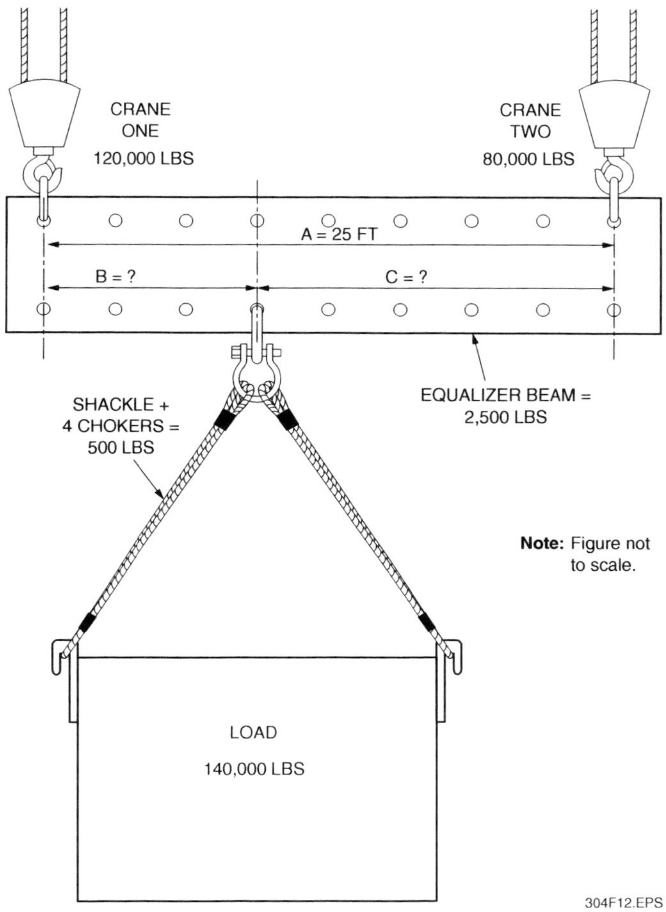

Figure 12 ◈ Multiple-crane lift with equalizer beam, determining distance B and C.

Show Transparencies 13 and 14 (Figures 12 and 13). Review the procedure for ensuring that neither crane is loaded beyond its lowest rated net capacity.

Demonstrate how to perform multiple-crane lifting calculations.

Have trainees practice performing multiple-crane lifting calculations. See the suggested exercise for Performance Task 3 located in the Test Booklet. Note the proficiency of each trainee. This laboratory corresponds to Performance Task 3.

Have trainees review Sections 4.0.0–6.0.0.

Ensure that you have everything required for teaching this session.

Describe the three categories of lifts: ordinary, critical, and pre-engineered, and give examples of each.

See the Teaching Tip for Section 4.0.0 at the end of this module.

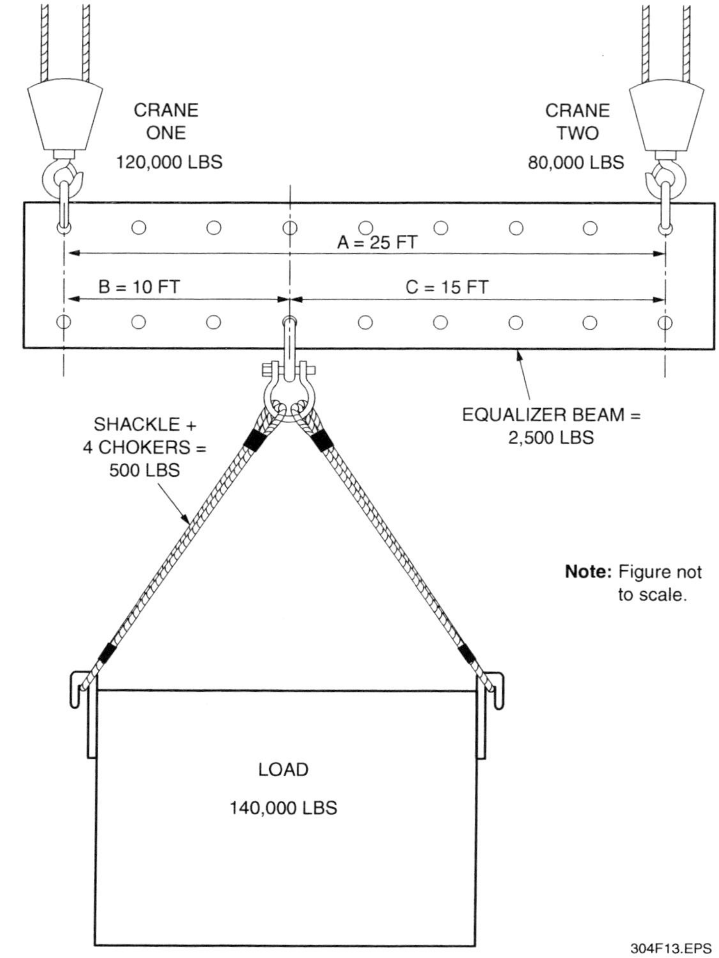

Figure 13 ◈ Multiple-crane lift with equalizer beam, distance B and C results.

Step 3 Determine the load that each crane will actually be carrying.

$$\text{Load on Crane One} = \frac{C}{A} \times (\text{total load weight})$$

$$= \frac{15}{25} \times 143{,}000 \text{ lbs}$$

$$= 85{,}800 \text{ lbs}$$

This number is less than the lowest rated net capacity and also meets the 75-percent capacity guideline (75 percent of 120,000 = 90,000 lbs).

$$\text{Load on Crane Two} = \frac{B}{A} \times (\text{total load weight})$$

$$= \frac{10}{25} \times 143{,}000 \text{ lbs}$$

$$= 57{,}200 \text{ lbs}$$

This number is less than the lowest rated net capacity and also meets the 75-percent capacity guideline (75 percent of 80,000 = 60,000 lbs).

4.0.0 ◈ CRITICAL LIFTS

A designated person must classify each lift into one of the following categories (ordinary, critical, or pre-engineered production) prior to planning the lift. A lift must be designated as a critical lift if

Instructor's Notes:

collision, upset, or dropping could result in any one of the following:

- Unacceptable risk of personnel injury or significant adverse health impact (on site or off site)
- Significant release of radioactive or other hazardous material or other undesirable conditions
- Undetectable damage that would jeopardize future operations or the safety of a facility
- Damage that would result in unacceptable delay to schedule or other significant program impact, such as loss of vital data

A lift should also be designated as critical if the load requires exceptional care in handling because of size, weight, close-tolerance installation, high susceptibility to damage, or other unusual factors.

For critical lifts, the operating organization must appoint a rigging superintendent or other qualified person who must be present at the lift site during the entire lifting operation. The rigging supervisor or other qualified person must ensure that a lift plan or procedure (see *Appendix A*) is prepared that defines the operation and includes the following:

- Identification of the items to be moved, the weight, dimensions, and center of gravity of the load, and any hazardous or toxic materials that are present
- Identification of operating equipment to be used by type and rated capacity
- Rigging sketches that include (as applicable):
  - Identification and rated capacity of slings, lifting bars, rigging accessories, and below-the-hook lifting devices
  - Load-indicating devices
  - Lifting points
  - Sling angles
  - Boom and swing angles
  - Methods of attachment
  - Crane orientations
  - Other factors affecting equipment capacity
- Operating procedures and special instructions to operators, including rigging precautions and safety measures to be followed as applicable

Experienced operators who have been trained and qualified to operate the specific equipment to be used must be assigned to make the lift.

Only designated, qualified signalers must give signals to the operator. However, the operator must obey a STOP signal at all times, no matter who gives the signal.

The procedure and rigging sketches must be reviewed and approved by the responsible manager (or designee) and the responsible oversight organization (such as safety, quality assurance, or quality control) before the lift is made.

5.0.0 ◆ ENGINEERING CONSIDERATIONS

The following **engineering** requirements need to be considered when planning a lift:

- Basic load data, such as the object to be lifted, its dimensions and general configuration (rectangle, square, sphere, or irregular), and its gross weight
- Location in which load is to be lifted and placed
- Maximum operating radius
- Lift attachment locations and descriptions
- Maximum distance between any two attachment locations
- Sling lengths needed to keep the horizontal angle greater than 60 degrees
- Specifications of the crane to be used (manufacturer, model)
- Counterweight attachments

Lattice boom cranes:

- Counterweights used on the lower and upper works
- Number of boom sections used and their sizes and lengths
- Number of jib sections used and their sizes and lengths
- Maximum capacity at the maximum lift radius
- Hoist block capacity
- Wire rope size and weight per foot
- Rated single line pull
- Parts of line reeved to the block
- The maximum capacity of the lift point for the boom and the jib

Ground support:

- Ground pressure allowed at the setup location and the ground pressure created by the crane and load at all radii out to the planned maximum load radius
- The maximum allowable crane load that will not exceed the ground support

Load calculations:

- Weight of the load, crane components below the lifting points, and wire rope

Administration:

- Pre-lift meeting
- Completion of lift plan

Discuss the factors that mandate designating a lift as critical. Explain that a lift must be designated as critical if collision, upset, or dropping could result in certain adverse situations.

Explain that a load must also be designated as critical if it requires exceptional care due to certain factors such as size or weight.

Discuss the role of the rigging supervisor and review the critical lift plan procedures in *Appendix A*.

Discuss engineering requirements when planning a lift.

Describe engineering considerations for lattice boom cranes, ground support, load calculations, and administration.

LIFT PLANNING 4.13

Discuss the best method to implement a lift plan.

Explain the benefits of a practice lift.

Summarize the major concepts presented in the module.

Have trainees complete the Review Questions, and go over the answers prior to administering the Module Examination.

6.0.0 ◆ LIFT PLAN IMPLEMENTATION

For a lift plan to be successful, the critical data must be communicated to all personnel involved. Holding a pre-lift meeting accomplishes this goal. The signal person(s), riggers, operators, management, and site supervisors should all be involved in this meeting. Officially identify which operator and signal person(s) will be assigned the job, and determine how communication is to be maintained during the lift (including the use of hand signals, radio, or both, or other methods). Also, determine whether tag lines will be used. Discuss all potential hazards, including weather, power lines, or mechanical failure. Review procedures in the event of an emergency. Finally, have the person responsible for the lift, the site superintendent, the site safety manager, and any others needed sign off on the lift plan.

If required by the critical lift procedure, a practice lift must be done before the critical lift. Conditions for a practice lift should closely simulate actual conditions, including weight, rigging selection and configuration, load movement path, and other relevant factors. The same crew, using the same lifting equipment, should do practice lifts.

Summary

Lift plans are just as important to the success of a lifting operation as proper load/capacity chart interpretation. The operator, supervisor, signal person, and all other personnel involved in the lift must understand what is required on a lift plan and how it affects them. Personnel must also understand mathematical calculations and the equipment's limitations.

Review Questions

1. A document that outlines all weight, rigging, and configuration requirements for a lift is a(n) _____.
   a. lift plan
   b. manufacturer's worksheet
   c. pre-lift qualification
   d. load chart

2. Lift plans are mandatory for _____.
   a. steel erection or multiple-crane lifts
   b. lifts using extensions or jibs
   c. luffing or derricking operations
   d. lifts using a load block

3. The type of hitch and number of shackles are criteria that fall under the category of _____.
   a. load weight
   b. sizing of slings
   c. wire ropes
   d. rope reeving

4. Foundation and location of electrical hazards are examples of _____ information.
   a. crane placement
   b. hoisting
   c. hydraulic
   d. pneumatic

5. The number of feet and the weight per foot of the _____ need to be considered as part of a lift plan.
   a. boom
   b. jib
   c. wire rope
   d. swing-away

6. Over-the-front, over-the-rear, and over-the-side are _____ considerations for crane placement.
   a. crawler configuration
   b. area of operation
   c. gantry position
   d. counterweight

7. Critical lifts always require _____.
   a. an equalizer beam
   b. beam clamps
   c. radio communication
   d. a lift plan

Instructor's Notes:

8. When lifting from the main boom of a lattice crane, an erected jib, whip line length below the jib, and the length of the line from the lift point to load block are calculations involving the _____.
   a. weight of crane components below the lifting point
   b. load block height
   c. rigging on the block
   d. effective weight of crane components past the lifting point

9. The load block weight is considered part of the weight of the _____.
   a. load to be lifted, not including rigging
   b. crane components below the lifting point
   c. crane components past the lifting point
   d. boom components

10. The effective weight of crane components past a main boom lifting point includes _____.
   a. an erected jib
   b. any parts of line considered part of the load
   c. the maximum wire rope weight to be experienced during the lift
   d. lifting beams, slings, and shackles

11. If the net capacity of a crane in a particular configuration is 95,000 lbs and the total weight of the deductions is 2,500 lbs, the gross capacity is _____ lbs.
   a. 92,500
   b. 95,000
   c. 97,500
   d. impossible to calculate without more information

12. If the gross capacity of a crane at 200 ft of boom length and 50 ft of load radius is 83,000 lbs, and the deductions are a 1,500-lb boom point, a 865-lb hook block, 335 lbs of rigging, and 1,900 lbs of rope, the net capacity is _____ lbs.
   a. 74,600
   b. 78,400
   c. 87,600
   d. impossible to calculate without more information

13. If the gross capacity of a crane is 90,000 lbs and its net capacity in a particular configuration is 83,425 lbs, the total weight of the deductions is _____ lbs.
   a. 2,875
   b. 4,932
   c. 6,575
   d. impossible to calculate without more information

14. By adjusting the crane and/or load attachment points on an equalizer beam, it is possible to _____.
   a. safely load a crane beyond its net capacity
   b. ensure that load rate exceeds the lowest rated net capacities of the cranes
   c. share the load between cranes in any proportion desired
   d. share the load between cranes equally

15. In a multiple-crane lift, the crane with the greatest net capacity must always be _____.
   a. equipped with single part hoist line
   b. loaded to at least 75 percent of its rated capacity
   c. farthest from the load's center of gravity
   d. closest to the load's center of gravity

16. When an equalizer beam is used, the load radius of each crane is measured to _____.
   a. the center of gravity of the load
   b. its pick-up point on the equalizer
   c. the center of the equalizer
   d. the attachment point of the load on the equalizer

17. In a multiple-crane lift using an equalizer, the weight of the equalizer is considered part of the load weight.
   a. True
   b. False

18. In a multiple-crane lift, if the total load weight is 100,000 lbs, and the total net capacity of Crane One is 80,000 lbs, the total net capacity of Crane Two must be no less than _____ lbs to meet the 75-percent capacity guidelines.
   a. 53,333
   b. 75,050
   c. 100,500
   d. 180,000

Administer the Module Examination. Record the results on Craft Training Report Form 200, and submit the results to the Training Program Sponsor.

Administer the Performance Test, and fill out Performance Profile Sheets for each trainee. If desired, trainee proficiency noted during laboratory sessions may be used to complete the Performance Test. Record the results on Craft Training Report Form 200, and submit the results to the Training Program Sponsor.

LIFT PLANNING 4.15

Refer to *Figure 1* for Questions 19 and 20.

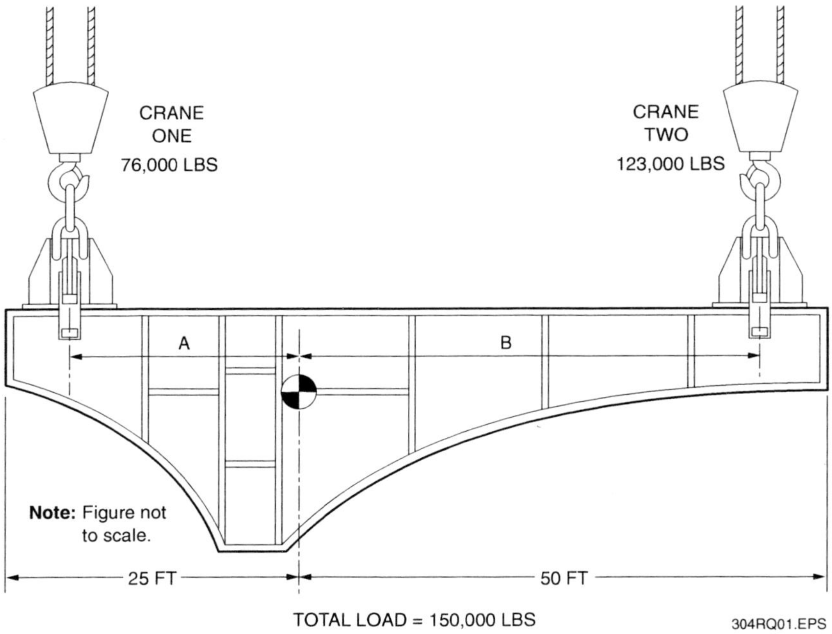

Figure 1

19. If A = 20 ft, then what is the value for B?

a. 10.6 ft
b. 12.4 ft
c. 30.1 ft
d. 37.6 ft

20. If A= 15 ft and B = 10 ft, then what is the actual load on Crane One?

a. 57,000 lbs
b. 60,000 lbs
c. 76,000 lbs
d. 150,000 lbs

21. In preparing for multiple-crane lifting, the load blocks of each crane have to be positioned relative to the load's center of gravity using the formula _____.

a. (net capacity Crane One) × A = (net capacity Crane Two) × B
b. (net capacity Crane One) + A = (net capacity Crane Two) + B
c. (net capacity Crane Two) × A = (net capacity Crane One) × B
d. (net capacity Crane One) – A = (net capacity Crane Two) – B

Instructor's Notes:

22. To determine the share of the load weight on Crane One that is positioned a distance of (A) feet from the load's center of gravity in a multiple-crane lift, the operator must use the formula _____.
    a. [B ÷ (A + B)] × load weight
    b. [B ÷ (A – B)] × load weight
    c. [A ÷ (A + B)] × load weight
    d. [A ÷ (A – B)] × load weight

23. Prior to planning a lift, a designated person must classify the lift into which of the following categories?
    a. Major, minor, critical
    b. Ordinary, critical, pre-engineered production
    c. Standard, critical
    d. General, special, critical

24. During critical lift operation, the operator must obey a _____ signal at all times.
    a. travel
    b. stop
    c. dog everything
    d. swing

25. A practice lift is recommended before _____.
    a. any lift
    b. a lift involving a luffing jib boom
    c. a lift involving a 100-ton hook block assembly
    d. a critical lift

LIFT PLANNING 4.17

Instructor's Notes:

GLOSSARY

Trade Terms Introduced in This Module

Engineering: Practical application of the knowledge of physics, leverage, and gravity as they relate to lifting operations.

Equalizer: Any of the various devices or appliances for equalizing stresses, strains, and pressures.

Lift plan: A written plan that outlines all weight, rigging, and configuration considerations for a lift.

LIFT PLANNING 4.19

Instructor's Notes:

APPENDIX A

Critical Lift Plan

APPENDIX A – CRITICAL LIFT PLAN

INDEX

1. OBJECTIVE

2. DEFINITIONS

2.1 Critical Lift
2.2 Minor Lift
2.3 Rigging
2.4 Lift Crane
2.5 Tail Crane

3. PROCEDURE

3.1 Responsibility for Critical Lift
3.2 Responsibility for Minor Lift
3.3 Special Lifting Requirements
3.4 Training

APPENDICES

Appendix 1 - Planned Lifting Procedure Form

PREPARED BY: ______________________
Engineering Representative

APPROVED BY: ______________________
Safety Representative

AUTHORIZED BY: ______________________
Project / Site Manager

DEVELOPED BY CRANE INSPECTION & CERTIFICATION BUREAU - 2004 1

304A01.EPS

LIFT PLANNING 4.21

CRITICAL LIFTS | ***NOTES***

1 PURPOSE

1.1 Define responsibilities for personnel when crane lifts are made.

- To ensure the rigging and lift crane or cranes have adequate capacity for handling the weight of the device to be lifted.
- To ensure that the personnel responsible for the lift are qualified and competent personnel.
- To protect personnel and property.

2 DEFINITIONS

2.1 **Critical Lift** – As an accepted "General Industry Standard" The combined weight of the load, rigging and crane configuration exceeds 75% - 85% of the capacity of the crane and or there are complications as a result of load configuration, clearance, danger to personnel, power transmission equipment, critical service equipment or hazardous materials.

- The weight of the load is 25 tons or greater and exceeds 50' in length, height, and width. (Detailed Lift Plan required)
- Lifts involving more than one crane to handle a common load.

2.2 **Minor Lift** - The combined weight of the device, rigging and crane configuration is less than 80% of the crane capacity and there are no compilations as above.

- The lift weight is less than 25 tons and 50' in length.
- Lift is single crane.

2.3 **Rigging** – Any component part used to connect the device to be lifted by the crane; i.e., choker, sling, shackles, jib, spreader bars, snatch block, etc.

2.4 **Lift Crane** – The primary crane used to ultimately support the total weight of the device.

2.5 **Tail Crane** - The equipment used to tail or maneuver the device while the lift crane supports the device.

DEVELOPED BY CRANE INSPECTION & CERTIFICATION BUREAU - 2004 2

304A02.EPS

Instructor's Notes:

CRITICAL LIFTS *NOTES*

3 PROCEDURE

3.1 Responsibility for Critical Lifts

3.1.1 The Rigging Superintendent making the lift is responsible for planning and executing the lift. The superintendent shall complete the attached lifting procedure and acquire all approval signatures prior to the lift. The original "approved" procedure is to be submitted to the organization's Engineering Representative and Safety director for record and one copy retained by the Construction Rigging Superintendent. The Rigging Superintendent must include the organization's Engineering Representative and Safety Director in developing the lift plan to ensure full consideration of existing conditions and circumstances. Any lifts exceeding 85% must have Project / Site Manager 's signature.

The planning should include:

- Determining the maximum weight of the device to be lifted.
- Calculating the combined weight of the device, rigging accessories, load blocks and crane jibs.
- Calculating the sling and rigging hardware capacities with consideration to load angle factors and stresses on all slings, hardware and lift points.
- Determining the required boom angle and swing radius.
- Selecting the proper crane for the lift and the tail crane when required.
- Inspecting the lift site to determine the soil condition to decide whether matting and / or soil stabilization is needed.
- Determine the area to be restricted during lift.

304A03.EPS

LIFT PLANNING 4.23

CRITICAL LIFTS | **NOTES**

3.1.2 After a careful review of the job site, the hoisting equipment and critical lift plan, the Rigging Supervisor will record the project data on the Planned Lifting Procedure Form. The crane operator will sign and date the completed form. In addition to his signature, the lifting procedure will be reviewed and signed by others as indicated on the lift form.

3.1.3 The crane operator has the authority and responsibility to ensure the lift is proper and in accordance with established local procedures, as well as OSHA and ANSI standards.

3.2 Responsibility for Minor Lifts

3.2.1 The Rigging Superintendent or competent person making the lift is responsible for completing the "Lifting Procedure" planning / executing the lift and submitting the lifting procedure to the Engineering Representative for record. Note: Repetitive minor lifts need only one "Lifting Procedure" completed per day of lifting activities; i.e., multiple pieces of structural steel and pipe installation. However, if the crane is repositioned, another "Lifting Procedure" should be completed before lifting is resumed.

- Should hoisting activities become routinely cycle duty procedures and located in remote locations the Engineering Representative or Project Superintendent may waive the completion of a lift form after review.

The Planning Should Include:

- Determining the maximum weight of the device to be lifted.
- Selecting the rigging based upon the maximum weight and calculate sling and hardware stresses.
- Calculating the combined weight of the device, rigging, load blocks and crane jibs.
- Determining the center of gravity of the load.

DEVELOPED BY CRANE INSPECTION & CERTIFICATION BUREAU - 2004 4

304A04.EPS

Instructor's Notes:

- Determining the required boom length, boom angle and swing radius at the pick point and the point the load is to be placed.
- Selecting the proper crane for the lift and the tail crane when required.
- Inspecting the lift site to determine the soil condition to decide whether matting and/or soil stabilization is needed.
- Identify the location of all utilities.
- Determine the area to be restricted to public access during lift.
- Determine the swing area.

3.2.2 After the careful review of the job site, hoisting equipment and the lift plan is completed, the Rigging Supervisor will record the project data on a Planned Lifting Procedure Form. The crane operator will sign and date the completed form. In addition to his signature, the Lifting Procedure will be reviewed and signed by the rigger.

3.3 Special Lifting Requirements

Whenever lifting devices which are smooth surfaced and lightweight; i.e., fiberglass or aluminum vessels, with canvas or nylon slings a strongback and / or bracing sling shall be used to prevent the lifting slings from rendering. Rendering (or slippage) of the main lifting slings can occur if the resistant force of friction between the sling and vessel wall is exceeded due to lift movement. Therefore, the strongback and / or bracing slings must be properly attached to prevent the main lifting slings from rendering in any position during the entire lift.

3.4 Training

3.4.1 Each person whose job requires them to operate cranes shall receive the following training:

- The recognition of the hazards associated with lifting equipment.

304A05.EPS

CRITICAL LIFTS | *NOTES*

- The fundamentals of rigging practices and the understanding of load capacity charts as they pertain to crane configurations.
- The method and means of calculating load capacities.
- Attend a training course in mobile crane operations and rigging considerations.

3.4.2 Each person who is involved in rigging shall be required to participate in a rigging training program and tested to become competent in the selection of rigging when a lift is performed.

3.4.3 All trades personnel and supervision involved in lifting shall be instructed about this procedure and the requirements.

DEVELOPED BY CRANE INSPECTION & CERTIFICATION BUREAU - 2004

6

304A06.EPS

Instructor's Notes:

3.4.4 PLANNED LIFTING PROCEDURE FORM

1. DATE OF FORM PREPARATION:	
2. PROJECT NUMBER:	
3. PROJECT TITLE / LOCATION:	
4. PERMIT REQUIRED: Yes / No	
5. PREPARED BY:	
6. LOAD DESCRIPTION / PHYSICAL SIZE:	
• Equipment Name	
• I.D. / Tag Number	
• Physical Dimensions	
• Equipment Orientation V or H	
• Lift Lug Orientation Degrees	
• Hoisted load / Device Age	
• Physical Condition	
• NDE Required Y / N	
7. LIFTING AND TAILING ATTACHMENTS APPROVED: Y / N	
8. LOAD ELEVATION:	
9. LOAD WEIGHT:	

304A07.EPS

LIFT PLANNING 4.27

CRITICAL LIFTS ***NOTES***

APPLICATION - LIFT CRANE

10. CRANE(S) TYPE:

11. CRANE CONFIGURATION:

12. PARTS OF WIRE ROPE: Wire Rope Breaking Strength

- Single Line Pull Wire Rope SWL

13. MAXIMUM RADIUS:

14. BOOM ANGLE:

15. BOOM CLEARANCE CHECK REQUIRED: Yes No

16. TAIL SWING CLEARANCE VERIFIED: Yes No

17. CHART CAPACITY:

18. CRANE RIGGING:	SIZE	CAPACITY	CONDITION
• Choker(s)			
• Sling(s)			
• Shackle(s)			
• Spreader Bar(s)			
• Other			

19. CRANE RIGGING COMPONENT WEIGHT:

- Jib Weight
- Ball Weight
- Load Block Weight
- Suspended Load Line Weight
- Shackle(s) Weight
- Choker(s) Weight
- Sling(s) Weight
- Spreader Bar(s) Weight
- Other Lifting Attachment Weights
- Total Rigging Weight

20. GROSS LOAD:

21. % OF CHART:USED

DEVELOPED BY CRANE INSPECTION & CERTIFICATION BUREAU - 2004 8

304A08.EPS

Instructor's Notes:

CRITICAL LIFTS ***NOTES***

APPLICATION - TAIL CRANE

10. CRANE(S) TYPE:

11. CRANE CONFIGURATION:

12. PARTS OF WIRE ROPE: Wire Rope Breaking Strength

- Single Line Pull Wire Rope SWL

13. MAXIMUM RADIUS:

14. BOOM ANGLE:

15. BOOM CLEARANCE CHECK REQUIRED: Yes No

16. TAIL SWING CLEARANCE VERIFIED: Yes No

17. CHART CAPACITY

18. CRANE RIGGING	SIZE	CAPACITY	CONDITION
• Choker(s)			
• Sling(s)			
• Shackle(s)			
• Spreader Bar(s)			
• Other			

19. CRANE RIGGING COMPONENT WEIGHT:

- Jib Weight
- Ball Weight
- Load Block Weight
- Suspended Load Line Weight
- Shackle(s) Weight
- Choker(s) Weight
- Sling(s) Weight
- Spreader Bar(s) Weight
- Other Lifting Attachment Weights
- Total Rigging Weight

20. GROSS LOAD:

21. % OF CHART USED:

22. RIGGING SAFETY DESIGN FACTOR 5 to 1 Yes No

304A09.EPS

CRITICAL LIFTS | **NOTES**

Engineering Certification Spreader Bar (Required)

- Proof Test Certification of Chokers 1 1/4" & Greater (Required)

23. SOIL TYPE / CONDITION:

24. MATS / CRIBBING Yes No

25. RIGGING INSPECTED: Yes No

26. TAG LINE:

27. HAZARDS

28. WEATHER:

29. DATE: TIME:

30. LIFT MEETING CONDUCTED: Yes No

31. APPROVALS:

Rigging Supervisor: ____________________________

Crane Operator. ____________________________

Rigger / Signalman. ____________________________

Client Representative (If Required):

Safety Representative (If Required):

Engineering Representative (If Required)

Project Manager (If Required)

Communications: What form of communications will be used and by whom?

DEVELOPED BY CRANE INSPECTION & CERTIFICATION BUREAU - 2004 10

304A10.EPS

Instructor's Notes:

INSTRUCTIONS FOR FILLING OUT PLANNED LIFTING PROCEDURE FORM

1. **Date of form preparation.**
2. **PROJECT NUMBER:** Enter the project number that has been assigned to this work activity.
3. **PROJECT TITLE/LOCATION:** Enter the name of the project, plant site or construction site of the lift.
4. **PERMIT REQUIRED (y/n):** Indicate if a permit will be required. Certain plant safe work procedures require a permit when making certain lifts over in-service facilities or lifts made near power lines or guy lines or exceeding 75% of load crane capacity chart.
5. **PREPARED BY:** Enter name of person responsible for research and preparation of this form.
6. **LOAD DESCRIPTION / PHYSICAL SIZE:** Describe the load to be lifted including name of the equipment and item number and or tag number. Enter physical dimensions of load to be lifted. Identify equipment orientation with a "V" or "H", "V" for vertical and "H" for horizontal. Identify age of existing equipment / device to be lifted, physical condition and / or if NDE is required on item to be lifted prior to lift.
7. **LIFTING / TAILING ATTACHMENTS APPROVED (y/n):** Indicate if lifting attachments are approved by, the equipment engineering firm.
8. **LOAD ELEVATION:** Indicate the maximum and minimum elevation in feet that the load will travel Enter the maximum radius the load will achieve during the lift cycle of pick, swing & set.
9. **LOAD WEIGHT:** Determine the weight of the load utilizing certified engineer drawings, vendor data or operations manual. Only use load weight information, which has been documented and certified, or calculated by a Mechanical or Vessel P.E.
10. **CRANE(S) TYPE:** Enter the type, model and rated capacity of lift equipment to be used for lifting
11. **CRANE CONFIGURATION:** Indicate the length of the main boom and the length of jib (if equipped) that will be in the lift & tail crane(s) at time of lift.
12. **PARTS OF WIRE ROPE / SINGLE LINE PULL:** Enter the number of parts in the load handling line during the lift. Enter the allowable winch single line pull in lbs.

304A11.EPS

LIFT PLANNING 4.31

CRITICAL LIFTS | *NOTES*

13. **MAXIMUM RADIUS:** Enter the maximum working radius the crane / load will achieve during the lift cycle of pick, swing & set.

14. **BOOM ANGLE:** Indicate the cranes' boom angle at the initial pick point / radius and final set point/radius.

15. **BOOM ANGLE:** Indicate the cranes' boom angle at the initial pick point / radius and final set point / radius.

16. **TAIL SWING CLEARANCE VERIFIED (y/n):** Indicate if the tail swing clearance has been verified. When it is required, the sketch / diagram verifying the tail swing clearance should be attached to the PCR Planned Crane Lifting Procedure Form.

17. **CHART CAPACITY:** Indicate the capacity of the crane(s) from the manufacturer's capacity chart with the parameters as indicated in items (8) thru (9).

18. **CRANE RIGGING:** Enter the size of chokers and shackles used for the lift and their physical condition.

19. **RIGGING COMPONENT WEIGHTS:** Enter the manufacturer's effective weight for the jib, load block, suspended load line, ball, suspended whip / load line, shackles, slings, spreader bars and any other lifting attachments suspended under the boom point.

20. **GROSS LOAD:** Indicate the total weight of the load by adding (9) and (19).

21. **% OF CAPACITY CHART USED:** Indicate the percentage of the crane's chart capacity to be utilized for the lift. If load / capacity percentage exceeds 75%, the lift should not be made without review of PCR Engineering Representative and Project / Site Manager. "Net load divided by net capacity = % Capacity Used "

22. **RIGGING SAFETY FACTOR 5 TO 1 (y/n):** Verify that the rigging equipment (shackles, chokers, etc.) used to perform the lift have a 5 to 1 Safety Factor.

- Engineering Certification on Spreader Bars
- Proof Test Certifications of Shackles and Slings

23. **SOIL CONDITION:** Indicate the type of soil in the area of the lift; loose, compact, or undisturbed earth. Consideration should be given to any adjacent excavation and its distance from the outriggers or crawler tracks.

24. **MATS REQUIRED:** Depending on the soil conditions determined above, indicate if any matting is required to properly protect any underground facilities and to secure the affected area under the

DEVELOPED BY CRANE INSPECTION & CERTIFICATION BUREAU - 2004

12

304A12.EPS

Instructor's Notes:

CRITICAL LIFTS | *NOTES*

lifting equipment. Matting location, configuration and the number of mats should be identified on the rigging plans and diagrams.

25. **RIGGING REVIEWED (y/n):** Indicate if the rigging has been reviewed. This includes chokers, shackles, spreaders, load line, lifting lugs and any other lifting attachments required. Depending on the nature of the lift, it may be required that a lift plan and diagram should be developed and attached to the "Planned Crane Lifting Procedure Form." This may be required at the discretion of the Client Representative or the Rigging Supervisor.

26. **TAG LINE (y/n)** Indicate if a tag lines will be required.

27. **HAZARDS:** Indicate if there are any existing power lines or overhead obstructions in the area of the lift. In addition, indicate where there are any existing underground utilities or items buried in the crane set-up area. Identify the hazard (water, sewage, drainage, electrical, etc.) and its depth.

28. **WEATHER:** Indicate the weather conditions prior to and during the lift with emphasis on wind velocity and direction, rain probability, etc.

29. **DATE: TIME:** Indicate the date and approximate time of day that the lift will be made.

30. **LIFT MEETING CONDUCTED (y/n):** Indicate if a pre-lift meeting will be conducted with all involved parties. The client representative may request such a meeting.

31. **APPROVALS:** A minimum of four (4) signatures will be required on the lift form. These include the Crane Operator, Rigging Supervisor, Safety, Client Representative. The Client Representative, at his option, may request additional signatures including Equipment Supervisors, Safety Representatives, and the organization's Engineering Representative.

DEVELOPED BY CRANE INSPECTION & CERTIFICATION BUREAU - 2004 13

304A13.EPS

Instructor's Notes:

Figure Credits

Manitowoc Crane Group	304F01–304F03
Crane Inspection & Certification Bureau	Appendix A

LIFT PLANNING 4.35

MODULE 21304-05 — TEACHING TIPS

The following are suggested activities or instructional methods to help you teach the material in this module.

General

When you call on someone to answer a question, the rest of the class relaxes or even tunes out because they expect that the question and answer will take place only between you and the trainee you called on. Instead, use this technique to involve more trainees in answering questions and to keep them on their toes.

1. Ask trainees to define a term or explain a concept.
2. After one trainee has answered, ask a trainee seated nearby if the answer is right. Then ask whether a trainee in the back of the room agrees.
3. Ask trainees to explain why they think an answer is right or wrong.
4. Use the session to clear up incorrect ideas and encourage trainees to learn from their mistakes.

Section 3.2.0 *Multiple-Crane Lifts*

Multiple-crane lifts must be carefully planned and executed. Trainees will need appropriate personal protective equipment, pencils, and paper. Arrange for an opportunity to observe a multiple-crane lift. Alternatively, have a manufacturer's representative give a presentation. Allow 60 to 90 minutes for this exercise.

1. Before the presentation discuss important lift considerations with trainees.
2. If possible, meet with the site safety supervisor, and review the lift plan with trainees.
3. Have trainees explain how the hook blocks were positioned based on the load's center of gravity and the cranes' net capacities.
4. Observe the lift. Point out safety considerations, signal people, and other factors. Have trainees take notes and write down questions during the tour.
5. After the lift answer any questions the trainees may have.

Section 4.0.0 *Critical Lifts*

Critical lifts must be carefully planned and executed. Trainees will need appropriate personal protective equipment, pencils, and paper. Arrange for an opportunity to observe a critical lift. Alternatively, have a manufacturer's representative give a presentation. Allow 60 to 90 minutes for this exercise.

1. Before the presentation discuss important lift considerations with trainees.
2. If possible, meet with the site safety supervisor, and review the lift plan with trainees.
3. Observe the lift. Point out safety considerations, signal people, and other factors. Have trainees take notes and write down questions during the tour.
4. After the lift answer any questions the trainees may have.

MODULE 21304-05 — ANSWERS TO REVIEW QUESTIONS

Answer	Section
1. a	1.0.0
2. a	1.0.0
3. b	2.0.0
4. a	2.0.0
5. c	2.0.0
6. b	2.0.0
7. d	2.0.0
8. d	3.0.0
9. b	3.0.0
10. a	3.0.0
11. c*	3.1.0
12. b*	3.1.0
13. c*	3.1.0
14. c	3.2.0
15. d	3.2.0
16. b	3.2.0
17. a	3.2.0
18. a*	3.2.0
19. b*	3.2.0
20. b*	3.2.0
21. a	3.2.0
22. a	3.2.0
23. b	4.0.0
24. b	4.0.0
25. d	6.0.0

*Please see the following page for math calculations.

MODULE 21304-05 — REVIEW QUESTION CALCULATIONS

11. Gross capacity = net capacity + deductions
Gross capacity = 95,000 + 2,500
Gross capacity = 97,500 lbs

12. Net capacity = gross capacity – deductions
Net capacity = 83,000 – (1,500 + 865 + 335 + 1,900)
Net capacity = 83,000 – 4,600
Net capacity = 78,400 lbs

13. Deductions = gross capacity – net capacity
Deductions = 90,000 – 83, 425
Deductions = 6,575 lbs

18. Total load weight = 100,000 lbs
Total net capacity of Crane One = 80,000 lbs
The sum of the lowest rated net capacities of the cranes must exceed 100,000 lbs.
To meet the 75-percent capacity guidelines, 75 percent of the sum of the lowest rated net capacities of the cranes must be greater than or equal to 100,000.

$0.75(80{,}000 + x) = 100{,}000$
$60{,}000 + 0.75x = 100{,}000$
$0.75x = 40{,}000$
$x = 40{,}000 \div 0.75$
$x = 53{,}333$ lbs

19. B = (net capacity of Crane One ÷ net capacity of Crane Two) × A
$B = 76{,}000 \div 123{,}000 \times 20$
$B = 12.36$, round to 12.4

20. Load on Crane One = B ÷ (A+B) × load weight
Load on Crane One = 10 ÷ (15 + 10) × 150,000
Load on Crane One = 10 ÷ 25 × 150,000
Load on Crane One = 60,000 lbs

CONTREN® LEARNING SERIES — USER UPDATE

The NCCER makes every effort to keep these textbooks up-to-date and free of technical errors. We appreciate your help in this process. If you have an idea for improving this textbook, or if you find an error, a typographical mistake, or an inaccuracy in NCCER's Contren® textbooks, please write us, using this form or a photocopy. Be sure to include the exact module number, page number, a detailed description, and the correction, if applicable. Your input will be brought to the attention of the Technical Review Committee. Thank you for your assistance.

Instructors – If you found that additional materials were necessary in order to teach this module effectively, please let us know so that we may include them in the Equipment/Materials list in the Annotated Instructor's Guide.

Write: Product Development and Revision
National Center for Construction Education and Research
P.O. Box 141104, Gainesville, FL 32614-1104

Fax: 352-334-0932

E-mail: curriculum@nccer.org

Craft ______________________ Module Name ______________________

Copyright Date ______________ Module Number ______________ Page Number(s) ______________

Description

____

____

____

____

(Optional) Correction

____

____

____

(Optional) Your Name and Address

____

____

____

Module 21305-05

Hoisting Personnel

NCCER STANDARDIZED CRAFT TRAINING PROGRAM

The National Center for Construction Education and Research (NCCER) provides a standardized national program of accredited craft training. Key features of the program include instructor certification, competency-based training, and performance testing. The program provides trainees, instructors, and companies with a standard form of recognition through a National Craft Training Registry. The program is described in full in the *Guidelines for Accreditation*, published by the NCCER. For more information on standardized craft training, contact the NCCER by writing us at P.O. Box 141104, Gainesville, FL 32614-1104; calling 352-334-0911; or e-mailing info@nccer.org. More information may be found at our Web site, www.nccer.org.

HOW TO USE THIS ANNOTATED INSTRUCTOR'S GUIDE

Each page presents two sections of information. The larger section displays each page exactly as it appears in the Trainee Module. The narrow column ties suggested trainee and instructor actions to each page and provides icons (detailed below) to call your attention to material, safety, audiovisual, or testing requirements. The bottom of each page includes space for your notes.

The **Audiovisual** icon indicates an appropriate time to show a transparency or other audiovisual aid.

The **Classroom** icon prompts you to define a term, stress a point, ask trainees to explain a concept, or give examples.

The **Demonstration** icon directs you to show trainees how to perform tasks.

The **Examination** icon tells you to administer the written module examination.

The **Homework** icon is placed where you may wish to assign reading for the next class, to assign a project, or to advise trainees to prepare for an examination.

The **Laboratory** icon is used when trainees are to practice performing tasks.

The **Materials** icon is a reminder for you to gather materials needed for classes, labs, and testing.

The **Performance Testing** icon tells you to administer a performance test or a portion thereof.

The **Safety** icon is used to emphasize safety issues. It is often keyed to *Caution* and *Warning* statements in the Trainee Module.

The **Teaching Tip** icon indicates additional guidance is available, such as how to conduct an exercise, get the most educational value from a field trip, or encourage class participation. Teaching Tips may expand on a feature (*Think About It*, *Did You Know?*) or provide Quick Quizzes or similar exercises. You will be referred to the Teaching Tips section at the back of the module if there is additional material.

The **Combination** icon indicates that the laboratory listed corresponds with a performance task. If desired, you can note the proficiency of the trainees during the laboratory and use it to satisfy performance testing requirements.

PREPARATION

Before teaching this module, you should review the Objectives, Performance Tasks, Materials and Equipment List, and the Module Outline. Be sure to allow ample time to prepare your own training or lesson plan and gather all required materials and equipment.

MODULE OVERVIEW

This module provides an in-depth discussion of the *ASME B30.23* and *29 CFR 1926.550(g)* requirements while presenting advanced operating techniques for hoisting personnel.

PREREQUISITES

Prior to training with this module, it is recommended that the trainee shall have successfully completed *Core Curriculum; Mobile Crane Operations Level One; Mobile Crane Operations Level Two;* and *Mobile Crane Operations Level Three,* Modules 21301-05 through 21304-05.

OBJECTIVES

Upon completion of this module, the trainee will be able to do the following:

1. Identify which federal regulations apply to hoisting personnel.
2. Identify which consensus standards apply to hoisting personnel.
3. Visually inspect the platform, suspension system, and attachment points.
4. Perform a test lift following appropriate safety procedures and regulations, including the use of fall protection.
5. Define operation techniques for hoisting personnel near power lines.

PERFORMANCE TASKS

Under the supervision of the instructor, the trainee should be able to do the following:

1. Visually inspect the platform, suspension system, and attachment points.
2. Perform a test lift following appropriate safety procedures and regulations, including the use of fall protection.
3. Perform operations for hoisting personnel.

MATERIALS AND EQUIPMENT LIST

Overhead projector and screen

Transparencies

Blank acetate sheets

Transparency pens

Whiteboard/chalkboard

Markers/chalk

Pencils and scratch paper

OSHA 29 CFR 1926.104 1926.106, 1926.550(g), and 1926.753

ASME B30.23

Fall protection equipment

Fall protection safety video or DVD

TV and VCR or DVD player

OSHA document: *Crane or Derrick Suspended Personnel Platforms*

Personnel platform

Personnel Platform Pre-lift Inspection Form

Crane and rigging hardware for personnel platform

Copies of the Quick Quiz*

Module Examinations**

Performance Profile Sheets**

*Located in the back of this module.

**Located in the Test Booklet.

SAFETY CONSIDERATIONS

Ensure that the trainees are equipped with appropriate personal protective equipment and know how to use it properly. This module requires trainees to operate a crane. Review site safety procedures and site evacuation procedures. Ensure that all trainees are familiar with hand signals and other site communication procedures. Brief trainees on pinching and crushing hazards attendant with crane operations. This module may require trainees to visit construction sites. Brief trainees on site safety procedures. This module may require trainees to be hoisted on a platform by a crane. Ensure that personnel have fall protection equipment and know how to use it properly.

TEACHING TIME FOR THIS MODULE

An outline for use in developing your lesson plan is presented below. Note that each Roman numeral in the outline equates to one session of instruction. Each session has a suggested time period of 2½ hours. This includes 10 minutes at the beginning of each session for administrative tasks and one 10-minute break during the session. Approximately 20 hours are suggested to cover *Hoisting Personnel*. You will need to adjust the time required for hands-on activity and testing based on your class size and resources. Because laboratories often correspond to Performance Tasks, the proficiency of the trainees may be noted during these exercises for Performance Testing purposes.

Topic	Planned Time
Sessions I through III. Introduction, Fall Protection, and Requirements	
A. Introduction	____________
B. Fall Protection	____________
C. Platform Requirements	____________
D. Boom-Attached Platform Requirements	____________
E. Crane and Operational Requirements	____________
Sessions II and III. Personnel Platform Inspection	
A. ASME-Prescribed Personnel Platform Inspection	____________
B. Laboratory – Trainees practice inspecting the platform, suspension system, and attachment points. This laboratory corresponds to Performance Task 1.	____________
Sessions IV and V. Trial Lift	
A. Proof Testing	____________
B. Laboratory – Trainees practice performing a test lift. This laboratory corresponds to Performance Task 2.	____________
Sessions VI and VII. Advanced Operations Techniques	
A. Advanced Operations Techniques for Hoisting Personnel	____________
B. Laboratory – Trainees practice hoisting personnel. This laboratory corresponds to Performance Task 3.	____________
Session VIII. Review, Module Examination, and Performance Testing	
A. Review	____________
B. Module Examination	____________
1. Trainees must score 70 percent or higher to receive recognition from NCCER.	
2. Record the testing results on Craft Training Report Form 200, and submit the results to the Training Program Sponsor.	
C. Performance Testing	____________
1. Trainees must perform each task to the satisfaction of the instructor to receive recognition from NCCER. If applicable, proficiency noted during laboratory exercises can be used to satisfy the Performance Testing requirements.	
2. Record the testing results on Craft Training Report Form 200, and submit the results to the Training Program Sponsor.	

Module 21305-05

Hoisting Personnel

Instructor's Notes:

COURSE MAP

This course map shows all of the modules in the third level of the *Mobile Crane Operations* curriculum. The suggested training order begins at the bottom and proceeds up. Skill levels increase as you advance on the course map. The local Training Program Sponsor may adjust the training order.

Assign reading of Module 21305-05.

MOBILE CRANE OPERATIONS LEVEL THREE

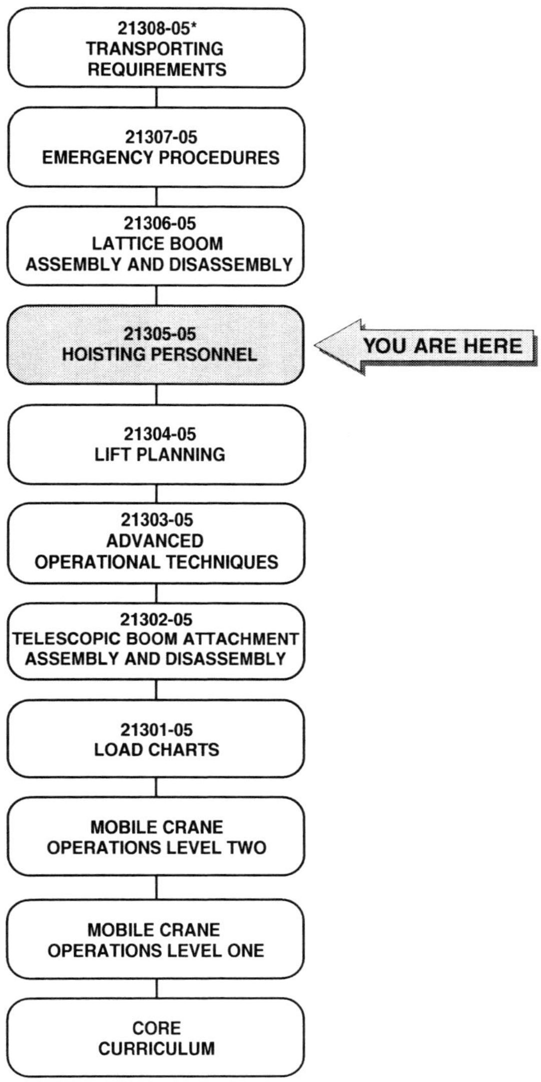

*Please note that 21308-05 is an elective module. It is not necessary to complete this module in order to advance in your training.

305CMAP.EPS

Copyright © 2005 National Center for Construction Education and Research, Gainesville, FL 32614-1104. All rights reserved. No part of this work may be reproduced in any form or by any means, including photocopying, without written permission of the publisher.

HOISTING PERSONNEL 5.iii

Instructor's Notes:

MODULE 21305-05 CONTENTS

HOISTING PERSONNEL 5.v

Figures

Tables

Instructor's Notes:

MODULE 21305-05

Hoisting Personnel

Objectives

When you have completed this module, you will be able to do the following:

1. Identify which federal regulations apply to hoisting personnel.
2. Identify which consensus standards apply to hoisting personnel.
3. Visually inspect the platform, suspension system, and attachment points.
4. Perform a test lift following appropriate safety procedures and regulations, including the use of fall protection.
5. Define operation techniques for hoisting personnel near power lines.

Prerequisites

Before you begin this module, it is recommended that you successfully complete *Core Curriculum*; *Mobile Crane Operations Level One*; *Mobile Crane Operations Level Two*; and *Mobile Crane Operations Level Three*, Modules 21301-05 through 21304-05

Required Trainee Materials

1. Pencil and paper
2. Appropriate personal protective equipment
3. Copies of OSHA and ASME regulations

1.0.0 ◆ INTRODUCTION

Like every other lift operation, hoisting personnel requires careful attention to detail and safety. This lifting practice and aspects related to it are covered under *OSHA 29 CFR 1926.104, 1926.106, 1926.550(g), 1926.753*, and *ASME B30.23*. The purpose of this module is to cover in-depth the guidelines of both the federal regulations and the consensus standard.

The intent of *29 CFR 1926.550* is to establish minimum design criteria, equipment features, and operational procedures when using material hoisting equipment to lift personnel.

Although using cranes to lift people was common in the past, OSHA regulations, as spelled out in *29 CFR 1926.550(g)(2)*, now prohibit the practice unless no other method is suitable. However, for steel erection, *29 CFR 1926.753(c)(4)* specifically exempts the requirement of *29 CFR 1926.550(g)(2)* and allows lifting of personnel. When lifting is permitted, certain controls must be in place, including the following:

- Depending on the type of rope, the load rope must be capable of supporting at least seven or ten times the maximum intended load.
- Cranes must be equipped with a boom angle indicator and, if applicable, a boom length indicator.
- The hoist drum must have a controlled load-lowering system or device other than the load hoist brake. Free fall is prohibited.
- Crane brakes and locking devices such as pawls or dogs must be engaged when the occupied suspended personnel platform is in a stationary position.
- The crane must be level within one percent of level grade and located on firm footing. Any outriggers must be fully deployed.
- The crane capacity is cut in half for the operating radius used.
- Anti-two-blocking devices are required.
- The suspended personnel platform must be specifically designed for lifting personnel.
- Before the lift is used, it must be tested with a comparable weight and then inspected.
- Every intended lift must undergo a trial run with weights rather than people.

Ensure that you have everything required to teach the course. Check the Materials and Equipment list at the front of this module.

See the general Teaching Tip at the end of this module.

Explain that terms shown in bold are defined in the Glossary lat the back of this module.

Show Transparency 1, Objectives, and Transparency 2, Performance Tasks. Review the goals of the module, and explain what will be expected of trainees.

Provide copies of *OSHA 29 CFR 1926.104, 1926.106, 1926.550(g), 1926.753*, and *ASME B30.23*. Note that platform authorization and inspection forms are provided in *Appendixes A* and *B*.

Identify the regulations and standards that cover hoisting personnel. Review the OSHA regulations regarding hoisting personnel. Discuss required controls.

Reiterate that every personnel lift must have a trial lift with weights.

Describe fall protection equipment and explain how it is used. Discuss the OSHA training requirements for fall protection.

Provide trainees with fall protection equipment, and have them practice putting it on.

See the Teaching Tip for Section 2.0.0 at the end of this module.

Review the requirements of *OSHA 29 CFR 1926.550* for platforms. Describe the important features of a manbasket, emphasizing the guardrail system.

Emphasize that the personnel platform must support its own weight and at least five times the maximum intended load.

See the Teaching Tip for Section 3.0.0 at the end of this module.

2.0.0 ◆ FALL PROTECTION

Fall protection equipment and guidelines for hoisting personnel are covered under *OSHA 29 CFR 1926.104, Full-Body Harnesses, Lifelines, and Lanyards*. These guidelines are as follows:

- Lifelines, full-body harnesses, and **lanyards** shall be used only for employee safeguarding. Any lifeline, full-body harness, or lanyard actually subjected to in-service loading, as distinguished from static load testing, shall be immediately removed from service and shall not be used again for employee safeguarding.
- Lifelines shall be secured above the point of operation to an **anchorage** or structural member capable of supporting a minimum dead weight of 5,400 pounds.
- Lifelines used on rock-scaling operations, or in areas where the lifeline may be subjected to cutting or abrasion, shall be a minimum of ⅞-inch wire core manila rope. For all other lifeline applications, a minimum of ¾-inch manila or equivalent with a minimum breaking strength of 5,400 pounds shall be used.
- Full-body harness lanyards shall be a minimum of ½-inch nylon or equivalent, with a maximum length to provide for a fall of no greater than six feet. The lanyard shall have a nominal breaking strength of 5,400 pounds.
- All full-body harness and lanyard hardware shall be drop-forged or pressed steel, cadmium-plated in accordance with Type 1, Class B plating specified in Federal Specification *QQ-P-416*. The surface shall be smooth and free of sharp edges.
- All full-body harness and lanyard hardware, except rivets, shall be capable of withstanding a tensile loading of 4,000 pounds without cracking, breaking, or taking a permanent deformation.

Employees who may be exposed to fall hazards must be trained to recognize the hazards to which they may be exposed. They must also be able to demonstrate proficiency in the use of fall protection systems. The following are among the training requirements:

- Understanding the general requirements of the OSHA fall protection standard
- Knowing the nature of fall hazards in the work area
- Being able to perform the correct procedures for erecting, maintaining, disassembling, inspecting, and using fall protection systems

When applicable, training must also be provided in the following areas:

- The role of each employee in the safety monitoring system
- Limitations on the use of mechanical equipment during the performance of elevated work
- Correct procedures for the handling and storage of equipment and materials
- The role of employees in fall protection plans

3.0.0 ◆ PLATFORM REQUIREMENTS

In compliance with *OSHA 29 CFR 1926.550*, platform requirements are broken down into four categories: design criteria [*29 CFR 1926.550(g)(4)(i)*], platform specifications [*29 CFR 1926.550(g)(4)(ii)*], platform loading [*29 CFR 1926.550(g)(4)(iii)*], and rigging [*29 CFR 1926.550(g)(4)(iv)*]. These requirements are detailed in the following sections.

3.1.0 Design Criteria

The suspended personnel platform, also called a manbasket or workbasket (*Figure 1*), shall be designed by a qualified engineer or a competent person qualified in structural design.

The suspension system shall be designed to minimize tipping of the platform due to movement of employees occupying the platform.

The personnel platform itself, except the **guardrail** system and personal fall-arrest system anchorages, shall be capable of supporting, without failure, its own weight and at least five times the maximum intended load.

> **NOTE**
> In the following sections taken from OSHA and ASME standards, a suspended personnel platform may be called a platform or a personnel platform.

3.2.0 Platform Specifications

Each personnel platform shall be equipped with a guardrail system and shall be enclosed at least from the toeboard to mid-rail with either solid construction or expanded metal having openings no greater than ½ inch (1.27 cm). Other platform specifications are as follows:

- A grab rail shall be installed inside the entire perimeter of the personnel platform.
- Access gates, if installed, shall not swing outward during hoisting.
- Access gates, including sliding or folding gates, shall be equipped with a restraining device to prevent accidental opening.

Instructor's Notes:

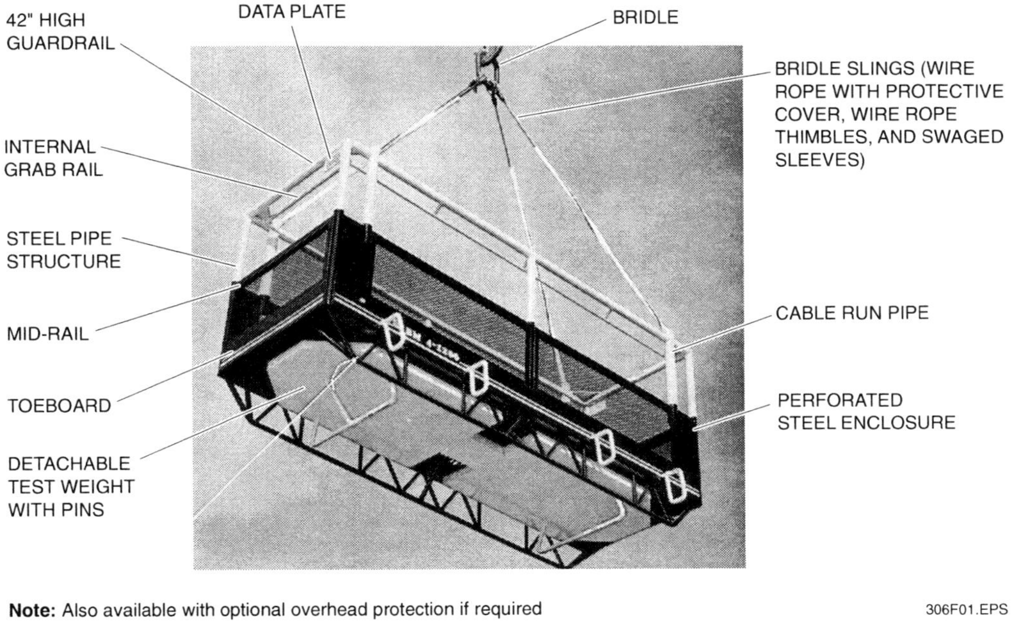

Figure 1 ◆ Suspended personnel platform.

- Head room shall be provided which allows employees to stand upright in the platform.
- In addition to the use of hard hats, employees shall be guarded by overhead protection on the personnel platform when employees are exposed to falling objects.
- All rough edges exposed to contact by employees shall be surfaced or smoothed in order to prevent injury to employees from punctures or lacerations.
- All welding of the personnel platform and its components shall be performed by a qualified welder familiar with the weld grades, types, and material specified in the platform design.
- The personnel platform shall be conspicuously posted with a plate or other permanent marking which indicates the weight of the platform and its rated load capacity or maximum intended load. The maximum intended load includes personnel, tools, accessory equipment, and work materials.

3.3.0 Personnel Platform Loading

The personnel platform shall not be loaded in excess of its rated load capacity. When a personnel platform does not have a rated load capacity, the personnel platform shall not be loaded in excess of its maximum intended load.

The number of employees, along with material, occupying the personnel platform shall not exceed the limit established for the platform and the rated load capacity or the maximum intended load.

Personnel platforms shall be used only for employees, their tools, and the materials necessary to do their work, and shall not be used to hoist only materials or tools when not hoisting personnel.

Materials and tools for use during a personnel lift shall be secured to prevent displacement. These items shall be evenly distributed within the confines of the platform while the platform is suspended.

3.4.0 Rigging

When a wire-rope **bridle sling** is used to connect the personnel platform to the load line, each bridle leg shall be connected to a master link or **shackle** in such a manner as to ensure that the load is evenly divided among the bridle legs (*Figures* 2 and 3).

Hooks on headache ball assemblies, lower load blocks, or other attachment assemblies shall be of a type that can be closed and locked, eliminating the hook throat opening. Alternatively, an alloy anchor type shackle with a bolt, nut, and retaining pin may be used.

Discuss the limitations and requirements for loading a personnel platform.

Explain that the platform must be rigged so that the weight is evenly divided among the bridle legs. Describe the requirements for the hook and attachment assembly.

Explain that all tools and materials on the platform must be secured.

Provide a locking hook and bridle sling for the trainees to examine.

HOISTING PERSONNEL 5.3

Explain that rigging hardware must be capable of supporting at least five times the maximum intended load. Discuss other requirements for rigging hardware.

Review the design criteria in *ASME B30.23* for boom-attached platforms.

Discuss the special features required for boom-attached platforms.

See the Teaching Tip for Sections 2.0.0–4.0.0 at the end of this module.

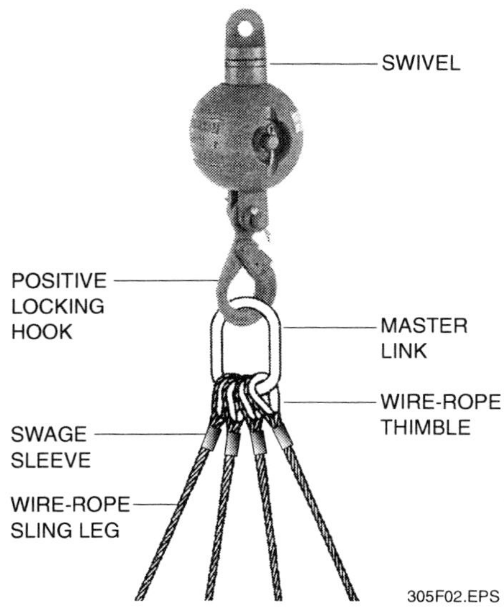

Figure 2 ◈ Bridle sling using a master link.

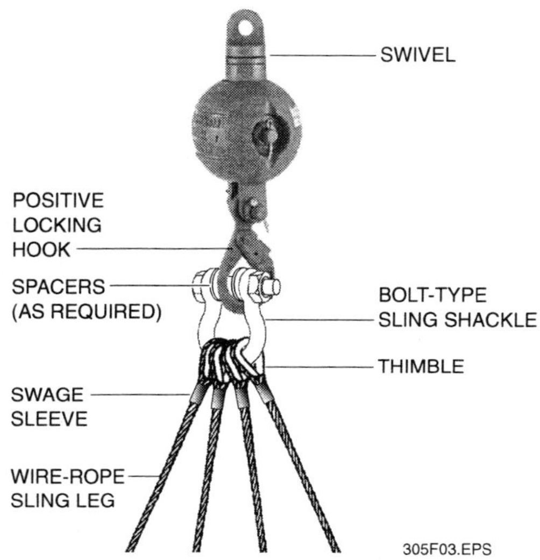

Figure 3 ◈ Bridle sling using a shackle.

Wire rope, shackles, rings, master links, and other rigging hardware must be capable of supporting, without failure, at least five times the maximum intended load applied or transmitted to that component. Where rotation-resistant rope is used, the slings shall be capable of supporting without failure at least ten times the maximum intended load. All eyes in wire rope slings shall be fabricated with thimbles.

Bridles and associated rigging for attaching the personnel platform to the hoist line shall be used only for the platform and the necessary employees, their tools, and the materials necessary to do their work, and they shall not be used for any other purpose when not hoisting personnel.

4.0.0 ◈ BOOM-ATTACHED PLATFORM REQUIREMENTS

Design criteria (*ASME B20.23*) for boom-attached platforms include the following:

- Based on the minimum yield strength of the material, the structural components of the platform responsible for supporting the load shall have a minimum design factor of two.
- Based on the minimum ultimate strength of the materials used, the structural components of the platform that are made of nonductile material and supporting the load shall have a minimum design factor of five.

Mandatory guidelines for specific features of personnel lifting platforms are as follows:

- A minimum platform rating of 300 lbs (136 kg), including tools and materials
- Guardrail protection consisting of a top rail, intermediate rail, toe board, and lower barrier with the following characteristics:
  - *Top rail:* Not less than 39 inches nor more than 45 inches in height with respect to the platform floor
  - *Toe board:* Around the entire perimeter of the platform and a minimum of $3\frac{1}{2}$ inches in height from the top of the platform floor
  - *Lower barrier:* Spanning the distance between the toe board and intermediate rail; it will also be of solid construction or expanded metal, having no opening that would allow a ½-inch sphere to pass through
- The top rail, toe board, and lower barrier must also be capable of withstanding a load of 300 pounds applied at any point in an outward or downward direction. This must occur without the rail, board, or barrier reaching its **yield** point.
- Anchorage points within the platform or on the boom tip which are suitable for attaching lanyards and capable of withstanding the impact created by falling loads.
- Hand railings, with a clearance of not less than 1½ inches between the railing and any other structure, located within the platform.
- Flooring with a slip-resistant surface and provisions to promote the free drainage of liquids.
- A method to secure loose items within the platform.

Instructor's Notes:

- An identification plate stating:
  - Manufacturer name and address
  - Platform rating in terms of weight and personnel
  - Platform identification number
  - Suspension system description
  - Weight of the empty platform and its suspension system
  - Date of the platform manufacture
  - Certification of compliance with the design, construction, and testing requirements of the *ASME B30.23* volume
  - Listing of any unique operational environments for which the platform has been designed
- Access gates, if used, equipped with a positive-acting device to suppress the gate from accidental opening.
- Minimal tipping due to movement of personnel on the platform.
- Minimum characteristics for the suspension system on a platform are as follows:
  - Rope sling systems with mechanically spliced flemish eyes designed with thimbles in the eyes
  - No wire-rope clips, wedge sockets or knots, synthetic webbing, or natural- or synthetic-fiber rope slings
  - Rope sling systems with poured socket end connections designed in accordance with the manufacturer's or qualified person's application instructions
  - Sling systems that utilize a master link or safety-type shackle for attachment to the hoisting equipment hook
  - For rope slings systems, each leg of the system permanently marked with the rated load of the leg
  - The master link in the system permanently marked with the suspension system's rated load and identification as a personnel lifting platform suspension component
  - Shackles that are categorized as a safety type
  - Overhead protection that allows view of hoist equipment components positioned above the personnel
  - Accessories for platform use in special environments included as specified by the manufacturer
  - Welds in accordance with the American Welding Society (AWS) or ASME
  - Bolted connections in accordance with the American Institute of Steel Construction, Inc. (AISC) *Specification for Structural Joints*
  - Weatherproof compartment for storage of the operator's manual and other documents

If motion controls are installed, they must be clearly marked, accessible to the operator, and protected from inadvertent actuation, and they must return to their neutral position and stop all motion when released. Boom motion controls, if installed, must be able to override platform controls. They also must have an emergency stop control and a control that is continuously activated for the controls to be operational.

Hoisting equipment used with personnel platforms shall meet these conditions:

- Have been removed for the period of personnel lifting if it has live boom or live load line capabilities
- Have an anti-two-block device, boom length indicator, positively locking hooks, and boom angle indicator
- Have automatic brakes such that when the equipment operating controls are released, the motions are brought to rest
- In the case of a holding device, be provided with hydraulic or pneumatic systems to prevent uncontrolled movement of the hoisting equipment in the event of a system failure
- Have a means to prevent the retraction of any hydraulically or pneumatically activated outriggers or stabilizers in the event a hydraulic or pneumatic line fails

5.0.0 ◆ CRANE AND OPERATIONAL REQUIREMENTS

According to *OSHA 29 CFR 1926.550(g)*, crane and operational requirements are broken down into operational criteria [*29 CFR 1926.550(g)(3)(i)*]; instruments and components [*29 CFR 1926.550(g)(3)(ii)*]; work practices [*29 CFR 1926.550(g)(6)*]; and traveling [*29 CFR 1926.550(g)(7)*]. These requirements are given in detail in the following sections.

5.1.0 Operational Criteria

Hoisting of the personnel platform shall be performed in a slow, controlled, cautious manner with no sudden movements of the crane or derrick or the platform.

Load lines shall be capable of supporting, without failure, at least seven times the maximum intended load. Where rotation-resistant rope is used, the lines shall be capable of supporting, without failure, at least ten times the maximum intended load. The required design factor is achieved by taking the current safety factor of 3.5, which is required under *OSHA 1926.550(b)(2)*, and applying the 50-percent derating of the crane capacity which is required by *OSHA 1926.550(g)(3)(i)(f)*.

Have trainees describe the requirements for an identification plate on a personnel platform.

Review the minimum requirements for a suspension system and motion controls.

Explain the criteria for hoisting equipment used with personnel platforms.

See the Teaching Tip for Section 5.0.0 at the end of this module.

Review the OSHA operational criteria for hoisting personnel.

Emphasize that hoisting should consist of slow, deliberate movements.

HOISTING PERSONNEL 5.5

Remind trainees that personnel should never be hoisted by machines with live booms.

Discuss specific instruments and components needed when hoisting personnel.

Review the work practices that must be observed when using a suspended personnel platform.

Explain specific considerations and precautions necessary when working over or near water.

Load and boom hoist drum brakes, swing brakes, and locking devices such as pawls or dogs shall be engaged when the occupied personnel platform is in a stationary position.

The crane shall be uniformly level within one percent of level grade and located on firm footing. Cranes equipped with outriggers shall have them all fully deployed following manufacturer's specifications, insofar as applicable, when hoisting personnel.

The total weight of the loaded personnel platform and related rigging shall not exceed 50 percent of the rated capacity for the radius and configuration of the crane or derrick.

The use of machines having live booms (booms in which lowering is controlled by a brake without aid from other devices which slow the lowering speeds) is prohibited.

5.2.0 Instruments and Components

Cranes and derricks with variable-angle booms shall be equipped with a boom angle indicator, readily visible to the operator.

Cranes with telescoping booms shall be equipped with a device to indicate clearly to the operator, at all times, the boom's extended length, or an accurate determination of the load radius to be used during the lift shall be made prior to hoisting personnel.

A positive-acting device shall be used which prevents contact between the load block or overhaul ball and the boom tip (anti-two-block device), or a system shall be used which deactivates the hoisting action before damage occurs in the event of a two-blocking situation (two-block damage prevention feature).

The load-line hoist drum shall have a system or device on the power train, other than the load hoist brake, which regulates the lowering rate of speed of the hoist mechanism (controlled load lowering). Free fall is prohibited.

5.3.0 Work Practices

The following work practices must be observed when using a suspended personnel platform. As defined in *29 CFR 1926.550(g)(3)(i)*, these requirements are detailed as follows:

- Personnel shall keep all parts of the body inside the platform during raising, lowering, and positioning. This provision does not apply to an occupant of the platform performing the duties of a signal person.
- Before workers exit or enter a hoisted personnel platform that is not landed, the platform shall be secured to the structure where the work is to be performed, unless securing to the structure creates an unsafe situation.
- Tag lines shall be used unless their use creates an unsafe situation.
- The crane or derrick operator shall remain at the controls at all times when the crane engine is running and the platform is occupied.
- Hoisting of personnel shall be promptly discontinued upon indication of any dangerous weather conditions or other impending danger.
- Personnel being hoisted shall remain in continuous sight of, and in direct communication with, the operator or signal person. In those situations where direct visual contact with the operator is not possible, and the use of a signal person would create a great hazard for the person, direct communication alone (such as by radio) may be used.
- Except over water, workers occupying the personnel platform shall use a body belt/harness system with lanyard appropriately attached to the lower load block or headache ball, or to a structural member within the personnel platform capable of supporting a fall impact for personnel using the anchorage. When working over water, the requirements of *OSHA 1926.106* shall apply.
- No lifts shall be made on another of the crane's or derrick's loadlines while personnel are suspended on a platform.

5.3.1 Working Over or Near Water – OSHA 1926.106

- Employees working over or near water where the danger of drowning exists shall be provided with U.S. Coast Guard-approved life jackets or buoyant work vests.
- Prior to and after each use, the buoyant work vests or life preservers shall be inspected for defects which would alter their strength or buoyancy. Defective units shall not be used.
- Ring buoys with at least 90 feet of line shall be provided and readily available for emergency rescue operations. Distance between ring buoys shall not exceed 200 feet.
- At least one lifesaving skiff shall be immediately available at locations where personnel are working over (or adjacent to) water.

Instructor's Notes:

5.4.0 Traveling

Hoisting of personnel while the crane is traveling is prohibited, except for portal, tower, and locomotive cranes, or where the supervisor demonstrates that there is no less hazardous way to perform the work.

Under any circumstances where a crane would travel while hoisting personnel, the employer or supervisor shall implement the following procedures to safeguard personnel:

- Crane travel shall be restricted to a fixed track or runway
- Travel shall be limited to the load radius of the boom used during the lift
- The boom must be parallel to the direction of travel
- A complete trial run shall be performed to test the route of travel before employees are allowed to occupy the platform.
- If travel is done with a rubber-tired carrier, the condition and air pressure of the tires shall be checked. The chart capacity for personnel lifts on rubber shall be used for application of the 50-percent reduction of rated capacity.
- Crane travel speed must be as slow as possible, and sudden stops and starts should be avoided.

6.0.0 ◆ PERSONNEL PLATFORM INSPECTION

According to *OSHA 29 CFR 1925.550(g)* guidelines, a visual inspection of the crane or derrick, rigging, personnel platform, and the crane or derrick base support or ground shall be conducted by a competent person immediately after the trial lift with the platform suspended several inches. This inspection is made to determine whether the testing has exposed any defect or produced any adverse effect upon any component or structure. Personnel shall not be hoisted unless the following conditions are determined to exist:

- Hoist ropes shall be free of kinks.
- Multiple-part lines shall not be twisted around each other.
- The primary attachment shall be centered over the platform.
- Fall protection equipment must be free of damage and defects.
- The hoisting system shall be inspected if the load rope is slack to ensure that all ropes are properly seated on drums and in sheaves.

Any defects found during inspections which create a safety hazard shall be corrected before hoisting personnel.

6.1.0 ASME-Prescribed Personnel Platform Inspection

ASME B30.23 guidelines break down inspections into three categories: initial, frequent, and periodic. Prior to use at each new job site, all platforms shall be inspected by a qualified person to verify compliance with the provisions of *ASME B30.23*. This is the initial inspection.

During a frequent inspection, the platform, suspension system, and attachment points shall be inspected by a designated person. This inspection is made to identify conditions that have been specifically indicated by the platform manufacturer or a qualified person designated by the employer as potentially creating a hazardous operating condition. The frequent inspection shall address at least the following items:

- Markings
- Structure
- Attachment mechanisms
- Special purpose items

Periodic inspections of the same items are to be conducted at least once every 12 months. Platforms that have been out of service for 12 or more consecutive months shall receive a periodic inspection prior to use.

Dated records of frequent inspections must be made and kept by the platform user for the duration of the personnel lift operation. Dated records of periodic inspections must be made by the platform owner. The last inspection shall be kept with the platform and be available for review.

7.0.0 ◆ TRIAL LIFT

The requirements for conducting a trial lift under *OSHA 29 CFR 1926.550(g)(5)* are as follows:

- A trial lift with the unoccupied personnel platform loaded at least to the anticipated liftweight shall be made from ground level or any other location where personnel will enter the platform, to each location at which the personnel platform is to be hoisted and positioned. This trial lift shall be performed immediately prior to placing personnel on the platform. The operator shall determine that all systems, controls, and safety devices are activated and functioning properly; that no interferences exist; and that all configurations necessary to reach those work locations will allow the operator to remain under the 50-percent limit of the hoist's rated capacity. A single trial lift may be performed at one time for all locations that are to be reached from a single setup position.

Discuss the safeguards required for hoisting personnel in rare cases while the crane is traveling.

Have trainees review Sections 6.0.0 and 6.1.0.

Ensure that you have everything required for teaching this session.

Explain how to perform a personnel platform inspection pursuant to OSHA guidelines.

Demonstrate how to perform a personnel platform inspection.

Have trainees practice performing a personnel platform inspection. Note the proficiency of each trainee. This laboratory corresponds to Performance Task 1.

Have trainees review Section 7.0.0.

Discuss the OSHA requirements for a trial lift.

Explain how to perform a proof test pursuant to *ASME B30.23*.

Demonstrate how to perform a test lift.

Have trainees practice performing a test lift. Note the proficiency of each trainee. This laboratory corresponds to Performance Task 2.

Have trainees review Section 8.0.0.

Ensure that you have everything required for teaching this session.

Show Transparencies 3 and 4 (Table 1 and Figure 4). Discuss the dangers of hoisting personnel near power lines and the required safety precautions.

Familiarize trainees with a power line awareness checklist (*Appendix C*).

- The trial lift shall be repeated prior to hoisting personnel whenever the crane or derrick is moved and set up in a new location or returned to a previously used location. Additionally, the trial lift shall be repeated when the lift route is changed, unless the operator determines that the route change is not significant.
- At each job site, prior to hoisting personnel, the platform and rigging shall be proof-tested to 125 percent of the platform's rated capacity. This is accomplished by holding the platform in a suspended position for five minutes with the test load evenly distributed on the platform. Proof testing can be accomplished concurrently with the trial lift. After proof testing the platform, a competent person shall inspect the platform and rigging. If any deficiencies are found, they shall be corrected and another test conducted. Personnel hoisting shall not take place until the proof testing requirements are satisfied.

7.1.0 Proof Testing

In proof testing per *ASME B30.23*, the platform and rigging shall be proof-tested to 125 percent of the platform's rating. For this test, the following guidelines will be met:

- The load will be evenly distributed.
- The platform shall be hoisted and held in a suspended position for not less than five minutes.
- Inspection will follow after the proof testing.
- If damage is found, it shall be corrected and another proof test conducted.
- The hoisting equipment that will be used during the personnel lifts should be the equipment used to perform the proof tests.

After any structural repair or modification, the platform shall be tested at 150 percent of its rating. The platform shall be raised to a height, then lowered at a speed of 100 ft/min. Once 100 ft/min is reached, the platform descent will be halted by applying the hoisting equipment brakes. The platform shall be held for no less than 5 minutes. After the test, a qualified person will inspect the platform. If necessary, any damage revealed shall be corrected and the testing done again. The most recent record of proof testing shall be maintained at the job site.

8.0.0 ◆ ADVANCED OPERATIONS TECHNIQUES FOR HOISTING PERSONNEL

Lifting personnel near power lines is not allowed unless there is no less hazardous way to do the job. There are two options when hoisting personnel near power lines. The first option is best and involves de-energized and grounded power lines. In this situation, a representative from the power company or the owner of the power lines will ensure that the lines have been de-energized before hoisting takes place. Even with the power lines de-energized, signs must be installed at the operator's station, and on the outside of the crane, warning of electrocution or bodily injury unless minimum clearance (*Table 1*) is maintained.

> **WARNING!**
> Even if operator aids such as high-voltage proximity warning devices are being used, they do not negate the need for power-line shutdown.

The second, less desirable situation is when the power lines are energized, with the equipment outside the prohibited zone (*Figure 4*). In this situation, the possibility exists that the equipment or platform will become energized. A power line awareness checklist can be performed (see *Appendix C*). A pre-lift meeting must be held between the job-site supervisor and a representative of the electric company. Considerations must

Table 1 High-Voltage Power Line Clearances

CRANE IN OPERATION (1)	
POWER LINE (kV)	BOOM OR MAST MINIMUM CLEARANCES (feet)
0 to 50	10
50 to 200	15
200 to 350	20
350 to 500	25
500 to 750	35
750 to 1000	45
CRANE IN TRANSIT (with no load and the boom or mast lowered) (2)	
POWER LINE (kV)	BOOM OR MAST MINIMUM CLEARANCES (feet)
0 to 0.75	4
0.75 to 50	6
50 to 345	10
345 to 7500	16
750 to 1000	20

Note 1: For voltages over 50kV, clearance increases 5 feet for every 150 kV.
Note 2: Environmental conditions such as fog, smoke, or precipitation may require increased clearances.

305T01.EPS

Instructor's Notes:

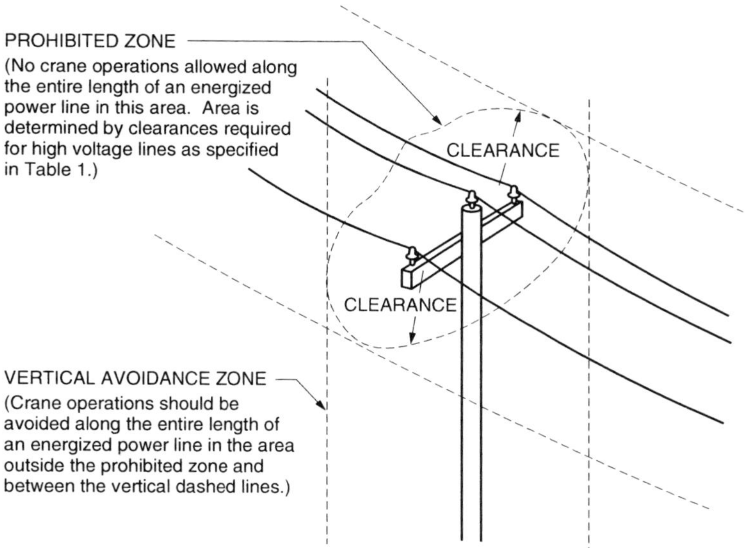

305F04.EPS

Figure 4 ◆ Prohibited zone and avoidance zone.

be given to possible power-line movement and platform restraint devices. Signs must be made available to warn of possible electrocution and bodily injury. These signs are to be posted at the operator station, on the outside of the equipment, and inside the personnel platform. In some cases, large high-voltage warning signs with upward pointing arrows placed under the lowest portion of the energized line may be desired or required.

The final situation is when the equipment must be moved under the prohibited zone and through the avoidance zone. In this situation, the boom must be lowered (*Figure 5*) and no personnel are permitted in the platform. During transit, the minimum clearances (*Table 1*) from power lines must be obeyed, and any speed and terrain effects on boom and equipment movement must be taken into consideration.

305F05.EPS

Figure 5 ◆ Moving a crane through an avoidance zone.

Summary

Although using cranes to lift personnel was common in the past, OSHA regulations prohibit the practice unless no other method is suitable. When lifting is permitted, certain controls must be in place. These controls are covered under *OSHA 29 CFR 1926.104, 1926.106, 1926.550(g), 1926.753*, and *ASME B30.23*.

The purpose of this module is to cover in-depth the guidelines of both the federal regulations and the consensus standard. The module discusses fall protection, platform requirements, crane operational requirements, how to inspect personnel platforms, and how to hoist personnel near power lines.

Explain that operator aids do not negate the need for power-line shutdown.

Keep Transparencies 3 and 4 (Table 1 and Figure 4) showing. Discuss methods for safely moving the boom through an avoidance zone.

Demonstrate how to perform operations for hoisting personnel.

Have trainees practice performing operations for hoisting personnel. Note the proficiency of each trainee. This laboratory corresponds to Performance Task 3.

Summarize the major concepts presented in the module.

HOISTING PERSONNEL 5.9

Have trainees complete the Review Questions, and go over the answers prior to administering the Module Examination.

Administer the Module Examination. Record the results on Craft Training Report Form 200, and submit the results to the Training Program Sponsor.

Administer the Performance Test, and fill out Performance Profile Sheets for each trainee. If desired, trainee proficiency noted during laboratory sessions may be used to complete the Performance Test. Record the results on Craft Training Report Form 200, and submit the results to the Training Program Sponsor.

Review Questions

1. A small cord or rope used for securing or suspending is a _____.
   a. tag line
   b. swage sleeve
   c. rope bridle
   d. lanyard

2. The proper place for securing a lanyard is a(n) _____.
   a. shackle
   b. anchorage
   c. guardrail
   d. swage sleeve

3. A full-body harness lanyard must be a minimum of _____ rope or equivalent.
   a. ½-inch nylon
   b. ¾-inch wire-core manila
   c. ¾-inch nylon
   d. ⅞-inch wire-core manila

4. Lifelines used in areas where they may be subjected to cutting or abrasion must be a minimum of _____ rope.
   a. ½-inch nylon
   b. ¾-inch wire-core manila
   c. ¾-inch nylon
   d. ⅞-inch wire-core manila

5. A personnel platform and suspension system is also known as a manbasket.
   a. True
   b. False

6. A personnel platform must be capable of supporting its own weight and at least _____ times the maximum intended load.
   a. two
   b. three
   c. four
   d. five

7. A personnel platform must have a guardrail, be enclosed from toeboard to mid-rail, and have a _____.
   a. grab rail
   b. locking device
   c. elastic floor
   d. lightning device

8. The platform must have a permanent marking stating the _____.
   a. type of crane that can lift it
   b. weight of the platform and the rated load capacity
   c. year it was made
   d. wind speeds that the platform can withstand before operations should cease

9. A grab rail must be installed _____.
   a. on each corner of the platform
   b. inside the entire perimeter of the platform
   c. on the floorboard
   d. at the owner's discretion

10. According to *OSHA 1926.550(g)*, access gates must be equipped with a _____.
    a. restraining device
    b. 5-inch handle
    c. 10-inch handle, for easy gripping
    d. steel grip device

11. The number of employees occupying the platform must not exceed _____.
    a. the limit established for the platform
    b. two (the signal person and the operator)
    c. three
    d. four

12. According to *OSHA 1926.550(g)*, each bridle-sling leg must be connected to a _____ or shackle to ensure the load is evenly distributed.
    a. rod
    b. lanyard
    c. master link
    d. auxiliary link

13. According to *ASME B30.23*, the toeboard on the guardrail must be a minimum of _____ in height from the top of the platform floor.
    a. 3½ inches
    b. 6 inches
    c. 1 foot
    d 2 feet

14. Welds on the suspension system must be in accordance with ASME or _____.
    a. CMAA
    b. DOE
    c. NEC
    d. AWS

Instructor's Notes:

15. The top rail on a personnel platform must be not less than _____ in height with respect to the platform floor.

a. 24 inches nor more than 48 inches
b. 36 inches nor more than 60 inches
c. 39 inches nor more than 45 inches
d. 48 inches nor more than 66 inches

16. The lower barrier of the guardrail on a personnel platform spans the distance between the _____.

a. top rail and the toe board
b. grab rail and the intermediate rail
c. top rail and the intermediate rail
d. toe board and the intermediate rail

17. A personnel platform must have hand railings with a clearance of not less than _____ between the railing and any other structure.

a. ½ inch
b. 1 inch
c. 1½ inches
d. 3 inches

18. Cranes having _____ must *not* be used for hoisting personnel.

a. crawlers for stabilizers
b. live booms
c. lattice booms
d. hydraulic booms

19. Load lines must be capable of supporting, without failure, at least _____ times the maximum intended load.

a. 4
b. 5
c. 6
d. 7

20. Controlled load lowering is described as lowering that _____.

a. uses positively locking hooks
b. uses tag lines
c. is regulated by a system or device on the power train, other than the load hoist brake
b. is controlled by a brake without aid from other devices that slow the lowering speed

21. Cranes and derricks with variable-angle booms must be equipped with a _____.

a. control lockout system
b. boom angle indicator
c. proximity warning device
d. hoist drum indicator

22. According to *OSHA 1926.550(g)*, when working over or near water the requirements of _____ must apply.

a. *1910.104*
b. *1910.114*
c. *1926.104*
d. *1926.106*

23. Dated records of _____ inspections must be made and kept by the platform user for the duration of the personnel lift operation.

a. initial
b. frequent
c. periodic
d. annual

24. Platforms which have been out of service for 12 or more consecutive months must receive a(n) _____ inspection prior to use.

a. initial
b. frequent
c. periodic

25. After any structural repair or modification during proof testing, the platform must be tested at _____ percent of its rating.

a. 120
b. 130
c. 140
d. 150

HOISTING PERSONNEL 5.11

Instructor's Notes:

GLOSSARY

Trade Terms Introduced in This Module

Anchorage: A place for properly securing a lanyard.

Bridle sling: A multi-leg, wire-rope sling.

Guardrail: A system of protective railings at waist height, mid-height, and toeboard that provides an added measure of safety around platforms and catwalks.

Lanyard: Any of various small cords or ropes used for securing or suspending.

Shackle: A fastening or coupling device.

Yield: To give way to force, pressure, etc., so as to move, bend, or collapse.

HOISTING PERSONNEL 5.13

Instructor's Notes:

APPENDIX A

Personnel Platform Lift Planning Authorization Form

Note: This Appendix is part of *ASME B30.23-1998*

1. Location: ____________________ Date: ____________________
2. Purpose of Lift: ____
3. Hoist Eqpt. Mfg: ____________ Model #: ____________ Serial #: ____________
4. Expected Radius: ____________ (maximum); ____________ (at work location)
5. (A) Rated Load at Radius: ____________ (B) Maximum Lift Load: ____________ [50% of 5(A)]
6. Platform ID: ____________________ Platform Rating: ____________
7. Platform Weight: ____________________ Type: (Pin On) ________ (Suspended)
8. (A) Number of Platform Occupants: ____________ (B) Approx. Wt. (With Equip): ____________
9. Total Lift Weight: ____________ ____________ [7 + 8(B)] [No more then 5(b) above]
10. Personnel Lift Supervisor: ________________________________
11. What Are The Alternatives To This Personnel Lift? ____________________
12. Why Are They Not Being Used? ________________________________
13. Pre-Lift Briefing Held: ____________ / ______ / ______ / ____________ AM/PM

    Attendees: ____
14. Anticipated Hazards (wind, weather, visibility, power lines): ____________________
15. Lift Accomplished Date: ____________________ Time: ____________
16. Remarks: ____

____

Personnel Lift Authorizer Signature   Date

HOISTING PERSONNEL 5.15

APPENDIX B

Personnel Platform Pre-Lift Inspection

Note: This Appendix is part of *ASME B30.23-1998*

Inspector: ______________________ Date: __________

Platform ID: ______________________

	SAT	UNSAT
1. Markings		
Platform (All Information Legible)	______	______
Suspension System	______	______
2. Structure		
Load Supporting Welds/Bolts	______	______
Load Supporting Members	______	______
Barrier From Toe Board to Intermediate Rail	______	______
Hand Rail	______	______
Fall Protection Device Anchorage Points	______	______
Gate Locking Mechanisms	______	______
Platform Flooring	______	______
Suspension Attachment Points	______	______
3. Attachment Mechanisms		
Pins/Ears/Bolt-Ups/Eyes (Circle)	______	______
Wire Rope/Chain/Rigid Leg (Circle)	______	______
Master Links	______	______
4. Special Purpose Items (Overhead Protection, Flotation, Platform Controls)		
List: (1) ______________	______	______
(2) ______________	______	______
(3) ______________	______	______

5. General Comments: ______________________

____

Personnel Lift Authorizer Signature   Date

Instructor's Notes:

____

____

____

APPENDIX C

Power Line Awareness Permit

POWER LINE AWARENESS PERMIT

Today's Date __________ **Job Number** ______________________________

Contractor Name			
Job Address			
Telephone Number		**Fax Number**	

Emergency Contact Number	

Survey

Before beginning any project, you must first survey your work area to find power lines at the job site. (See job-site sketch on reverse side.)

Identify

After finding all of the power lines at your site, identify the activities you'll be doing that may put you or your workers at risk. Mark one or more of the following:

- ☐ Cranes (mobile or truck mounted)
- ☐ Drilling rigs
- ☐ Backhoes/excavators
- ☐ Long-handled tools
- ☐ Other tools/high-reaching equipment
- ☐ Concrete pumper
- ☐ Aerial lifts
- ☐ Dump trucks
- ☐ Ladders
- ☐ Material handling & storage
- ☐ Scaffolding
- ☐ Other__________________

Eliminate or Control

After identifying the power line and high-risk activities on the job site, determine how to eliminate or control the risk of electrocution (a successful determination is often reached only after consultation with the utility).

Mark one or more of the following:

- ☐ Move the activity
- ☐ Change the activity
- ☐ Have the utility de-energize the power line
- ☐ Have the utility move the power line
- ☐ Use barrier protection (insulated sleeves)
- ☐ Use an observer
- ☐ Use warning lines with flags
- ☐ Use non-conductive tools
- ☐ Use a protective technology:
  - ☐ Insulated link
  - ☐ Boom cage guard
  - ☐ Proximity device

Always maintain your minimum safe clearance distance from the power line, except when the utility has de-energized and visibly grounded the power line.

Voltages	Distance from Power Line
Less than 50 kV	10 Feet
More than 50 kV	10'+(0.4")(# of kV over 50 kV)

WARNING!
It is unlawful to operate any piece of equipment within 10' of energized lines

CONSTRUCTION SAFETY COUNCIL

305A01.EPS

HOISTING PERSONNEL 5.17

Job-Site Sketch

(Draw in location of power lines and their proximity to construction site, including such things as proposed excavations, location of heavy equipment, scaffolding, material storage areas, etc.)

Completed by________________________________ **Date**____________

Title________________________________

Approved by________________________________ **Date**____________

Title________________________________

CONSTRUCTION SAFETY COUNCIL

305A02.EPS

Instructor's Notes:

Figure Credits

Lifting Technologies, Inc.	305F01
The Crosby Group, Inc.	305F02 (photo), 305F03 (photo)
Wire Rope Technical Board	305F02 (art), 305F03 (art)
Construction Safety Council	305F05, Appendix C
Reprinted from ASME *B30.23-1998* by permission of The American Society of Mechanical Engineers. All rights reserved.	Appendixes A and B

HOISTING PERSONNEL 5.19

MODULE 21305-05 — TEACHING TIPS

The following are suggested activities or instructional methods to help you teach the material in this module.

General

When you call on someone to answer a question, the rest of the class relaxes or even tunes out because they expect that the question and answer will take place only between you and the trainee you called on. Instead, use this technique to involve more trainees in answering questions and to keep them on their toes.

1. Ask trainees to define a term or explain a concept.
2. After one trainee has answered, ask a trainee seated nearby if the answer is right. Then ask whether a trainee in the back of the room agrees.
3. Ask trainees to explain why they think an answer is right or wrong.
4. Use the session to clear up incorrect ideas and encourage trainees to learn from their mistakes.

Section 2.0.0 *Fall Protection*

Falls are one of the leading causes of death on construction sites. Trainees will need pencils and paper. Obtain a fall protection safety video and/or arrange for a safety technician to do a presentation on fall protection. Allow 30 to 45 minutes for this exercise.

Fall protection videos are available from the following:

Fall Protection, videocassette, 21 minutes. www.safetyvideo.com, $99.95.

Fall Protection, Marcom Group Ltd., videocassette, 12 minutes. www.safetyvideodirect.com, $99.95.

Fall Protection, Compliance Consulting Services, videocassette. www.compliance-cs.com, $69.00.

1. Before the presentation discuss fall protection with trainees.
2. Show the video.
3. After the presentation answer any questions the trainees may have.

Section 3.0.0 *Platform Requirements*

OSHA requirements for personnel platforms are very specific. Trainees will need pencils and paper. Obtain a copy of the OSHA guidance on personnel platforms and/or arrange for an OSHA safety inspector to do a presentation on platform requirements. Allow 30 to 45 minutes for this exercise.

Crane or Derrick Suspended Personnel Platforms, OSHA 1996, available from the Construction Safety and Health Outreach Program, OSHA Office of Training and Education, U.S. Department of Labor.

1. Before the presentation discuss personnel platform requirements with trainees.
2. Review the guidance document and/or have the OSHA inspector give a presentation.
3. After the presentation answer any questions the trainees may have.

Sections 2.0.0–4.0.0

Terms and Definitions

This Quick Quiz will familiarize trainees with the terms and definitions that are commonly used in lifting personnel. You will need photocopies of the quiz provided on the following page. Trainees will need pencils. If you allow trainees to use the Trainee Module, decrease the amount of time you give them to complete the quiz.

1. Make a photocopy of the quiz for each trainee.
2. Give trainees between 5 and 10 minutes to complete the quiz.
3. Go over the answers to the quiz.
4. Ask trainees if they have questions.

Answers to Quick Quiz

1. e
2. b
3. d
4. c
5. a
6. f

Section 5.0.0

Crane and Operational Requirements

Hoisting personnel with a crane can be dangerous if not performed correctly. Trainees will need appropriate personal protective equipment, pencils, and paper. Arrange for an opportunity to observe personnel being hoisted with a crane. Alternatively, have a platform manufacturer's representative give a presentation on hoisting personnel safely. Allow 1 to 2 hours for this exercise.

1. Before the tour discuss the safety considerations with trainees.
2. Tour the site, and point out specific safety considerations. Have trainees take notes and write down questions during the tour.
3. If possible, arrange to have trainees lifted off the ground while wearing proper personal protective and fall protection equipment.
4. After the tour answer any questions the trainees may have.

Quick Quiz *Terms and Definitions*

For each description listed, identify the term that the text best describes. Write the corresponding letter in the blank provided.

_____ **1.** A fastening or coupling device
_____ **2.** A place for properly securing a lanyard
_____ **3.** A system of protective railings that provide an added measure of safety around platforms and catwalks
_____ **4.** To give way to force or pressure so as to move, bend, or collapse
_____ **5.** A multi-leg, wire-rope sling
_____ **6.** Any of various small chords or ropes used for securing or suspending

a. Bridle sling
b. Anchorage
c. Yield
d. Guardrails
e. Shackle
f. Lanyard

Copyright © 2005 National Center for Construction Education and Research. Permission is granted to reproduce this page provided that copies are for local use only and that each copy contains this notice.

MODULE 21305-05 — ANSWERS TO REVIEW QUESTIONS

Answer	Section
1. d	2.0.0
2. b	2.0.0
3. a	2.0.0
4. d	2.0.0
5. a	3.1.0
6. d	3.1.0
7. a	3.2.0
8. b	3.2.0
9. b	3.2.0
10. a	3.2.0
11. a	3.3.0
12. c	3.4.0
13. a	4.0.0
14. d	4.0.0
15. c	4.0.0
16. d	4.0.0
17. c	4.0.0
18. b	5.1.0
19. d	5.1.0
20. c	5.2.0
21. b	5.2.0
22. d	5.3.0
23. b	6.1.0
24. c	6.1.0
25. d	7.1.0

CONTREN® LEARNING SERIES — USER UPDATE

The NCCER makes every effort to keep these textbooks up-to-date and free of technical errors. We appreciate your help in this process. If you have an idea for improving this textbook, or if you find an error, a typographical mistake, or an inaccuracy in NCCER's Contren® textbooks, please write us, using this form or a photocopy. Be sure to include the exact module number, page number, a detailed description, and the correction, if applicable. Your input will be brought to the attention of the Technical Review Committee. Thank you for your assistance.

Instructors – If you found that additional materials were necessary in order to teach this module effectively, please let us know so that we may include them in the Equipment/Materials list in the Annotated Instructor's Guide.

Write: Product Development and Revision
National Center for Construction Education and Research
P.O. Box 141104, Gainesville, FL 32614-1104

Fax: 352-334-0932

E-mail: curriculum@nccer.org

Craft ______________________ Module Name ______________________

Copyright Date __________ Module Number __________ Page Number(s) __________

Description

____

____

____

(Optional) Correction

____

____

(Optional) Your Name and Address

____

____

Module 21306-05

Lattice Boom Assembly and Disassembly

NCCER STANDARDIZED CRAFT TRAINING PROGRAM

The National Center for Construction Education and Research (NCCER) provides a standardized national program of accredited craft training. Key features of the program include instructor certification, competency-based training, and performance testing. The program provides trainees, instructors, and companies with a standard form of recognition through a National Craft Training Registry. The program is described in full in the *Guidelines for Accreditation*, published by the NCCER. For more information on standardized craft training, contact the NCCER by writing us at P.O. Box 141104, Gainesville, FL 32614-1104; calling 352-334-0911; or e-mailing info@nccer.org. More information may be found at our Web site, www.nccer.org.

HOW TO USE THIS ANNOTATED INSTRUCTOR'S GUIDE

Each page presents two sections of information. The larger section displays each page exactly as it appears in the Trainee Module. The narrow column ties suggested trainee and instructor actions to each page and provides icons (detailed below) to call your attention to material, safety, audiovisual, or testing requirements. The bottom of each page includes space for your notes.

The **Audiovisual** icon indicates an appropriate time to show a transparency or other audiovisual aid.

The **Classroom** icon prompts you to define a term, stress a point, ask trainees to explain a concept, or give examples.

The **Demonstration** icon directs you to show trainees how to perform tasks.

The **Examination** icon tells you to administer the written module examination.

The **Homework** icon is placed where you may wish to assign reading for the next class, to assign a project, or to advise trainees to prepare for an examination.

The **Laboratory** icon is used when trainees are to practice performing tasks.

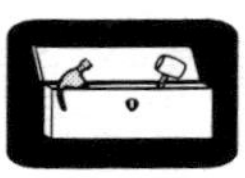

The **Materials** icon is a reminder for you to gather materials needed for classes, labs, and testing.

The **Performance Testing** icon tells you to administer a performance test or a portion thereof.

The **Safety** icon is used to emphasize safety issues. It is often keyed to *Caution* and *Warning* statements in the Trainee Module.

The **Teaching Tip** icon indicates additional guidance is available, such as how to conduct an exercise, get the most educational value from a field trip, or encourage class participation. Teaching Tips may expand on a feature (*Think About It, Did You Know?*) or provide Quick Quizzes or similar exercises. You will be referred to the Teaching Tips section at the back of the module if there is additional material.

The **Combination** icon indicates that the laboratory listed corresponds with a performance task. If desired, you can note the proficiency of the trainees during the laboratory and use it to satisfy performance testing requirements.

PREPARATION

Before teaching this module, you should review the Objectives, Performance Tasks, Materials and Equipment List, and the Module Outline. Be sure to allow ample time to prepare your own training or lesson plan and gather all required materials and equipment.

MODULE OVERVIEW

This module provides a step-by-step look at short and long lattice boom assembly and disassembly.

PREREQUISITES

Prior to training with this module, it is recommended that the trainee shall have successfully completed *Core Curriculum; Mobile Crane Operations Level One; Mobile Crane Operations Level Two;* and *Mobile Crane Operations Level Three,* Modules 21301-05 through 21305-05.

OBJECTIVES

Upon completion of this module, the trainee will be able to do the following:

1. Determine if there is adequate space and resources for crane assembly and disassembly.
2. Identify lattice boom components.
3. Define the relationship of the counterweight to the assembly and disassembly of the boom.
4. Assemble and disassemble a short lattice boom.
5. Assemble and disassemble a long lattice boom.
6. Assemble and disassemble a jib at a lattice boom top.
7. Define and evaluate foundation requirements unique to long lattice boom erection.

PERFORMANCE TASKS

Under the supervision of the instructor, the trainee should be able to do the following:

1. Determine the amount of counterweight required for lattice boom erection.
2. Identify lattice boom components.
3. Assemble and disassemble a short lattice boom.
4. Assemble and disassemble a long lattice boom.
5. Assemble a jib at the lattice boom top.

MATERIALS AND EQUIPMENT LIST

Overhead projector and screen

Transparencies

Blank acetate sheets

Transparency pens

Whiteboard/chalkboard

Markers/chalk

Pencils and scratch paper

Lattice boom crane manufacturer's assembly and disassembly instructions

Jib manufacturer's assembly instructions

Manufacturer's assembly instructions for a short lattice boom

Manufacturer's assembly instructions for a long lattice boom

Lattice boom crane and counterweights

Short and long lattice boom sections

Jib and rigging hardware

Blocking

Tools and rigging hardware for attachment of boom sections

Copies of the Quick Quiz*

Module Examinations**

Performance Profile Sheets**

*Located in the back of this module.

**Located in the Test Booklet.

SAFETY CONSIDERATIONS

Ensure that the trainees are equipped with appropriate personal protective equipment and know how to use it properly. This module requires trainees to assemble and disassemble cranes. Review site safety procedures and site evacuation procedures. Ensure that all trainees are familiar with hand signals and other site communication procedures. Brief trainees on pinching and crushing hazards attendant with assembling and disassembling cranes.

TEACHING TIME FOR THIS MODULE

An outline for use in developing your lesson plan is presented below. Note that each Roman numeral in the outline equates to one session of instruction. Each session has a suggested time period of 2½ hours. This includes 10 minutes at the beginning of each session for administrative tasks and one 10-minute break during the session. Approximately 25 hours are suggested to cover *Lattice Boom Assembly and Disassembly*. You will need to adjust the time required for hands-on activity and testing based on your class size and resources. Because laboratories often correspond to Performance Tasks, the proficiency of the trainees may be noted during these exercises for Performance Testing purposes.

Topic	**Planned Time**
Sessions I and II. Introduction, Assembly and Counterweights Considerations, and Boom Parts	
A. Introduction	____________
B. Pre- and Post-Assembly Considerations	____________
C. Counterweight Considerations	____________
D. Laboratory – Trainees practice determining the amount of counterweight required for lattice boom erection. This laboratory corresponds to Performance Task 1.	____________
E. Boom Parts	____________
F. Laboratory – Trainees practice identifying lattice boom components. This laboratory corresponds to Performance Task 2.	____________
Sessions III through VII. Lattice Boom Assembly and Disassembly	
A. Assembling Short Lattice Booms	____________
B. Laboratory – Trainees practice assembling and disassembling a short lattice boom. This laboratory corresponds to Performance Task 3.	____________
C. Assembling Long Lattice Booms	____________
D. Laboratory – Trainees practice assembling and disassembling a long lattice boom. This laboratory corresponds to Performance Task 4.	____________
Sessions VIII and IX. Jib Assembly	
A. Jib Assembly	____________
B. Laboratory – Trainees practice assembling a jib at the lattice boom top. This laboratory corresponds to Performance Task 5.	____________
Session X. Review, Module Examination, and Performance Testing	
A. Review	____________
B. Module Examination	____________
1. Trainees must score 70 percent or higher to receive recognition from NCCER.	
2. Record the testing results on Craft Training Report Form 200, and submit the results to the Training Program Sponsor.	
C. Performance Testing	____________
1. Trainees must perform each task to the satisfaction of the instructor to receive recognition from NCCER. If applicable, proficiency noted during laboratory exercises can be used to satisfy the Performance Testing requirements.	
2. Record the testing results on Craft Training Report Form 200, and submit the results to the Training Program Sponsor.	

Module 21306-05

Lattice Boom Assembly and Disassembly

Instructor's Notes:

COURSE MAP

This course map shows all of the modules in the third level of the *Mobile Crane Operations* curriculum. The suggested training order begins at the bottom and proceeds up. Skill levels increase as you advance on the course map. The local Training Program Sponsor may adjust the training order.

Assign reading of Module 21306-05.

MOBILE CRANE OPERATIONS LEVEL THREE

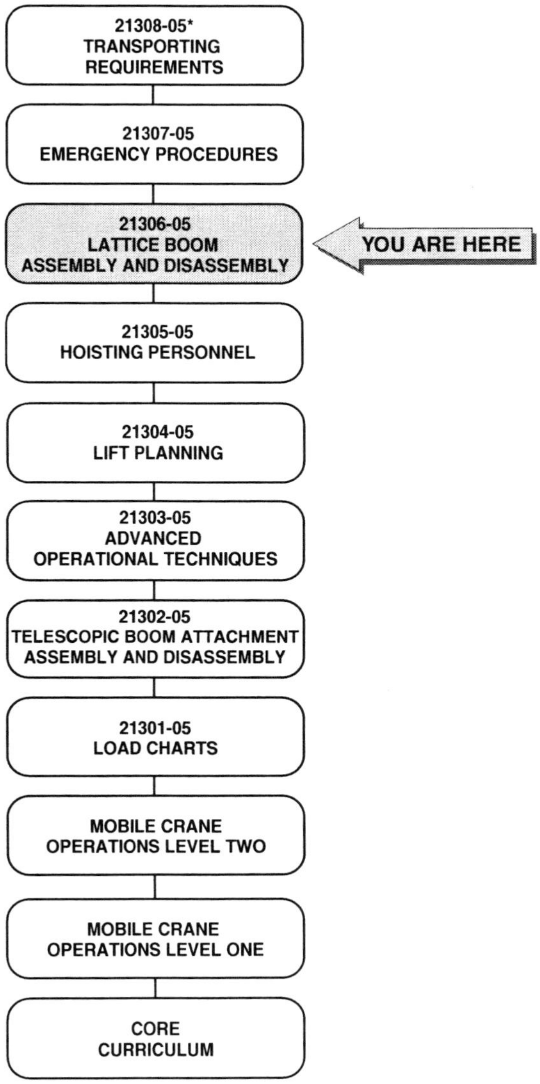

*Please note that 21308-05 is an elective module. It is not necessary to complete this module in order to advance in your training.

306CMAP.EPS

Copyright © 2005 National Center for Construction Education and Research, Gainesville, FL 32614-1104. All rights reserved. No part of this work may be reproduced in any form or by any means, including photocopying, without written permission of the publisher.

LATTICE BOOM ASSEMBLY AND DISASSEMBLY 6.iii

Instructor's Notes:

MODULE 21306-05 CONTENTS

LATTICE BOOM ASSEMBLY AND DISASSEMBLY 6.v

Figures

Instructor's Notes:

MODULE 21306-05

Lattice Boom Assembly and Disassembly

Objectives

When you have completed this module, you will be able to do the following:

1. Determine if there is adequate space and resources for crane assembly and disassembly.
2. Identify lattice boom components.
3. Define the relationship of the counterweight to the assembly and disassembly of the boom.
4. Assemble and disassemble a short lattice boom.
5. Assemble and disassemble a long lattice boom.
6. Assemble and disassemble a jib at a lattice boom top.
7. Define and evaluate foundation requirements unique to long lattice boom erection.

Prerequisites

Before you begin this module, it is recommended that you successfully complete *Core Curriculum; Mobile Crane Operations Level One; Mobile Crane Operations Level Two;* and *Mobile Crane Operations Level Three*, Modules 21301-05 through 21305-05.

Required Trainee Materials

1. Pencil and paper
2. Appropriate personal protective equipment

1.0.0 ◆ INTRODUCTION

Depending on the crane and the overall weight, cranes must be disassembled for transportation and reassembled on site. For most lattice boom cranes, the boom and counterweights are removed. For larger cranes the crawler treads are also removed. Once the crane arrives at the site, it must be properly unloaded and assembled (*Figure 1*). After the lifts are completed, it must be disassembled and loaded for transportation.

Over half of all mobile crane accidents are the result of unskilled work performed when the crane is assembled and placed to make lifts. Operators need to ensure that the crane is assembled and disassembled correctly. The project engineer, site superintendent, contractor, or whoever is causing the crane to be used are also responsible for safety.

Specific guidelines and instructions for assembling the crane and installing various boom attachments and components are contained in the operator's safety manual for each crane. Some manufacturers provide a separate assembly manual.

A typical set of assembly procedures is shown in *Figure 2*. This module will give an overview of the assembly and disassembly procedures relevant to lattice boom cranes.

Ensure that you have everything required to teach the course. Check the Materials and Equipment list at the front of this module.

See the general Teaching Tip at the end of this module.

Explain that terms shown in bold are defined in the Glossary at the back of this module.

Show Transparency 1, Objectives, and Transparency 2, Performance Tasks. Review the goals of the module, and explain what will be expected of trainees.

Explain that over half of all crane accidents could be avoided if cranes were properly placed and assembled.

Show Transparencies 3 and 4 (Figures 1 and 2). Explain the general procedure for unloading and assembling a crane.

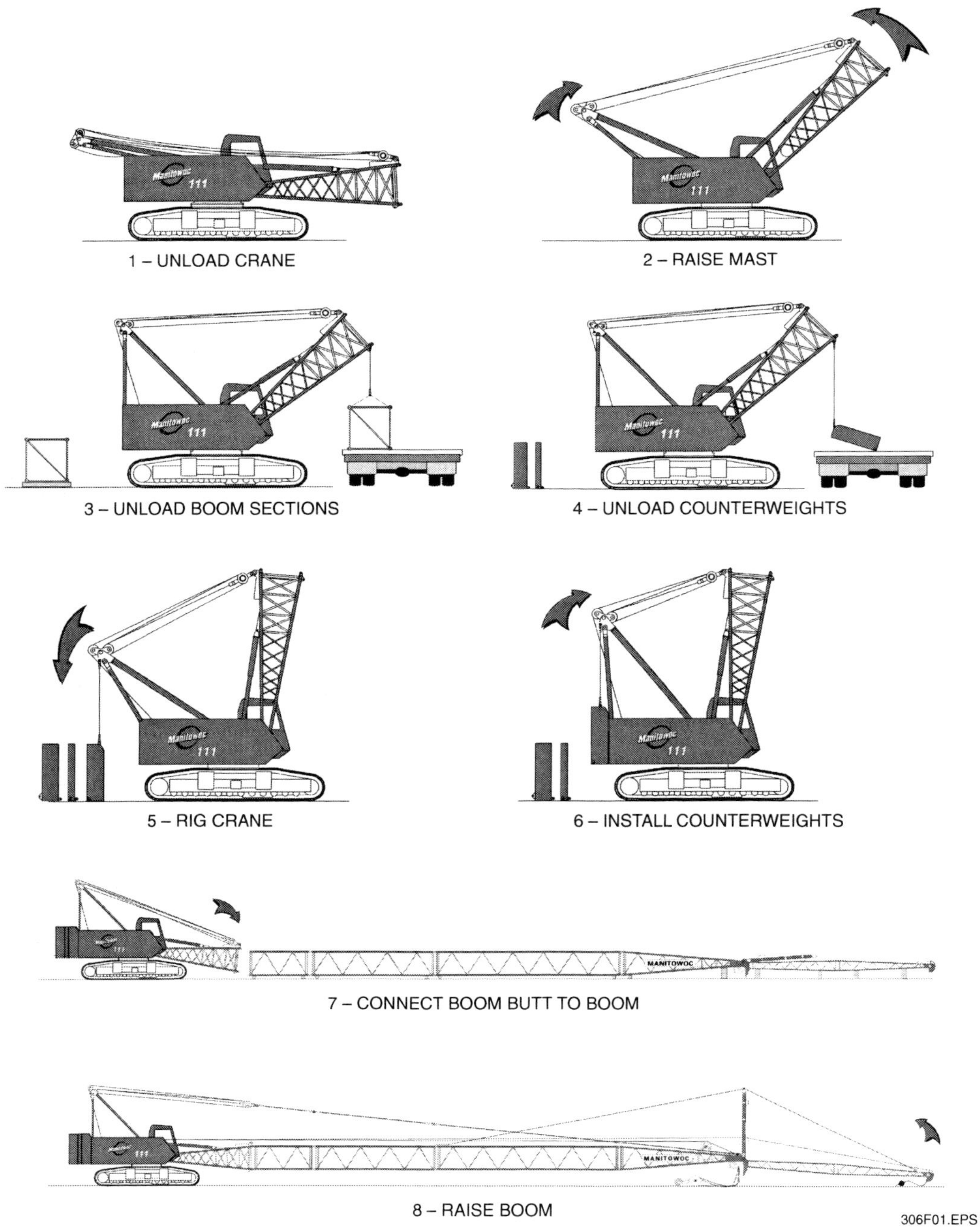

Figure 1 ◆ Crane assembly.

Instructor's Notes:

CRANE ASSEMBLY AND DISASSEMBLY GUIDE
MODEL 999

Table of Contents

Component Identification

NOTE: Except for procedures specific to #82 boom, mast for #22E boom is shown throughout this folio. Mast for #22EL and #82 booms is identical except that pendants (#22EL boom) or straps (#82 and #82LR boom) are suspended from mast.

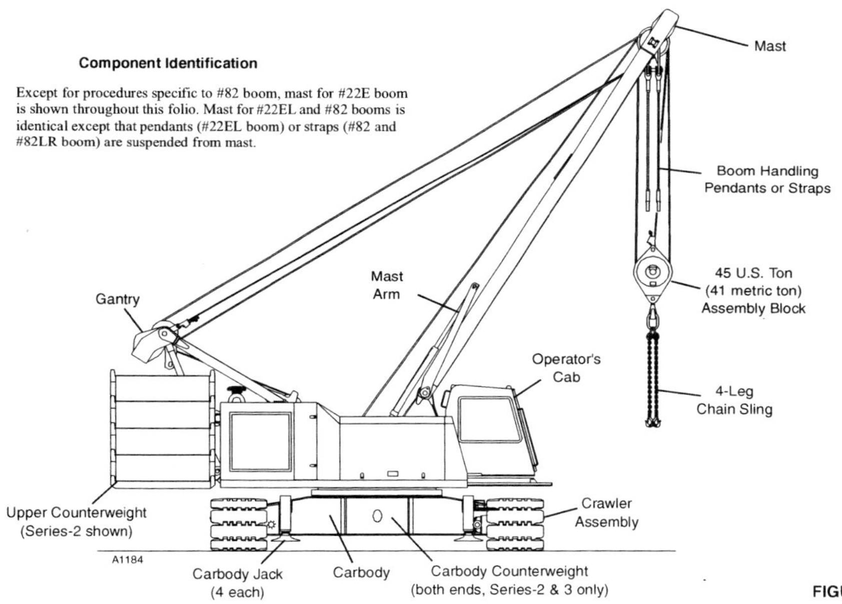

FIGURE 1

© 2002 Manitowoc Cranes, Inc.

Rev. 02-13-2002 **Folio 1864-1**

306F02.EPS

Figure 2 ◈ Crane assembly procedures.

LATTICE BOOM ASSEMBLY AND DISASSEMBLY 6.3

Provide a copy of the manufacturer's assembly and disassembly instructions for the trainees to examine.

See the Teaching Tip for Section 2.0.0 at the end of this module.

Explain that the first consideration is selecting a good site to assemble and disassemble the crane. Discuss space and movement considerations.

Discuss the counterweight considerations for boom assembly.

Explain what personal protective equipment and work practices are needed to safely assemble and disassemble boom sections.

Demonstrate how to determine the amount of counterweight required for lattice boom erection.

Emphasize the need to install counterweights in the proper sequence according to the operator's manual.

Have trainees practice determining the amount of counterweight required for lattice boom erection. Note the proficiency of each trainee. This laboratory corresponds to Performance Task 1.

2.0.0 ◆ PRE- AND POST-ASSEMBLY CONSIDERATIONS

The first consideration for crane assembly is to select a good location to assemble the crane. The assembly/disassembly area should have a firm, level, uniformly supporting surface. Make sure that the area is large enough to accommodate the crane, the selected boom and jib lengths, and the movement of the trailer. An erection area should be barricaded off, as shown in *Figure 3*. It should be large enough to allow the components to be stacked and handled. It should also allow a crane to be assembled without interfering with or posing a hazard to site personnel. As the crane swings, there needs to be at least 2 feet of clearance between the counterweight and the nearest obstacle. The operator and the person responsible for the lift must consider the following:

- Is there enough room for the crane to maneuver into position on the site?
- Is there sufficient room in which to erect the boom?
- Can the trucks hauling boom sections get into place and be unloaded safely?
- Is there enough timber blocking to support the boom when it is being assembled and disassembled?
- Is there a good system for the crane operator to communicate with the assemblers?

306F03.EPS

Figure 3 ◆ Barricaded crane assembly area.

3.0.0 ◆ COUNTERWEIGHT CONSIDERATIONS

Crane operators and any personnel involved in the assembly and disassembly of lattice boom cranes must be aware of counterweight requirements. Do not begin assembling the boom until you find out what the correct counterweight is, and that it is sufficient to get the boom off the ground.

Never add or remove a counterweight from a carrier-mounted crane without first locating the machine on firm, level ground and extending all outrigger beams and jacks, as shown in *Figure 4*. In order to lift some long booms off the ground, it may be necessary to install extra counterweights beyond those which are required for normal operation. On some models of cranes, this extra weight must be removed before operations begin or when shorter booms are used. If it is not, the crane's backward stability will be reduced.

306F04.EPS

Figure 4 ◆ Install counterweights.

The swing lock should be engaged whenever any work is being performed on the counterweights and whenever the boom is being assembled, lifted, or lowered. Moving counterweights can cause the wire rope to break or the mast to collapse if the load exceeds the manufacturer's specifications.

> **WARNING!**
> Assembly and disassembly personnel must always wear the appropriate personal protective equipment and observe safe working practices when working with lattice boom sections.

Instructor's Notes:

> **WARNING!**
> To avoid tipping the crane over, install counterweights in the exact sequence given in the operator's manual. Lifting lugs may fail if the load exceeds the manufacturer's specifications. Falling counterweights can cause death, serious injury, or property damage.

4.0.0 ◆ BOOM PARTS

After the boom sections are unloaded they should be inspected and inventoried. The boom sections should be inspected for damage or wear. Damaged or worn gear can cause serious accidents or injuries if they are used. They should be repaired or replaced pursuant to manufacturer's specifications.

Check the boom sections for bent **lacings**, bent or damaged main **chords**, or any broken or cracked **welds**. If the fittings are bent or damaged the boom sections cannot be fit together easily. Check that the suspension ropes, **pendants**, and boom pendant pins are all in good condition. Pins must be retained by cotter pins, as shown in *Figure 5*.

You should make sure that you have the correct size and quantity of nuts, bolts, and pins needed to secure the boom sections. Use only the fasteners specified in the operator's manual.

Before putting the boom together, determine the weight of the loads and rigging and the height of the lift and the radius of operation. Find the shortest length that will accomplish the job. Check the operator's manual to ensure that the load weight and radius are within the tolerances listed.

Figure 5 ◆ Pin secured with cotter pins.

Boom sections should be handled with care to prevent damage to chords, lacing, and connectors. It is very important to pay close attention to the type of sling being used for assembly. Slings made of nylon are best for handling boom sections, although wire rope or chain slings can be used. When using wire or chain slings, stay clear of lacing to prevent damage. Some boom sections have welded lifting lugs as shown in *Figure 6*. If the boom section has lifting lugs, they should be used for lifting the boom sections.

Figure 6 ◆ Welded lifting lugs.

5.0.0 ◆ LATTICE BOOM ASSEMBLY

The information in this module is presented as a general guideline. Always refer to the manufacturer's manual for the specific lattice boom crane before any assembly work is done.

A boom rigging drawing is furnished with each machine for the particular boom ordered. Important rigging information, including boom assembly, **boom gantry** and back hitch assembly, boom point assemblies, and pendants, are contained within the rigging drawing. The operator should read, study, and understand all the content before assembly.

> **WARNING!**
> Boom section connecting points can be hazardous. Do not perform maintenance directly under a boom. Collapse could occur, resulting in serious injury or death.

See the Teaching Tip for Section 4.0.0 at the end of this module.

Explain the hazards in assembling boom sections. Stress the importance of following the operator's manual. Emphasize that maintenance should never be performed directly under a boom.

Identify lattice boom components and explain how to inspect them for damage.

Demonstrate how to identify lattice boom components.

Have trainees practice identifying lattice boom components. Note the proficiency of each trainee. This laboratory corresponds to Performance Task 2.

Have trainees review Sections 5.0.0–5.2.0.

Ensure that you have everything required for teaching this session.

Discuss the importance of following the manufacturer's assembly and disassembly manual.

See the Teaching Tip for Section 5.0.0 at the end of this module.

Show Transparencies 5 through 7 (Figures 7, 8, and 10). Explain the procedure for assembling a short lattice boom.

Provide a copy of the manufacturer's assembly instructions for a short boom for the trainees to examine.

Discuss the importance of ensuring that boom sections are installed in the proper sequence. Explain pinching and crushing hazards inherent in assembling boom sections. Warn trainees against standing inside the boom while installing connector pins.

5.1.0 Assembling Short Lattice Booms

This section presents a general guideline for assembling short lattice booms. Always consult the manufacturer's recommendations before any assembly or disassembly takes place.

Step 1 Verify the amount of counterweight required in the load/capacity chart for the lifts to be made and the length of boom being installed. Make sure enough counterweight is installed to lift the boom off the ground.

Step 2 If so equipped, the extendable counterweight must be properly extended to manufacturer requirements.

Step 3 Rotate the upperworks to face the direction recommended by the manufacturer, or in the direction of maximum stability.

Step 4 Set the swing lock.

Step 5 Fully extend the gantry (see *Figure 7*). If the machine has a live **boom mast**, check the load/capacity chart to see if it must be used. Never attempt to raise a boom until this equipment has been erected. Crane manufacturers will specify the maximum length of boom which can be raised with standard equipment using the crane machinery, and which lengths require the use of special equipment.

Step 6 Install the boom base (*Figure 8*) section, and attach the supporting pendants.

WARNING!
Boom sections must be installed using the sequence and methods specified by the manufacturer. A boom can collapse if not properly assembled. Do not mix and match boom sections.

Step 7 Assemble the boom sections in the order specified by the manufacturer's manual. The boom number may be stamped on the side of the boom joint connector on two diagonal chords of each boom section. The boom may also have an identification plate, as shown in *Figure 9*. Boom identification is done to prevent mixing boom sections of one type with those of another. Boom identification numbers can exist as either a boom number (20), a boom component part number (42597-2), or a date-manufactured number (12-10-92).

WARNING!
Do not stand inside the boom while installing connector pins. Stand outside the boom.

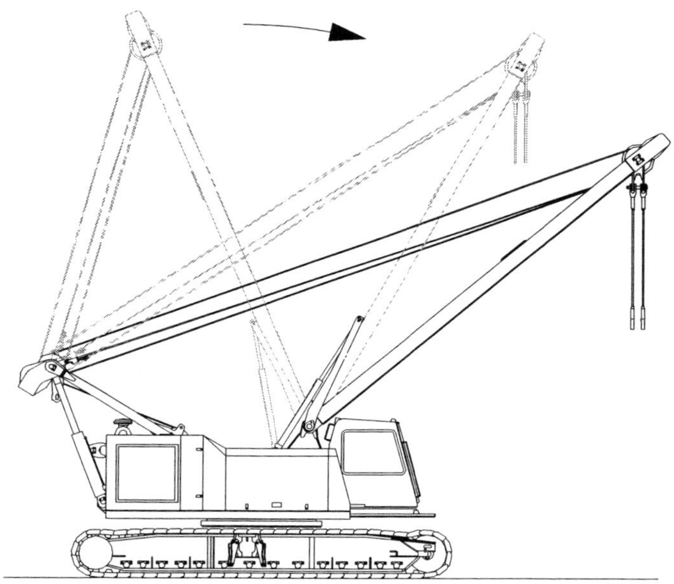

Figure 7 ◆ Extend the gantry.

Instructor's Notes:

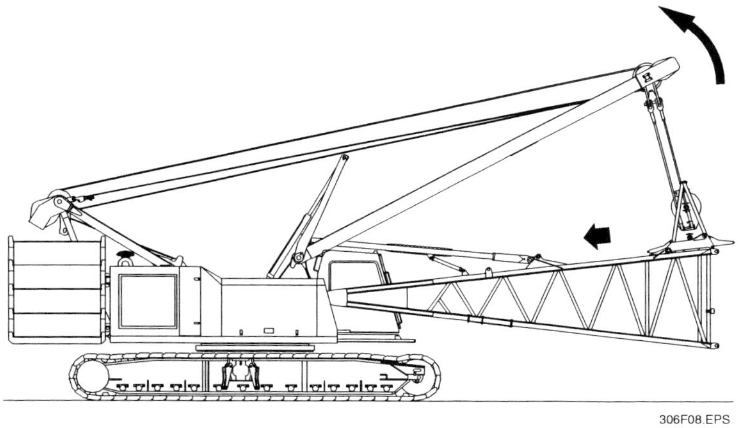

Figure 8 ◈ Install the boom base.

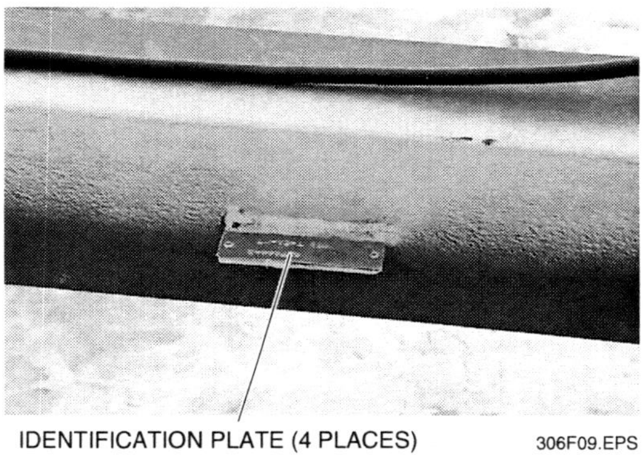

Figure 9 ◈ Boom identification plate.

Step 8 Insert the top pins first.

> **WARNING!**
> Ensure that all personnel are clear before operating the crane.

Step 9 Keep the pendants attached behind the boom sections being connected. Line up the two upper pin connection holes between the two sections being pinned together. Insert the two top boom pins and install the pin retainers (*Figure 10*).

Step 10 Lay out the remaining sections, draw them together, and insert all top pins.

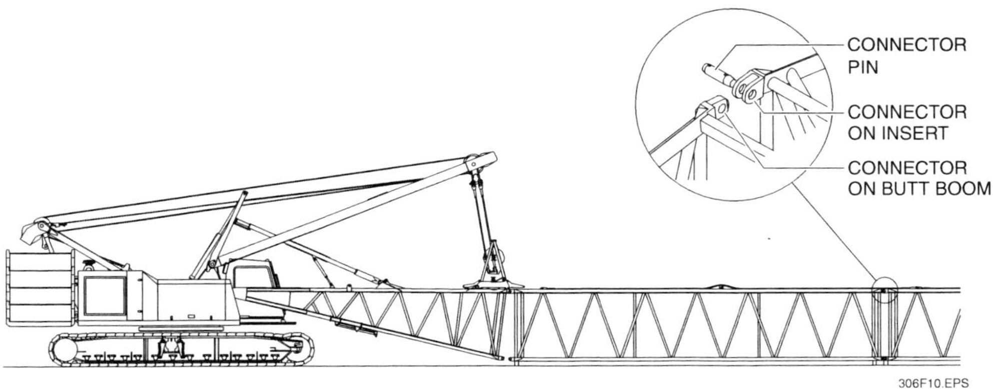

Figure 10 ◈ Connect top pins.

Discuss the importance of ensuring that all personnel are clear before operating the crane. Explain the communications procedure used on the yard.

LATTICE BOOM ASSEMBLY AND DISASSEMBLY 6.7

Explain the importance of checking the load capacity chart before raising the boom. Explain how the crawlers can be blocked to maintain stability. Discuss the dangers of using pendants that are different lengths.

Demonstrate how to assemble and disassemble a short lattice boom.

Have trainees practice assembling and disassembling a short lattice boom. Note the proficiency of each trainee. This laboratory corresponds to Performance Task 3.

> **CAUTION**
> When the boom is assembled, check that the pendants are the proper length for the boom length.

Step 11 If the pendants are slightly long on a machine with a short boom, the weight of the live mast will pull an unloaded boom into the backstops when the minimum radius is approached. The boom will have to be pulled away from the backstops by hanging a load on the hook.

> **CAUTION**
> Make sure that the suspension ropes and pendants do not interfere with the boom pins and/or clips that maintain the pin connections.

> **WARNING!**
> Check the load capacity charts before raising the boom. It may be necessary to block the crawlers to maintain stability.
> It is very important to have the pendants exactly the same length. If they aren't, the boom will be twisted, and the load will be altered. This creates the potential for structural failure.

Step 12 Boom up slightly until the bottom pin connection points line up. Engage the boom hoist pawl. Install the pins and retainers (*Figure 11*).

Step 13 Lower the boom onto blocking, and extend the pendants out to the boom tip (*Figure 12*). Pendants come in matched sets of either two or four. Identification numbers are stamped into both ends of each pendant to aid in keeping them in sets. Always install pendants in sequence as noted on the rigging drawing for the particular boom length being used.

Figure 12 ◆ Lower the boom onto the blocking.

Step 14 Rig the main hoist line (and auxiliary or whip line, if necessary).

Step 15 Check that the boom angle indicator is properly installed and adjusted.

Step 16 Boom up slightly, engage the boom hoist pawl, and reeve the block and/or ball.

Step 17 Make sure that the boom hoist wire rope is spooled tightly onto the drum and engaged with the proper sheaves, and that the wire rope is securely anchored to the wedge socket at the gantry.

Step 18 Do not lift the load block or ball when raising the boom. Pay out the hoist line to keep the block on the ground. Raise the boom smoothly in one continuous lift until the minimum boom angle is reached.

> **CAUTION**
> Always follow manufacturer recommendations when raising the boom to avoid damaging the equipment.

Figure 11 ◆ Install lower pins.

Instructor's Notes:

Step 19 Once the boom is in the air, check that the boom hoist limiting device (kickout) is installed and in working order.

5.2.0 Assembling Long Lattice Booms

This assembly method applies when the boom length exceeds the manufacturer's maximum **cantilever** length (*Figure 13*). As with short lattice booms, this is a guideline, and any manufacturer's recommendations must be followed.

Step 1 Check the amount of counterweight required in the load/capacity chart for the lifts to be made and the length of boom being installed. Check to ensure that enough counterweight is installed to lift the boom off the ground (*Figure 14*).

Step 2 Check to see if a front bumper counterweight is required.

Step 3 If so equipped, extend the extendible counterweight.

> **WARNING!**
> The appropriate counterweights must be used, or the crane may tip.

Step 4 Rotate the upperworks to face the direction recommended by the manufacturer or the direction of maximum stability.

Step 5 Set the swing lock.

Step 6 Fully extend the gantry. If the machine has a live mast, check the load/capacity chart to see if it must be used. Never attempt to raise a boom until this equipment has been fitted. Crane manufacturers will specify the maximum length of boom which can be raised with standard equipment using the crane machinery and which lengths require the use of special equipment.

Step 7 Certain crane manufacturers recommend blocking the front idlers on the crawler ends (*Figure 15*). Consult the operator's manual.

Step 8 Install the base or foot section of the boom, and attach the supporting pendants to the ends.

Step 9 Assemble the boom base sections in the order specified by the manufacturer.

Show Transparencies 8 through 10 (Figures 13, 18, and 20). Explain the procedure for assembling a long lattice boom.

Provide a copy of the manufacturer's assembly instructions for a long boom for the trainees to examine.

Discuss the importance of using the appropriate counterweights.

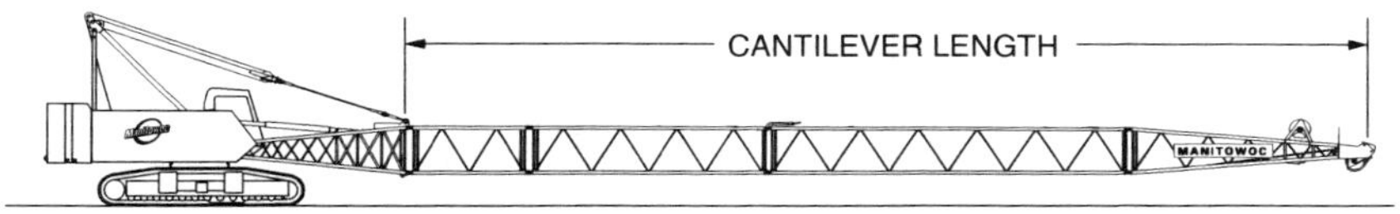

306F13.EPS

Figure 13 ◆ Cantilever length.

306F14.EPS

Figure 14 ◆ Install counterweights.

306F15.EPS

Figure 15 ◆ Blocking the front idlers on crawler ends.

LATTICE BOOM ASSEMBLY AND DISASSEMBLY 6.9

Discuss the dangers of using pendants that are too short.

Step 10 Draw some, but not all, of the sections together. The total length of the inserts to assemble at this stage must not exceed the maximum cantilever length specified by the manufacturer. Line up the upper pin connection points. Insert the top pins only, and secure them with pin retainers.

Step 11 Boom up slightly until the bottom pin connection points line up. Engage the boom hoist pawl. Install the pins and pin retainers (*Figure 16*).

Step 12 Lower the boom onto blocking until the pendants are slack. Use pendant **spreader bars** (*Figure 17*) as required by the manufacturer.

Step 13 Move the pendants out to the end of the last insert (*Figure 18*). Remember to watch for twisting of the boom.

Step 14 Draw the remaining boom sections together. Ensure that their length does not go over the maximum cantilever length.

Step 15 Line up the upper pin connection points. Insert the top pins only, and secure them with retainer pins.

Step 16 Boom up slightly until the bottom pin connection points line up. Engage the boom hoist pawl. Install pins and cotters.

Step 17 Lower the boom onto blocking until the pendants are slack (*Figure 19*).

WARNING!
If the pendants are too short on a machine with a longer boom, it will overload the boom, mast, pendants, and boom hoist systems when handling near-capacity loads.

Figure 16 ◆ Secure lower pins.

Figure 17 ◆ Pendant spreader bar.

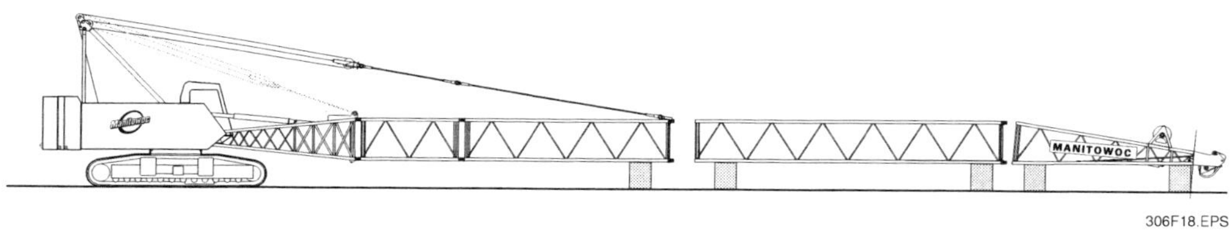

Figure 18 ◆ Move the pendants to the end of the boom section.

Instructor's Notes:

306F19.EPS

Figure 19 ◈ Lower the boom onto blocking.

Step 18 Move the pendants out to the end of the boom (*Figure 20*).

Step 19 Rig the main hoist line, and if necessary, the whip line.

> **NOTE**
> Some manufacturers require boom suspension for long booms.

Step 20 Boom up slightly, engage the boom hoist pawl, and reeve the block.

Step 21 To gain extra stability, do not lift the load block when raising the boom. Pay out the hoist line to keep the block on the ground. Raise the boom slowly.

Some manufacturers recommend certain checks be performed before the boom is raised. An example of these checks is shown in *Figure 21*.

Pre-Raising Checks

Make following checks and correct any defects before raising attachment:

- ❑ Crane is on firm, level surface.
- ❑ Boom hinge pins are fully engaged and locking cover is pinned in up position.
- ❑ Crawler connecting pins are engaged and locking pins are installed.
- ❑ Carbody jack pads are stored.
- ❑ Carbody jacks are fully retracted and pinned in stored position.
- ❑ Boom and jib inserts are installed in proper sequence per rigging drawing
- ❑ Spreader (22EL boom only) is installed at proper location and spread to proper position.
- ❑ Intermediate suspension installed if required (#82 Long Reach boom only).
- ❑ Links properly connected between all straps.
- ❑ All insert and pendant connector pins installed. Cotter pins installed and spread.
- ❑ Gantry fully raised.
- ❑ Mast arms fully lowered.
- ❑ Mast arms bypass handle turned off.
- ❑ Setup mode turned off (desired operating mode is selected and confirmed).
- ❑ Boom hoist wire rope spooled tightly onto drum and engaged with proper sheaves. Wire rope securely anchored to wedge socket at gantry.
- ❑ Load lines spooled tightly onto drums and engaged with proper sheaves. Load lines securely anchored to wedge sockets at boom and jib point or at load block and weight ball.
- ❑ All blocking, tools, and other items removed from boom and jib.
- ❑ Automatic boom stop properly installed and adjusted (refer to Boom Stop Adjustment folio in Operator's Manual).
- ❑ Cable from crane control system connected to cable reel on boom butt.
- ❑ Block-up limit control properly installed and operational.
- ❑ LMI properly installed and operational (refer to Block-Up Limit Adjustment folio in Operator's Manual).
- ❑ Crane and attachment properly lubricated (refer to Lubrication Guide in your Operator's Manual).
- ❑ Crawlers blocked if required per capacity chart.
- ❑ Wind within allowable limits for operation given on capacity chart.

306F21.EPS

Figure 21 ◈ Pre-raising checklist chart.

Show Transparency 11 (Figure 21). Discuss recommended pre-raising checks.

Demonstrate how to assemble and disassemble a long lattice boom.

Have trainees practice assembling and disassembling a long lattice boom. Note the proficiency of each trainee. This laboratory corresponds to Performance Task 4.

Have trainees review Sections 6.0.0–7.0.0.

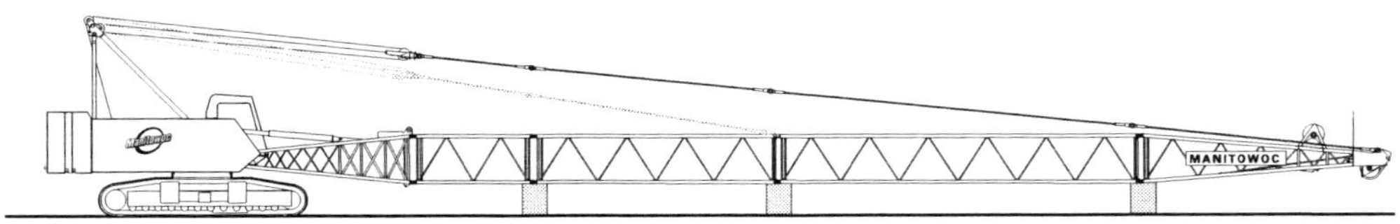

306F20.EPS

Figure 20 ◈ Move the pendant to the end of the boom.

LATTICE BOOM ASSEMBLY AND DISASSEMBLY 6.11

Ensure that you have everything required for teaching this session.

Show Transparencies 12 and 13 (Figures 23 and 24). Explain the procedures for assembling a jib onto a lattice boom tip.

Provide a copy of the jib manufacturer's assembly instructions for the trainees to examine.

Demonstrate how to assemble a jib at the lattice boom top.

Have trainees practice assembling a jib at the lattice boom top. Note the proficiency of each trainee. This laboratory corresponds to Performance Task 5.

Discuss the procedures for disassembly of lattice booms.

Reiterate the dangers of performing maintenance directly under a boom section.

6.0.0 ◆ JIB ASSEMBLY

Once the boom is assembled, the jib can be assembled and attached to the boom. The following steps are general guidelines for installing a jib onto a lattice boom tip:

Step 1 Lay the boom tip on blocking on the ground. Some boom tips have welded brackets to attach the jib, as shown in *Figure 22*. If a jib adapter is required for the boom, install the jib adapter to the boom tip.

Step 2 Assemble the jib on blocking on the ground, pinning the jib base to the boom tip.

Step 3 Pin the jib strut to the jib base (*Figure 23*).

Step 4 Pin the jib forestay and backstay pendants to the strut tip with the pins furnished with the pendants (*Figure 24*).

Step 5 Install a wire-rope roller guide on each jib insert used, if applicable.

Step 6 Pin the other end of the jib pendants to the jib point.

306F22.EPS

Figure 22 ◆ Boom tip with jib bracket.

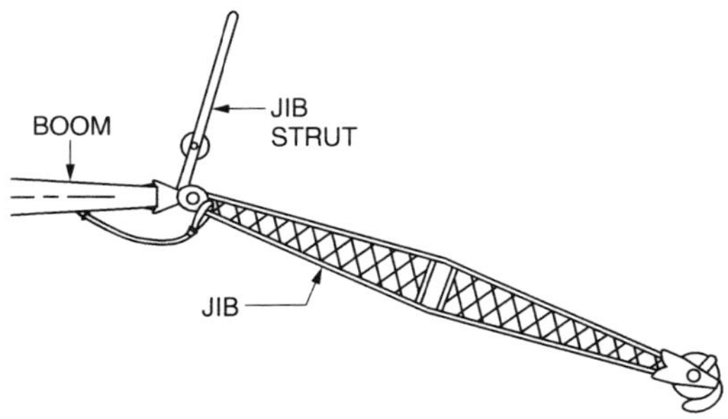

306F23.EPS

Figure 23 ◆ Jib strut to jib base.

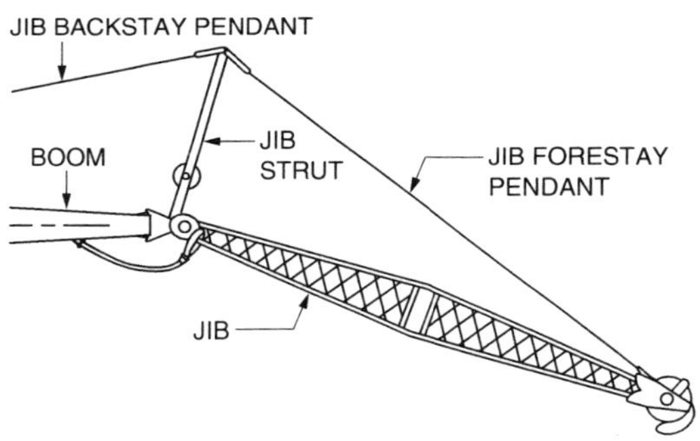

306F24.EPS

Figure 24 ◆ Jib forestay and backstay pendants.

7.0.0 ◆ DISASSEMBLY OF LATTICE BOOMS

The following sequence is a guideline for removing lattice boom sections. Always consult manufacturer requirements when performing any kind of disassembly.

> **WARNING!**
> Boom section connecting points can be hazardous. Do not perform maintenance directly under a boom. Collapse could occur, resulting in serious injury or death.

Step 1 Move the crane onto a level area that is cleared of personnel.

Step 2 Set the outriggers and level the machine, or block idlers on a crawler.

Step 3 Check that the counterweight is correct.

Step 4 Set the swing lock.

Instructor's Notes:

> **CAUTION**
> A thorough inspection of the boom hoist cable should be made before booming down. Cable damaged during crane operations can break under maximum load. Collapse could occur, resulting in serious equipment damage.

Step 5 Lower the boom over the rear or over the area of maximum stability.

Step 6 Place blocking under the boom tip and under both sides of every joint, if possible.

Step 7 Lower the boom onto this blocking until the pendants go slack.

Step 8 Move the pendants back by removing a length and re-anchoring them at a lower boom section.

Step 9 Elevate the boom slightly. Do not elevate it so far that the boom peak is lifted off its blocking.

Step 10 Drive out the lower pins on the sections ahead of the pendants.

Step 11 Lower the boom back down onto the blocking. Make sure every section is supported.

Step 12 Drive out the upper pins on the sections ahead of the pendants.

Step 13 Repeat this procedure until the boom is disassembled.

Summary

Lifting operations depend on proper assembly of the crane's boom. On lattice boom cranes, this involves the proper attachment of the boom sections. It is imperative that assembly and disassembly personnel be aware of clearance requirements on the job site, and that they also follow the appropriate safety requirements associated with the work.

Review Questions

1. For each crane, guidelines and instructions for installing various boom attachments and components are _____.

a. contained in the operator's safety manual
b. listed on the manufacturer's web site
c. the same regardless of serial number, as long as the crane is manufactured by the same company
d. the same regardless of model number, as long as the crane is manufactured by the same company

2. Cranes must be assembled on site because _____.

a. the configuration depends on site conditions
b. each part must be separately inspected
c. new wire ropes must be installed
d. they are too heavy to transport intact

3. Over half of all mobile crane accidents are the result of _____.

a. poorly-trained operators
b. mechanical and equipment failures
c. unskilled work performed when the crane is assembled
d. boom failure due to overweight load

4. Disassembly of most lattice boom cranes involves removal of the _____.

a. carbody and counterweights
b. outriggers and boom
c. base mounting and counterweights
d. boom and counterweights

5. The first concern for crane assembly is to _____.

a. select a good location
b. make sure you have all the pins and bolts
c. calculate the counterweight needed
d. inspect all the boom sections

6. The crane assembly area must be _____.

a. close to or at the lift location
b. on a hill to make it easier to lift the boom
c. a firm and level surface
d. close to the main access road

Point out the importance of inspecting cable before booming down.

Summarize the major concepts presented in the module.

Have trainees complete the Review Questions, and go over the answers prior to administering the Module Examination.

Administer the Module Examination. Record the results on Craft Training Report Form 200, and submit the results to the Training Program Sponsor.

Administer the Performance Test, and fill out Performance Profile Sheets for each trainee. If desired, trainee proficiency noted during laboratory sessions may be used to complete the Performance Test. Record the results on Craft Training Report Form 200, and submit the results to the Training Program Sponsor.

LATTICE BOOM ASSEMBLY AND DISASSEMBLY 6.13

7. Barricades should be placed _____.
   a. completely around the assembly area
   b. at a distance of 5 feet around the crane body
   c. at the swing radius of the crane
   d. around boom sections

8. The operator should check if there is enough timber to _____.
   a. chock the wheels of delivery trucks
   b. brace the crawlers on the crane
   c. block the boom sections
   d. build barricades

9. The crane operator and the assemblers must have _____.
   a. the same instruction book
   b. a way to communicate
   c. crane operation training
   d. commercial driver's licenses

10. In order to lift a longer boom off the ground it may be necessary to install _____ on the crane.
    a. a boom extension tip
    b. pendant spreaders
    c. a luffing jib
    d. additional counterweights

11. Whenever any work is being performed on the counterweights the ____ should be engaged.
    a. swing lock
    b. load moment indicator
    c. anti-two-blocking device
    d. emergency brake

12. Before adding or removing counterweights you must _____.
    a. locate the crane on a firm level surface
    b. retract the outriggers
    c. chock the crawlers
    d. lockout forward motion control

13. Before adding or removing counterweight from a carrier-mounted crane, the _____ must be extended.
    a. hoist rope
    b. telescopic boom sections
    c. pendant spreader bars
    d. outrigger beams and jacks

14. If the fittings on boom sections are bent, _____.
    a. the pins should be replaced
    b. use larger pins to fill the gap
    c. the boom sections will not fit together properly
    d. do not use them to lift the boom

15. After the boom parts are unloaded they should be _____.
    a. placed for assembly
    b. properly lubricated
    c. connected with pins
    d. inspected for damage

16. All pins should be secured with _____.
    a. cotter pins
    b. wire rope
    c. bolts
    d. nuts

17. The boom sections should be inspected for _____ before assembly.
    a. proper length
    b. correct order
    c. bent lacings
    d. manufacturer's identification

18. Before assembly, suspension ropes and pendants should be in good condition and _____.
    a. well lubricated
    b. reeved into the block
    c. coiled on the drum
    d. retained by cotter pins

19. When you determine the weight of the loads, the height of the lift, and the radius of operation, the boom length should be the shortest length that will accomplish the job.
    a. True
    b. False

20. When handling the boom, to prevent damage you should _____.
    a. always use chain slings
    b. use a choker hitch around the lacing
    c. attach the sling to the lifting lugs if present
    d. use a basket sling around boom section

21. Any pins and fasteners can be used to assemble the boom as long as they are secured by cotter pins.
    a. True
    b. False

22. A rope or bar of specified length with fixed end connections is called a _____.
    a. cantilever
    b. pendant
    c. placard bar
    d. jib adapter

Instructor's Notes:

23. Information on boom assembly, gantry and back hitch assembly, boom point assemblies, and pendants is available _____.
 a. in the manufacturer's inspection manual
 b. on the boom rigging drawing
 c. at the manufacturer's web site
 d. on the range diagram

24. The operator must make sure that the _____ interfere with the boom pins and/or clips that maintain the pin connections.
 a. suspension ropes do not
 b. anti-two-block device does not
 c. load cell does not
 d. gantry does not

25. Pendants may come in matched sets of _____.
 a. four
 b. six
 c. eight
 d. ten

26. After rigging the main hoist line during boom assembly, the operator must make sure the _____ is properly installed and adjusted.
 a. boom angle indicator
 b. boom length indicator
 c. hoist drum rotation indicator
 d. selected controls system

27. The amount of counterweight needed to lift the boom can be verified by _____.
 a. asking the project manager
 b. using the load chart
 c. using all the weights on hand
 d. checking the project specifications

28. The boom identification number is used _____.
 a. by the manufacturer for tracking
 b. to date-stamp each section
 c. only for inventory control
 d. to prevent mixing boom sections of different types

29. When connecting boom sections, the bottom pins should be secured first.
 a. True
 b. False

30. To aid in keeping pendant sets together, both ends of each pendant are _____.
 a. color coded
 b. stamped with identification numbers
 c. clipped together
 d. the same size

31. When raising the boom, the block must be _____.
 a. on the ground
 b. at the boom tip
 c. level with the boom base
 d. exactly half the distance to the ground

32. A frame hinged at or near the boom hinge for use in supporting a boom is called a boom _____.
 a. gantry
 b. pendant
 c. chord
 d. mast

33. When assembling long lattice booms, each group raised _____.
 a. cannot exceed 20 feet
 b. must be a minimum of two boom sections
 c. cannot exceed the maximum cantilever length
 d. are a maximum of three boom sections

34. A thorough inspection of _____ should be made before booming down.
 a. the hydraulic system
 b. outrigger positioning
 c. boom hoist cable
 d. the pendants

35. In the jib assembly procedure discussed in this module, the jib is assembled and attached to the lattice boom tip after the _____.
 a. boom is assembled
 b. wire rope roller guide is installed on each jib insert
 c. jib strut is pinned to the jib boom
 d. jib forestay is pinned to the strut tip

LATTICE BOOM ASSEMBLY AND DISASSEMBLY 6.15

Instructor's Notes:

GLOSSARY

Trade Terms Introduced in This Module

Boom gantry: A structural frame, extending above the crane superstructure, to which the boom or boom mast support ropes are reeved.

Boom mast: A frame hinged at or near the boom hinge for use in supporting a boom. The head of the mast is usually supported and raised or lowered by the boom hoist ropes.

Cantilever: A rigid structural member projecting from a vertical support so that the upper part is in tension and the lower part is in compression.

Chord: A load-bearing member of a lattice boom.

Lacing: A lattice strut member of a boom structure that is used to add structural support to the boom and keep the chords in their proper position.

Pendant: A rope or bar of specified length with fixed end connections.

Spreader bars: Adjustable bars used with long boom lengths to keep the pendants from scuffing on top of the boom.

Weld: A localized body of metals or nonmetals produced either by heating the materials to the melting temperature, with or without the application of pressure; or by the application of pressure alone, and with or without the use of filler material.

LATTICE BOOM ASSEMBLY AND DISASSEMBLY 6.17

Instructor's Notes:

Figure Credits

Manitowoc Crane Group	306F01, 306F02, 306F05–306F11, 306F13, 306F21
Topaz Publications, Inc.	306F03, 306F12, 306F16, 306F19
Link-Belt Construction Equipment Company	306F04, 306F14, 306F15, 306F17, 306F22

LATTICE BOOM ASSEMBLY AND DISASSEMBLY 6.19

MODULE 21306-05 — TEACHING TIPS

The following are suggested activities or instructional methods to help you teach the material in this module.

General

When you call on someone to answer a question, the rest of the class relaxes or even tunes out because they expect that the question and answer will take place only between you and the trainee you called on. Instead, use this technique to involve more trainees in answering questions and to keep them on their toes.

1. Ask trainees to define a term or explain a concept.
2. After one trainee has answered, ask a trainee seated nearby if the answer is right. Then ask whether a trainee in the back of the room agrees.
3. Ask trainees to explain why they think an answer is right or wrong.
4. Use the session to clear up incorrect ideas and encourage trainees to learn from their mistakes.

Section 2.0.0 *Pre- and Post-Assembly Considerations*

Selecting an area to assemble and disassemble a crane is critical to safety. Trainees will need appropriate personal protective equipment, pencils, and paper. Arrange for an opportunity to observe several sites where cranes are being or have been erected. Alternatively, have a manufacturer's representative give a presentation on site and staging considerations for different types of cranes. Allow 1 to 2 hours for this exercise.

1. Before the tour discuss the pre- and post-assembly considerations, including counterweight requirements with trainees.
2. Tour the site(s), and point out specific safety considerations. See that trainees can point out overhead power lines and other clearance issues, traffic and truck access, and unsafe ground conditions. Have trainees take notes and write down questions during the tour.
3. Have the trainees identify the best location to erect a crane in each location.
4. After the tour answer any questions the trainees may have.

Section 4.0.0 *Boom Parts*

This Quick Quiz will familiarize trainees with the terms and definitions that are commonly used in boom erection. You will need photocopies of the quiz provided in the following page. Trainees will need pencils. If you allow trainees to use the Trainee Module, decrease the amount of time you give them to complete the quiz.

1. Make a photocopy of the quiz for each trainee.
2. Give trainees between 5 and 10 minutes to complete the quiz.
3. Go over the answers to the quiz.
4. Ask trainees if they have questions.

Answers to Quick Quiz

1. c
2. e
3. f
4. d
5. a
6. b

Quick Quiz *Terms and Definitions*

For each description listed, identify the term that the text best describes. Write the corresponding letter in the blank provided.

_____ **1.** A loadbearing member of a lattice boom

_____ **2.** A rigid structural member projecting from a vertical support so that the upper part is in tension and the lower part is in compression

_____ **3.** Adjustable bars used with long boom lengths to keep pendants from scuffing the top of the boom

_____ **4.** A lattice strut member of a boom structure that is used to add structural support to the boom and keep the chords in proper position

_____ **5.** A localized body of metals or nonmetals produced by heating the materials to the melting temperature

_____ **6.** A rope or bar of specified length with fixed connections

a. Weld
b. Pendant
c. Chord
d. Lacing
e. Cantilever
f. Spreader bars

Copyright © 2005 National Center for Construction Education and Research. Permission is granted to reproduce this page provided that copies are for local use only and that each copy contains this notice.

Section 5.0.0 *Lattice Boom Assembly*

Lattice booms from different models of crane or manufacturers are not assembled in the same way. Trainees will need appropriate personal protective equipment, pencils, and paper. Arrange for an opportunity to observe several different types of lattice booms being assembled. Alternatively, have a manufacturer's representative give a presentation on the various types of lattice booms and jibs for their cranes. Allow 1 to 2 hours for this exercise.

1. Before the tour discuss at least one method for assembling a lattice boom with trainees.
2. Ask the site safety technician to explain how to assemble a lattice boom or jib. Have trainees take notes and write down questions during the tour.
3. Have the trainees identify differences in how the various booms are assembled.
4. After the tour answer any questions the trainees may have.

MODULE 21306-05 — ANSWERS TO REVIEW QUESTIONS

Answer	Section
1. a	1.0.0
2. d	1.0.0
3. c	1.0.0
4. d	1.0.0
5. a	2.0.0
6. c	2.0.0
7. a	2.0.0
8. c	2.0.0
9. b	2.0.0
10. d	3.0.0
11. a	3.0.0
12. a	3.0.0
13. d	3.0.0
14. c	4.0.0
15. d	4.0.0
16. a	4.0.0
17. c	4.0.0
18. d	4.0.0
19. a	4.0.0
20. c	4.0.0
21. b	4.0.0
22. b	4.0.0; Glossary
23. b	5.0.0
24. a	5.1.0
25. a	5.1.0
26. a	5.1.0
27. b	5.1.0
28. d	5.1.0
29. b	5.1.0
30. b	5.1.0
31. a	5.1.0
32. d	5.1.0; Glossary
33. c	5.2.0
34. c	7.0.0
35. a	6.0.0

CONTREN® LEARNING SERIES — USER UPDATE

The NCCER makes every effort to keep these textbooks up-to-date and free of technical errors. We appreciate your help in this process. If you have an idea for improving this textbook, or if you find an error, a typographical mistake, or an inaccuracy in NCCER's Contren® textbooks, please write us, using this form or a photocopy. Be sure to include the exact module number, page number, a detailed description, and the correction, if applicable. Your input will be brought to the attention of the Technical Review Committee. Thank you for your assistance.

Instructors – If you found that additional materials were necessary in order to teach this module effectively, please let us know so that we may include them in the Equipment/Materials list in the Annotated Instructor's Guide.

Write: Product Development and Revision
National Center for Construction Education and Research
P.O. Box 141104, Gainesville, FL 32614-1104

Fax: 352-334-0932

E-mail: curriculum@nccer.org

Craft ______ Module Name ______

Copyright Date ______ Module Number ______ Page Number(s) ______

Description

______

______

______

(Optional) Correction

______

______

(Optional) Your Name and Address

______

______

Module 21307-05

Emergency Procedures

NCCER STANDARDIZED CRAFT TRAINING PROGRAM

The National Center for Construction Education and Research (NCCER) provides a standardized national program of accredited craft training. Key features of the program include instructor certification, competency-based training, and performance testing. The program provides trainees, instructors, and companies with a standard form of recognition through a National Craft Training Registry. The program is described in full in the *Guidelines for Accreditation*, published by the NCCER. For more information on standardized craft training, contact the NCCER by writing us at P.O. Box 141104, Gainesville, FL 32614-1104; calling 352-334-0911; or e-mailing info@nccer.org. More information may be found at our Web site, www.nccer.org.

HOW TO USE THIS ANNOTATED INSTRUCTOR'S GUIDE

Each page presents two sections of information. The larger section displays each page exactly as it appears in the Trainee Module. The narrow column ties suggested trainee and instructor actions to each page and provides icons (detailed below) to call your attention to material, safety, audiovisual, or testing requirements. The bottom of each page includes space for your notes.

The **Audiovisual** icon indicates an appropriate time to show a transparency or other audiovisual aid.

The **Classroom** icon prompts you to define a term, stress a point, ask trainees to explain a concept, or give examples.

The **Demonstration** icon directs you to show trainees how to perform tasks.

The **Examination** icon tells you to administer the written module examination.

The **Homework** icon is placed where you may wish to assign reading for the next class, to assign a project, or to advise trainees to prepare for an examination.

The **Laboratory** icon is used when trainees are to practice performing tasks.

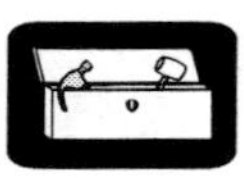

The **Materials** icon is a reminder for you to gather materials needed for classes, labs, and testing.

The **Performance Testing** icon tells you to administer a performance test or a portion thereof.

The **Safety** icon is used to emphasize safety issues. It is often keyed to *Caution* and *Warning* statements in the Trainee Module.

The **Teaching Tip** icon indicates additional guidance is available, such as how to conduct an exercise, get the most educational value from a field trip, or encourage class participation. Teaching Tips may expand on a feature (*Think About It*, *Did You Know?*) or provide Quick Quizzes or similar exercises. You will be referred to the Teaching Tips section at the back of the module if there is additional material.

The **Combination** icon indicates that the laboratory listed corresponds with a performance task. If desired, you can note the proficiency of the trainees during the laboratory and use it to satisfy performance testing requirements.

PREPARATION

Before teaching this module, you should review the Objectives, Performance Tasks, Materials and Equipment List, and the Module Outline. Be sure to allow ample time to prepare your own training or lesson plan and gather all required materials and equipment.

MODULE OVERVIEW

This module covers accident prevention and investigation, the hazards of power line contact, and various failures that may occur during lifting operations.

PREREQUISITES

Prior to training with this module, it is recommended that the trainee shall have successfully completed *Core Curriculum; Mobile Crane Operations Level One; Mobile Crane Operations Level Two;* and *Mobile Crane Operations Level Three,* Modules 21301-05 through 21306-05.

OBJECTIVES

Upon completion of this module, the trainee will be able to do the following:

1. Explain legal responsibilities associated with crane operations, including accident prevention and investigation guidelines.
2. Describe the fire prevention standard for mobile crane operations.
3. Describe operational guidelines for lifting around power lines.
4. Recognize various failures in lifting and how to avoid them.

PERFORMANCE TASKS

There are no performance tasks for this module.

MATERIALS AND EQUIPMENT LIST

Overhead projector and screen
Transparencies
Blank acetate sheets
Transparency pens
Whiteboard/chalkboard
Markers/chalk
Pencils and scratch paper
ASME B30.5
Copies of sample accident investigation
Proximity warning device
Crane safety DVD or video
TV and VCR or DVD player
Copies of site safety procedures
Mobile crane with rigging and lifting equipment
Crane manufacturer's operator's manual
Copies of the Quick Quiz*
Module Examinations**

*Located in the back of this module

**Located in the Test Booklet.

SAFETY CONSIDERATIONS

Ensure that the trainees are equipped with appropriate personal protective equipment and know how to use it properly. Review site safety procedures and site evacuation procedures. Ensure that all trainees are familiar with hand signals and other site communication procedures. Brief trainees on pinching and crushing hazards attendant with crane operations. This module may require trainees to visit construction sites. Brief trainees on site safety procedures.

TEACHING TIME FOR THIS MODULE

An outline for use in developing your lesson plan is presented below. Note that each Roman numeral in the outline equates to one session of instruction. Each session has a suggested time period of 2½ hours. This includes 10 minutes at the beginning of each session for administrative tasks and one 10-minute break during the session. Approximately 15 hours are suggested to cover *Emergency Procedures*. You will need to adjust the time required for hands-on activity and testing based on your class size and resources.

Topic	Planned Time
Session I. Introduction and Accident Prevention and Investigation	
A. Introduction	____________
B. Accident Prevention and Investigation	____________
Session II. Fire	
A. Fire Prevention	____________
B. Extinguishing Fires	____________
Sessions III and IV. Power Lines	
A. De-energized and Grounded Electric Power Lines	____________
B. Prohibited Zone	____________
C. Avoidance Zone	____________
D. Operations in Transit	____________
E. Power Line Contact	____________
Session V. Various Failures in Lifting	
A. Striking the Boom	____________
B. Backward Collapse of a Boom	____________
Session VI. Review and Module Examination	
A. Review	____________
B. Module Examination	____________

1. Trainees must score 70 percent or higher to receive recognition from NCCER.
2. Record the testing results on Craft Training Report Form 200, and submit the results to the Training Program Sponsor.

Module 21307-05

Emergency Procedures

Instructor's Notes:

COURSE MAP

This course map shows all of the modules in the third level of the *Mobile Crane Operations* curriculum. The suggested training order begins at the bottom and proceeds up. Skill levels increase as you advance on the course map. The local Training Program Sponsor may adjust the training order.

Assign reading of Module 21307-05.

MOBILE CRANE OPERATIONS LEVEL THREE

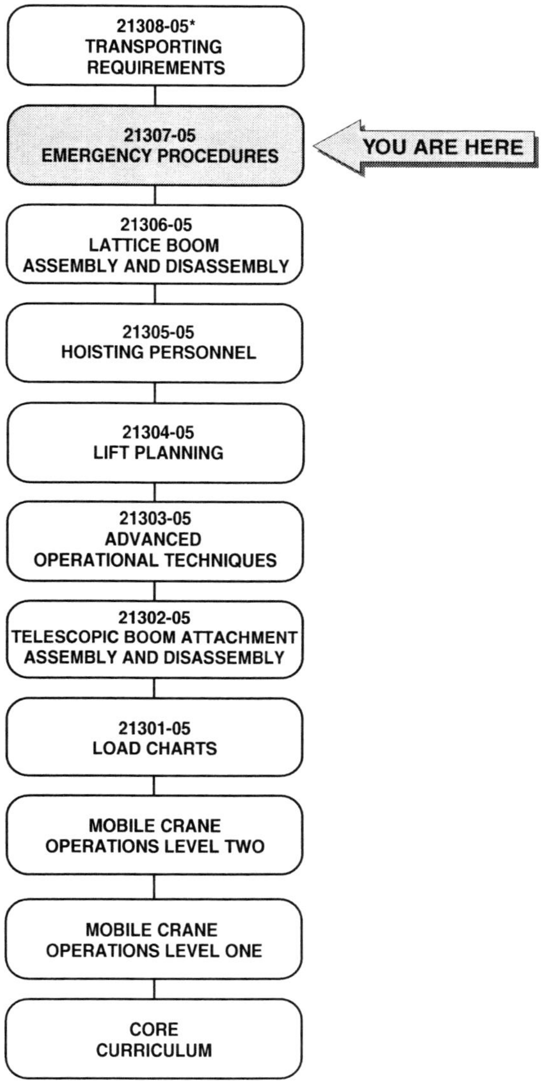

*Please note that 21308-05 is an elective module. It is not necessary to complete this module in order to advance in your training.

307CMAP.EPS

Copyright © 2005 National Center for Construction Education and Research, Gainesville, FL 32614-1104. All rights reserved. No part of this work may be reproduced in any form or by any means, including photocopying, without written permission of the publisher.

EMERGENCY PROCEDURES 7.iii

Instructor's Notes:

MODULE 21307-05 CONTENTS

Figures

Tables

EMERGENCY PROCEDURES 7.v

Instructor's Notes:

MODULE 21307-05

Emergency Procedures

Objectives

When you have completed this module, you will be able to do the following:

1. Explain legal responsibilities associated with crane operations, including accident prevention and investigation guidelines.
2. Describe the fire prevention standard for mobile crane operation.
3. Describe operational guidelines for lifting around power lines.
4. Recognize various failures in lifting and how to avoid them.

Prerequisites

Before you begin this module, it is recommended that you successfully complete *Core Curriculum; Mobile Crane Operations Level One; Mobile Crane Operations Level Two;* and *Mobile Crane Operations Level Three*, Modules 21301-05 through 21306-05.

Required Trainee Materials

1. Pencil and paper
2. Appropriate personal protective equipment
3. Copy of relevant OSHA regulations

1.0.0 ◆ INTRODUCTION

In crane accidents, equipment can be damaged, and people can be severely injured. Lives, careers, and companies can be lost whenever a crane accident occurs. This module gives an overview of accident situations that could possibly occur on the job site, and the steps that need to be taken to ensure the safety of personnel and the equipment.

2.0.0 ◆ ACCIDENT PREVENTION AND INVESTIGATION

A crane has the potential to be the most dangerous piece of equipment at a job site. Careful planning, quality inspection, and maintenance is the best way to prevent accidents. You can only satisfy your ethical and legal responsibility for crane safety if you have the right information. It is critical to verify the following:

- A complete operating manual is in the cab at all times.
- A copy of the operating and maintenance manuals are on file.
- The last frequent and periodic inspection forms and any deficiency correction statements are provided with the crane.
- The crane and its attachments meet the manufacturer's requirements for safe operation.

Before performing any lifting operation, an operator should do all of the following:

- Be familiar with the operator's manual for the crane.
- Review the load chart for the crane in use.
- Check that the crane is level.
- Check that the crane boom is assembled/extended correctly.
- Check that the required daily pre-startup inspection has been performed.
- Understand and walk through the lift plan if one exists.
- Determine the weight deduction items on the crane and their values.
- Determine the hoist line pull and the maximum permissible line pull.
- Check the wind speed limitations.

Ensure that you have everything required to teach the course. Check the Materials and Equipment list at the front of this module.

See the general Teaching Tip at the end of this module.

Explain that terms shown in bold are defined in the Glossary at the back of this module.

Show Transparency 1, Objectives. Review the goals of the module, and explain what will be expected of trainees.

Discuss accident prevention. List the important documents that crane operators must verify.

Describe the checks and activities an operator must perform before every lift.

See the Teaching Tip for Section 2.0.0 at the end of this module.

Provide copies of *ASME B30.5* for the trainees to examine.

Discuss the responsibilities of crane operators pursuant to *ASME B30.5*. Discuss accident investigation procedures.

If possible, provide copies of a sample accident investigation for trainees to examine.

Have trainees review Section 3.0.0.

Ensure that you have everything required for teaching this session.

See the Teaching Tips for Section 3.0.0 at the end of this module.

Explain that all cranes must be equipped with a 10 BC-rated portable fire extinguisher. Discuss what types of fires it can be used on and how to operate it.

Discuss safety precautions that should be used when refueling a crane.

Explain the fire triangle and discuss general fire prevention guidelines.

- Determine if the lifting operation is limited by stability or structure.
- Determine what procedure will be used if there is not enough room to extend tracks or outriggers.
- Determine how close the crane or load can come to power lines and still be considered safe.
- Determine the weight of the load.
- Check the load charts for the specific lift.
- Determine the options available if not comfortable with a lift.

Every operator must be qualified by *ASME B30.5* criteria (written, practical, and oral examinations). Each operator shall be held responsible for those operations under his or her direct control. Whenever there is any doubt as to safety, the operator shall consult with the supervisor before handling the loads.

Lifting operations must always be conducted in a safe manner to prevent accidents from happening. However, should an accident occur, the operator should adhere to emergency procedures and follow company policy for accident investigation. The operator should also have an understanding of the type of information that may be requested on a typical accident investigation checklist, including:

- Correct company name and mailing address
- Person receiving the report: name, title, and phone number
- Investigator's name, title, and company
- Date and time of investigation
- Location and exact time of accident
- Description of equipment involved, such as the manufacturer, model, serial number, age of machine, unit number, and configuration
- Synopsis of accident as told by witness(es)
- List of people who will make formal statements, including witnesses to the events from one hour before the accident to the end of the accident sequence, all personnel who perform maintenance on the involved equipment, and all personnel involved in determining the lift characteristics
- Sketches of the accident scene to scale with as much detail as possible
- Photographs from as many angles as possible
- Weather conditions at the time of the accident
- Ground conditions
- Boom length
- Load radius
- Actual load weight including load block, sling, and boom attachments
- Background information on crane operator – name, age, training, and experience
- Summary of accident
- Conclusions based on findings

Normally, a safety officer or inspector of the company responsible for the job site and/or crane will conduct the accident investigation and complete the required forms. After the investigation is completed, the report is filed by the responsible company with the proper state and federal authorities. If the accident results in one or more fatalities or immediate injuries to three or more persons, the accident must be reported to the Occupational Safety and Health Administration (OSHA) within eight hours. Then, investigators from OSHA and/or the National Institute for Occupational Safety and Health (NIOSH) will also conduct investigations of the accident.

3.0.0 ◆ FIRE

According to *ASME B30.5*, all mobile cranes are required to have a 10 BC-rated, portable fire extinguisher mounted in or near the cab. The operator is required to be trained to use the extinguisher. This type of extinguisher has a dry agent capable of combating Class B (flammable liquids) and Class C (electrical) fires.

The extinguisher is constructed so that the agent becomes pressurized when the pin is pulled and the handle is compressed. Compressing the lever on the handle activates a carbon dioxide cartridge in the bottle, pressurizing the canister and forcing the dry powder out of the nozzle. The effective range of this type of extinguisher is 5 to 20 feet, provided it is pointed at the base of the fire.

ASME standards also require that refueling of the crane must only be done when the engine is not running and in the absence of smoking or open flames. If refueling is being done with a portable container, the container must be a safety-type can equipped with an automatic closing cap and a flame **arrester**.

The best way to prevent a fire is to make sure that the three elements needed for fire—fuel, oxygen, and heat—are never present in the same place at the same time. Here are some basic safety guidelines for fire prevention:

- Always work in a well-ventilated area, especially when you are using flammable materials such as shellac, lacquer, paint stripper, or construction adhesives.
- Never smoke or light matches when you are working with flammable materials.
- Keep oily rags in approved, self-closing metal containers.
- Know where to find fire extinguishers, what kind of extinguisher to use for different kinds of fires, and how to use the extinguishers.
- Store combustible materials only in approved containers.

Instructor's Notes:

- Keep open fuel containers away from any sources of sparks, fire, or extreme heat.
- Make sure all extinguishers are fully charged. Never remove the tag from an extinguisher; it shows the date the extinguisher was last serviced and inspected.
- Do not fill a gasoline or diesel fuel container while it is resting on a truck bed liner or other ungrounded surface. The flow of fuel creates static electricity that can ignite the fuel if the container is not grounded.

4.0.0 ◈ POWER LINES

High-voltage contact is the largest single cause of fatalities associated with cranes; however, these fatalities can be prevented. Surrounding every energized (live) power line is an area referred to as the **prohibited zone** (see *Figure 1*). As defined in *ASME B30.5* and *OSHA 1910.333*, the prohibited zone is defined as the mandatory area around an energized power line that no part of a crane, person, or conductive object can enter. The prohibited area around an energized power line is shown in *Table 1* for various line voltages. The **avoidance zone** (*Figure 1*) is an unofficial area where crane operations should not be conducted, if possible, due the high probability of accidental power line contact.

Table 1 High-Voltage Power Line Clearances

CRANE IN OPERATION (1)	
POWER LINE (kV)	BOOM OR MAST MINIMUM CLEARANCES (feet)
0 to 50	10
50 to 200	15
200 to 350	20
350 to 500	25
500 to 750	35
750 to 1000	45
CRANE IN TRANSIT (with no load and the boom or mast lowered) (2)	
POWER LINE (kV)	BOOM OR MAST MINIMUM CLEARANCES (feet)
0 to 0.75	4
0.75 to 50	6
50 to 345	10
345 to 7500	16
750 to 1000	20

Note 1: For voltages over 50kV, clearance increases 5 feet for every 150 kV.
Note 2: Environmental conditions such as fog, smoke, or precipitation may require increased clearances.

307T01.EPS

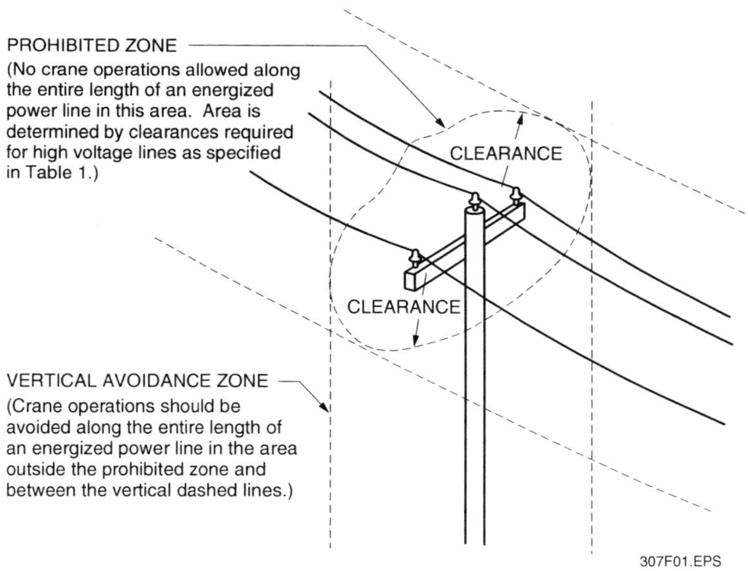

Figure 1 ◈ Prohibited zone and avoidance zone.

Have trainees review Sections 4.0.0–4.5.0.

Ensure that you have everything required for teaching this session.

See the Teaching Tips for Section 4.0.0 at the end of this module.

Discuss the dangers of working near power lines.

Emphasize that the largest cause of crane-related fatalities is high-voltage contact.

Show Transparency 2 (Figure 1). Describe prohibited zones and avoidance zones.

Show Transparency 3 (Table 1). Explain required clearances for high-voltage power lines.

EMERGENCY PROCEDURES 7.3

Discuss notification requirements for operations near power lines.

Have trainees describe the four conditions for operating near power lines.

Discuss crane operations near de-energized and grounded electric power lines. Review the steps that must be taken to ensure de-energization.

Describe proximity warning devices, and explain how they are used.

Provide a proximity warning device for the trainees to examine.

Discuss crane operations near the prohibited zone when the power lines are energized.

Explain the precautions that must be taken to minimize hazards.

Review the power line awareness checklist in *Appendix A*.

Show Transparency 3 (Table 1) again. Emphasize the importance of maintaining required clearances.

According to OSHA regulations and consensus standards, before operations are to begin near electrical lines, the owners of the lines or their authorized representative must be notified and provided with all pertinent information. In addition, the cooperation of the owner must be requested. Any overhead wire must be considered to be an energized line unless and until the owner or the electrical utility authorities indicate that it is not. Crane operators must not rely on the coverings of wires for their protection.

Operating mobile cranes where they can become electrified from electric power lines is an extremely hazardous practice. Work must be performed so that there is no possibility of the crane, load line, or load becoming a conductive path.

There are four conditions to consider when operating a mobile crane near electric power lines:

- Power lines de-energized and grounded
- Power lines energized, crane operating less than the erected/fully extended boom length away
- Power lines energized, crane components within the avoidance zone
- Crane in transit, no load, and boom lowered

4.1.0 Crane Operation Near De-energized and Grounded Electric Power Lines

This is the preferred condition under which the operation can be performed because the hazard of injury or death due to electrocution has been removed. The following steps must be taken to ensure de-energization of the power lines:

- The power company or owner of the power lines must de-energize the lines.
- The lines must be visibly grounded to avoid electrical feedback and be appropriately marked at the job-site location.
- A qualified representative of the owner of the lines or a designated representative of the electrical utility must be on the site to verify that the first two steps have been completed and that the lines are not energized.
- Durable signs must be installed warning that electrocution or serious bodily injury may occur unless minimum clearances are maintained between the crane or the load being handled and energized power lines (see *Table 1*). These warnings must be posted at the operator's station and on the outside of the crane.

If cage-type boom guards, insulating links, or **proximity warning devices** are used on cranes, such devices shall not be a substitute for the requirements of *ASME B30.5-3.4.5* and *OSHA 1910.333* for power line clearances, even if such devices are required by law or regulation. Because of the complex, invisible, and lethal nature of electrical hazards and to lessen the potential of false security, the crane operator, crew, and load handling personnel must understand instructions on the electrical hazard involved, operating conditions for the devices, limitations of the devices, and any testing requirements prescribed by the device manufacturer. Compliance with *ASME B30.5-3.4.5* is the recommended practice in determining permissible proximity of the crane and its protuberances (projections), including load, to electric power lines.

4.2.0 Crane Operation Near the Prohibited Zone, with the Power Lines Energized

Operation of the boom and load near electric power lines can be extremely dangerous, due to perception of distance and multiple contact points as viewed from the position of the operator and/or position of the signal person. The operator should avoid operating the crane, with or without a load, in such area. If it is necessary to operate in this area, the following steps must be taken to minimize the hazard of electrocution or serious injury as a result of contact between the energized power lines and the crane, load line, or load:

- An on-site meeting between project management and a qualified representative of the owner of the lines or a designated representative of the electrical utility must take place to establish the procedures to safely complete the operations. A power-line awareness checklist similar to the one shown in *Appendix A* should be prepared prior to the meeting and completed during the meeting.
- The specified clearance (stated in *Table 1*) between the power lines and the crane, load line, and load must be maintained at all times.
- Load control, when required, must utilize non-conductive tag lines.
- A qualified signal person(s) whose sole responsibility is to verify that the required clearance is maintained must be in constant contact with the crane operator.
- No one must be permitted to touch the crane or the load unless the signal person indicates it is safe to do so.

Instructor's Notes:

- The horizontal and vertical distance of movement of long-span power lines due to the wind must be added to the minimum clearance distance specified in *Table 1*. A qualified representative of the owner of the lines or a designated representative of the electrical utility must be consulted for specific distances.
- Devices such as ribbons and balls should be attached by a qualified person to the power lines to improve visibility, or equivalent means should be employed to aid in locating the prohibited zone.
- Durable signs must be installed at the operator's station, and on the outside of the crane, warning that electrocution or serious bodily injury may occur unless minimum clearances are maintained between the crane or the load being handled and energized power lines. In some cases, large, high-voltage warning signs with upward pointing arrows or a temporary non-conductive goal post-type guide placed under the lowest portion of the energized lines may be desired or required, especially for transit under power lines.
- All non-essential personnel must be kept from the work area for a distance of the full boom extension and swing plus 10 feet.
- The same stipulations regarding cage-type boom guards, insulating links, or proximity warning devices discussed previously apply.

4.3.0 Crane Operation Within the Avoidance Zone with the Power Lines Energized

Limited crane operations can be performed within the avoidance zone if absolutely necessary. The following steps must be taken to minimize the hazard of electrocution or serious injury as a result of contact between energized power lines and the crane, load line, or load:

- Before such operations take place, a qualified person together with a qualified representative of the utility or an engineer qualified in power line transmission must, after visiting the site, determine if this is the most feasible way to complete the operation and set minimum required clearances and procedures. These operations must be under their supervision. The following may be required:
  - Crane/load grounded to line neutral by the utility
  - Electrical system protective devices that automatically re-energize the circuit after a power line contact occurrence should be blocked or disengaged to inhibit this function
  - Insulated barriers, which are not a part of nor an attachment to the crane and which will not allow contact between the energized electric power lines and the crane, load lines, or load
  - Non-conductive barricades to restrict access to the crane work area
- Load control, when required, must utilize tag lines of a non-conductive type.
- A qualified signal person(s), whose sole responsibility is to verify that clearances are maintained, must be in constant contact with the crane operator.
- The person(s) responsible for the operation must alert and warn the crane operator and all persons working around or near the crane about the hazard of electrocution or serious injury and instruct them on how to avoid the hazard.
- All non-essential personnel must be removed from the crane work area for a distance of the full boom extension and swing plus 10 feet.
- No one must be permitted to touch the crane or the load unless the signal person indicates it is safe to do so.
- The same stipulations regarding warning signs, cage-type boom guards, insulating links, or proximity warning devices discussed previously apply.

4.4.0 Operations in Transit with No Load and Boom Lowered

While in transit with no load and the boom and boom support system lowered, minimum clearances as specified in *Table 1* must be maintained. When planning transit of the crane, the effect of speed and terrain on boom and crane movement must be considered.

WARNING!

Careful planning and supervision offer better protection than any known device. Insulated boom cages, proximity devices, and insulating links have limitations and can fail without warning. Insulated boom cages and links only protect part of the crane, and they can break down electrically when contaminated with dust and water. Operation of proximity warning devices can be affected by the different arrangement of multiple power lines, the movement of trucks, materials, the crane itself, and other influences. Relying completely on these devices could be dangerous because operators may think they are providing protection when they are not.

Explain that movement of power lines due to wind must be factored into the clearance values.

Describe how power lines can be marked, and discuss required signage.

Explain that limited crane operations may be performed within the avoidance zone if absolutely necessary.

Discuss precautions to minimize hazards in the avoidance zone.

Discuss how safety devices can fail. Explain that operators must always be careful and not rely solely on these devices.

Show Transparency 3 (Table 1) again. Explain that minimum clearance values must be maintained when the crane is in transit.

EMERGENCY PROCEDURES 7.5

Discuss what steps operators must take after contacting power lines. Remind trainees to stay in the cab if there is no danger of fire or explosion.

Show Transparency 4 (Figure 2). Explain how the ground is energized when a crane contacts a power line.

Explain the proper way to exit a crane after contact. Have trainees practice the movement of exiting with feet together.

Have trainees review Sections 5.0.0–5.2.0.

Ensure that you have everything required for teaching this session.

Describe various types of crane accidents. Discuss how the crane can be damaged if the boom strikes an object.

4.5.0 Power Line Contact

If power line contact does occur, first try to reverse the action that caused the contact. Otherwise, if there is no danger of fire or explosion, remain inside the cab. The hand controls inside the cab are insulated. Do not panic.

If you must get off the crane due to fire, jump off of it from the lowest point of the crane, land on the ground with your feet together, and make no further contact with the crane. Shuffle your feet in small increments (about six inches) to move away from the crane, or hop away with your feet together until you are a safe distance from the crane. High-voltage current transmitted from the power lines through the boom to the ground energizes the ground near the crane with high voltage (*Figure 2*). The ground farther away from the crane will have lower voltage. The percentage of voltage decrease varies depending on the soil **resistivity**. The voltage **differential** can be deadly. If you take large steps, it is possible for one foot to be in a high-voltage area and the other to be in a lower-voltage area.

> **WARNING!**
> Do not try to step down from the crane. Both feet must touch the ground at the same time. The second you make contact between the crane and the ground, you will be electrocuted and possibly killed.

Instruct all other personnel on the site to stay away from the crane, ropes, and load. The crane, along with the ground around it, will have electric potential.

307F02.EPS

Figure 2 ◆ Voltage radiated in the ground decreases with distance.

If you remain in the crane, wait until the electrical authorities de-energize the circuit and confirm that conditions are safe.

Completely inspect the machine for possible damage caused by the electrical contact. Wire rope should be replaced if it touches a line since the arc is usually of sufficient power to weld, melt, or badly pit the rope. The damaged section of rope will look like it was burnt by a torch.

Report every incident involving contact with power lines to the electrical authority and safety inspector. Inspections can be made to prevent damaged power lines from falling later.

5.0.0 ◆ VARIOUS FAILURES IN LIFTING

Carelessness by the operator can lead to accidents other than those associated with overloading the crane. These accidents include hitting the boom on some other object, backward collapse of the boom, and two-blocking.

5.1.0 Striking the Boom

The operator must never allow a crane boom to strike any structure or load (*Figure 3*). Contact will dent, bow, or slightly bend the lower boom chords and may cause a total boom collapse. If the boom touches or rests on any structure, the boom loading changes from compression to bending. The boom is very strong in compression but weak in bending. If the boom, mast, or gantry is struck or damaged, stop. The load on the boom increases as the boom is lowered. As a result, a damaged boom or boom suspension may collapse during lowering. A second crane may be required to help lower a damaged boom. Check with your supervisor.

307F03.EPS

Figure 3 ◆ Striking the boom.

Instructor's Notes:

5.2.0 Backward Collapse of a Boom

When operating near minimum radius with the boom at its highest angle, boom down as you set the load down. This will compensate for the tendency of the boom to move back against the boom stops when the load is released, resulting in a possible backward collapse of the boom. This action occurs because of the elasticity in the boom and boom hoist systems (*Figure 4*).

Another factor which may cause backward collapse of a boom or upset the crane is high wind speed (*Figure 5*). The manufacturer's manual should contain information and instructions regarding maximum wind speeds for operation or wind speeds that require boom stowage.

Other factors that may contribute to boom collapse include the following:

- Continual pull on the hoist line after two-blocking if the crane is not equipped with an anti-two-blocking device

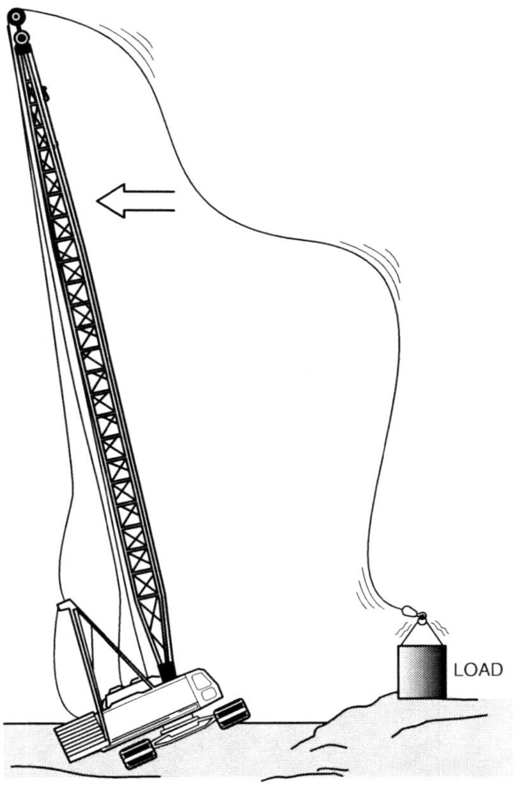

307F04.EPS

Figure 4 ◆ Sudden release of a load.

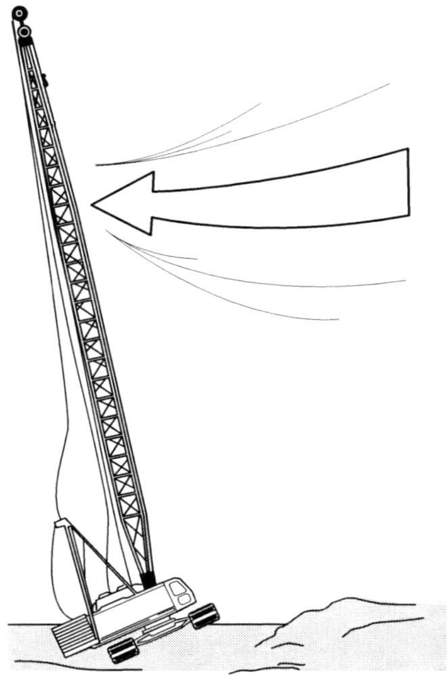

307F05.EPS

Figure 5 ◆ High wind speed.

- Starting or stopping a swing suddenly if the boom is at a high angle
- Sudden forward movement of a crane, which can collapse the boom over backward if it is carried at a high boom angle
- Snubbing the hook block to the boom foot, then pulling it up tight
- The crane becoming unstable in certain positions when on an incline

> **WARNING!**
> High boom angles when traveling on an incline can lead to backward collapse of the boom.

Summary

The purpose of this material is to inform operators of emergency situations and the proper steps that should be taken if they occur. The importance of this information cannot be emphasized enough. Not only can a failure to follow proper procedures result in extensive damage to the equipment, it can mean the difference between life and death for the operator and site personnel.

Show Transparencies 5 and 6 (Figures 4 and 5). Discuss situations when the boom can collapse backward. Explain how to avoid these accidents.

Summarize the major concepts presented in the module.

EMERGENCY PROCEDURES 7.7

Have trainees complete the Review Questions, and go over the answers prior to administering the Module Examination.

Administer the Module Examination. Record the results on Craft Training Report Form 200, and submit the results to the Training Program Sponsor.

Review Questions

1. Information that must be kept in the crane cab at all times includes _____.
   a. the complete operating manual
   b. a copy of *ASME B30.5*
   c. a copy of *OSHA 1910.333*
   d. the crane's maintenance records

2. A crane accident must be reported to OSHA within eight hours if the accident results in _____.
   a. injuries to two persons
   b. one or more fatalities
   c. fire within the crane cab
   d. property damage exceeding $5,000

3. What kind of fire extinguisher is required to be mounted in or around the cab of a mobile crane according to consensus standards?
   a. A fire extinguisher is not required in the cab as long as there is one somewhere on the job site
   b. 5 AC
   c. 10 BC
   d. 20 BC

4. ASME standards require that crane refueling must be done only _____.
   a. with a portable container
   b. in the presence of a supervisor
   c. when the engine is running
   d. when the engine is not running

5. The prohibited zone is the _____.
   a. quadrant the crane is not allowed to work in
   b. area near a power line that no part of the crane or load can enter, as specified in *ASME B30.5*
   c. length and radius at which the crane cannot operate if it is making a lift on uneven soil
   d. the radius of the crane's outriggers when extended

6. The single largest cause of fatalities associated with cranes is _____.
   a. crushing
   b. electrocution
   c. falling load
   d. two-blocking

7. The unofficial zone where limited crane operations can be performed if absolutely necessary is the _____ zone.
   a. allowance
   b. prohibited
   c. proximity
   d. avoidance

8. For lines rated 50 kV or below, the minimum clearance between the lines and any part of the crane or load, including handling appendages, must be _____ feet.
   a. 5
   b. 10
   c. 15
   d. 20

9. Crane operators can be warned of electrical hazards by a device called a(n) _____.
   a. proximity warning device
   b. LMI
   c. line rider
   d. anti-two-blocking device

10. For voltages over 50 kV, the minimum required high-voltage power line clearance increases _____ ft for every 150kV.
    a. 5
    b. 10
    c. 15
    d. 20

11. The preferred condition for crane operation near power lines is that power lines are _____.
    a. de-energized and grounded
    b. energized, but specified clearances maintained at all times
    c. energized, but with the crane operating less than the fully extended boom length away
    d. energized, but cage-type boom guards, insulating links, and proximity warning devices are used

12. While in transit with no load and boom or mast lowered, if the voltage is between 50 and 345 kV, the minimum required clearance is _____ ft.
    a. 4
    b. 6
    c. 10
    d. 16

Instructor's Notes:

13. During crane operation near energized power lines, the crane operator must be in constant contact with a(n) _____.

a. qualified signal person whose sole responsibility is to verify that the required clearance is maintained
b. designated representative of the electrical utility
c. engineer qualified in power line transmission
d. project manager

14. All non-essential personnel must be removed from the work area before a crane operation that takes place with _____

a. power lines energized and the crane operating less than the fully extended boom length away
b. power lines de-energized and grounded
c. power lines energized and the crane within the avoidance zone
d. the crane in transit with the boom lowered

15. If it becomes necessary to jump off a crane that has hit a power line, you should keep your feet _____.

a. together
b. about one foot apart, side-by-side
c. about one foot apart, one in front of the other
d. in a position where only your heels are touching the ground

16. The percentage of voltage decrease in the area of the ground around a crane after it has struck a power line varies depending on the _____.

a. height of the boom
b. weight of the machine
c. length of the outrigger extensions
d. soil resistivity

17. If a lattice boom touches or rests on any structure, the boom loading changes from _____.

a. bending to compression
b. compression to bending
c. bending to fracturing
d. fracturing to bending

18. If the boom, mast, or gantry is struck or damaged, the operator should _____.

a. lower the boom
b. raise the boom
c. stop operations
d. boom down while setting the load down

19. To prevent backward collapse of a crane, the operator should pull continually on the hoist line after two-blocking.

a. True
b. False

20. If the boom angle is high, the boom may move back against the boom stops when a load is released because of _____.

a. elasticity in the boom and boom hoist systems
b. friction in the boom extensions
c. operator error
d. ground stabilizing

EMERGENCY PROCEDURES 7.9

Instructor's Notes:

GLOSSARY

Trade Terms Introduced in This Module

Arrester: A device that stops flashback ignition and may have a release mechanism that vents pressure above a certain psi.

Avoidance zone: An unofficial zone where limited crane operations can be performed if absolutely necessary.

Differential: The quantitative difference between two or more forces, motions, etc.

Prohibited zone: The area around a power line that no part of the crane or load can enter as specified in *ASME B30.5*.

Proximity warning device: A device that warns crane operators of electrical hazards by detecting the electrostatic field of any AC power line.

Resistivity: Ratio of electric intensity to cross-sectional area; the opposite of conductivity.

EMERGENCY PROCEDURES 7.11

Instructor's Notes:

APPENDIX

Power Line Awareness Permit

POWER LINE AWARENESS PERMIT

Today's Date __________ **Job Number** ____________________________________

Contractor Name			
Job Address			
Telephone Number		**Fax Number**	

Emergency Contact Number	

Survey

Before beginning any project, you must first survey your work area to find power lines at the job site. (See job-site sketch on reverse side.)

Identify

After finding all of the power lines at your site, identify the activities you'll be doing that may put you or your workers at risk. Mark one or more of the following:

- ☐ Cranes (mobile or truck mounted)
- ☐ Drilling rigs
- ☐ Backhoes/excavators
- ☐ Long-handled tools
- ☐ Other tools/high-reaching equipment
- ☐ Concrete pumper
- ☐ Aerial lifts
- ☐ Dump trucks
- ☐ Ladders
- ☐ Material handling & storage
- ☐ Scaffolding
- ☐ Other___________________

Eliminate or Control

After identifying the power line and high-risk activities on the job site, determine how to eliminate or control the risk of electrocution (a successful determination is often reached only after consultation with the utility).

Mark one or more of the following:

- ☐ Move the activity
- ☐ Change the activity
- ☐ Have the utility de-energize the power line
- ☐ Have the utility move the power line
- ☐ Use barrier protection (insulated sleeves)
- ☐ Use an observer
- ☐ Use warning lines with flags
- ☐ Use non-conductive tools
- ☐ Use a protective technology:
  - ☐ Insulated link
  - ☐ Boom cage guard
  - ☐ Proximity device

Always maintain your minimum safe clearance distance from the power line, except when the utility has de-energized and visibly grounded the power line.

Voltages	Distance from Power Line
Less than 50 kV	10 Feet
More than 50 kV	10'+(0.4")(# of kV over 50 kV)

WARNING!
It is unlawful to operate any piece of equipment within 10' of energized lines

CONSTRUCTION SAFETY COUNCIL

307A01.EPS

EMERGENCY PROCEDURES 7.13

Job-Site Sketch

(Draw in location of power lines and their proximity to construction site, including such things as proposed excavations, location of heavy equipment, scaffolding, material storage areas, etc.)

Completed by______________________________ **Date**____________

Title______________________________

Approved by______________________________ **Date**____________

Title______________________________

CONSTRUCTION SAFETY COUNCIL

307A02.EPS

Instructor's Notes:

Figure Credits

Construction Safety Council	307F02, Appendix
Equipment Safety Services, Inc.	307F03

EMERGENCY PROCEDURES 7.15

MODULE 21307-05 — TEACHING TIPS

The following are suggested activities or instructional methods to help you teach the material in this module.

General

When you call on someone to answer a question, the rest of the class relaxes or even tunes out because they expect that the question and answer will take place only between you and the trainee you called on. Instead, use this technique to involve more trainees in answering questions and to keep them on their toes.

1. Ask trainees to define a term or explain a concept.
2. After one trainee has answered, ask a trainee seated nearby if the answer is right. Then ask whether a trainee in the back of the room agrees.
3. Ask trainees to explain why they think an answer is right or wrong.
4. Use the session to clear up incorrect ideas and encourage trainees to learn from their mistakes.

Section 2.0.0 *Accident Prevention and Investigation*

Accidents can be very costly and even fatal. Trainees will need pencils and paper. Invite an OSHA safety inspector or manufacturer's safety technician to give a presentation to the class on accident prevention and investigation. Allow 60 to 90 minutes for this exercise.

1. Before the presentation discuss accident prevention and investigations with trainees.
2. Have the inspector give a presentation on accident prevention and investigations, including required recordkeeping, permits, and accident reporting.
3. After the presentation ask the presenter to answer any questions the trainees may have.

Section 3.0.0 *Fire*

Crane accidents and fires can be deadly. Trainees will need pencils and paper. Obtain a crane safety video and/or arrange for a safety technician to do a presentation on crane accidents. Allow 1 to 1½ hours for this exercise. The following is an example of a video you can present:

Crane Accidents, Overland Park Fire Department, 2002, DVD, 60 minutes. Available from Media Manager, OPFD, 12401 Hemlock, Overland Park, KS 66213.

1. Before the presentation discuss various types of accidents with trainees.
2. Show the video.
3. After the presentation answer any questions the trainees may have.

Section 3.0.0 *Fire*

Some fires can be handled immediately using a fire extinguisher. Trainees will need pencils, paper, and appropriate personal protective equipment. Arrange for a representative from the fire department to do a presentation on the proper use of a fire extinguisher. Allow 45 to 60 minutes for this exercise.

1. Before the presentation discuss various types of fire extinguishers that are standard equipment on cranes with trainees.
2. Ask the presenter to cover different types of fire extinguishers, the fire triangle, and the PASS technique.
3. If possible, have the trainees practice using a fire extinguisher to put out a fire.
4. After the presentation answer any questions the trainees may have.

Section 4.0.0 *Power Lines*

Power lines pose specific hazards for crane operations. Trainees will need pencils and paper. Arrange for an opportunity to observe crane operations in an avoidance zone or an area that contains high voltage power lines. Alternatively, invite a safety technician from the utility company to give a presentation to the class on working safely around power lines. Allow 60 to 90 minutes for this exercise.

1. Before the tour/presentation discuss safety precautions with trainees.
2. Have the safety technician point out specific safety measures for around power lines, covering grounding, grounding equipment, and zones.
3. After the tour/presentation ask the presenter to answer any questions the trainees may have.

Section 4.0.0 *Power Lines Terms and Definitions*

This Quick Quiz will familiarize trainees with the terms and definitions that are commonly used in working near power lines. You will need photocopies of the quiz provided in the following page. Trainees will need pencils. If you allow trainees to use the Trainee Module, decrease the amount of time you give them to complete the quiz.

1. Make a photocopy of the quiz for each trainee.
2. Give trainees between 5 and 10 minutes to complete the quiz.
3. Go over the answers to the quiz.
4. Ask trainees if they have questions.

Answers to Quick Quiz

1. d
2. a
3. f
4. c
5. b
6. e

Quick Quiz *Power Lines Terms and Definitions*

For each description listed, identify the term that the text best describes. Write the corresponding letter in the blank provided.

_____ **1.** A device that tells crane operators of electrical hazards by detecting the electrostatic field of any AC power line

_____ **2.** The area around a power line that no part of the crane or load can enter as specified by *ASME B30.5*

_____ **3.** Ratio of electric intensity to cross-sectional area; the opposite of conductivity

_____ **4.** The quantitative difference between two or more forces or motions

_____ **5.** An unofficial zone where limited crane operations can be performed if absolutely necessary

_____ **6.** A device that stops flashback ignition and may have a release mechanism that vents pressure above a certain psi

a. Prohibited zone
b. Avoidance zone
c. Differential
d. Proximity warning device
e. Arrester
f. Resistivity

Copyright © 2005 National Center for Construction Education and Research. Permission is granted to reproduce this page provided that copies are for local use only and that each copy contains this notice.

MODULE 21307-05 — ANSWERS TO REVIEW QUESTIONS

Answer	Section
1. a	2.0.0
2. b	2.0.0
3. c	3.0.0
4. d	3.0.0
5. b	4.0.0
6. b	4.0.0
7. d	4.0.0
8. b	4.0.0; Table 1
9. a	4.1.0
10. a	4.0.0; Table 1
11. a	4.1.0
12. c	4.4.0; Table 1
13. a	4.2.0
14. c	4.3.0
15. a	4.5.0
16. d	4.5.0
17. b	5.1.0
18. c	5.1.0
19. b	5.2.0
20. a	5.2.0

CONTREN® LEARNING SERIES — USER UPDATE

The NCCER makes every effort to keep these textbooks up-to-date and free of technical errors. We appreciate your help in this process. If you have an idea for improving this textbook, or if you find an error, a typographical mistake, or an inaccuracy in NCCER's Contren® textbooks, please write us, using this form or a photocopy. Be sure to include the exact module number, page number, a detailed description, and the correction, if applicable. Your input will be brought to the attention of the Technical Review Committee. Thank you for your assistance.

Instructors – If you found that additional materials were necessary in order to teach this module effectively, please let us know so that we may include them in the Equipment/Materials list in the Annotated Instructor's Guide.

Write: Product Development and Revision
National Center for Construction Education and Research
P.O. Box 141104, Gainesville, FL 32614-1104

Fax: 352-334-0932

E-mail: curriculum@nccer.org

Craft ____________________ Module Name ____________________

Copyright Date ____________ Module Number ____________ Page Number(s) ____________

Description

____

____

____

(Optional) Correction

____

____

(Optional) Your Name and Address

____

____

Module 21308-05

Transporting Requirements

NCCER STANDARDIZED CRAFT TRAINING PROGRAM

The National Center for Construction Education and Research (NCCER) provides a standardized national program of accredited craft training. Key features of the program include instructor certification, competency-based training, and performance testing. The program provides trainees, instructors, and companies with a standard form of recognition through a National Craft Training Registry. The program is described in full in the *Guidelines for Accreditation*, published by the NCCER. For more information on standardized craft training, contact the NCCER by writing us at P.O. Box 141104, Gainesville, FL 32614-1104; calling 352-334-0911; or e-mailing info@nccer.org. More information may be found at our Web site, www.nccer.org.

HOW TO USE THIS ANNOTATED INSTRUCTOR'S GUIDE

Each page presents two sections of information. The larger section displays each page exactly as it appears in the Trainee Module. The narrow column ties suggested trainee and instructor actions to each page and provides icons (detailed below) to call your attention to material, safety, audiovisual, or testing requirements. The bottom of each page includes space for your notes.

The **Audiovisual** icon indicates an appropriate time to show a transparency or other audiovisual aid.

The **Classroom** icon prompts you to define a term, stress a point, ask trainees to explain a concept, or give examples.

The **Demonstration** icon directs you to show trainees how to perform tasks.

The **Examination** icon tells you to administer the written module examination.

The **Homework** icon is placed where you may wish to assign reading for the next class, to assign a project, or to advise trainees to prepare for an examination.

The **Laboratory** icon is used when trainees are to practice performing tasks.

The **Materials** icon is a reminder for you to gather materials needed for classes, labs, and testing.

The **Performance Testing** icon tells you to administer a performance test or a portion thereof.

The **Safety** icon is used to emphasize safety issues. It is often keyed to *Caution* and *Warning* statements in the Trainee Module.

The **Teaching Tip** icon indicates additional guidance is available, such as how to conduct an exercise, get the most educational value from a field trip, or encourage class participation. Teaching Tips may expand on a feature (*Think About It*, *Did You Know?*) or provide Quick Quizzes or similar exercises. You will be referred to the Teaching Tips section at the back of the module if there is additional material.

The **Combination** icon indicates that the laboratory listed corresponds with a performance task. If desired, you can note the proficiency of the trainees during the laboratory and use it to satisfy performance testing requirements.

PREPARATION

Before teaching this module, you should review the Objectives, Performance Tasks, Materials and Equipment List, and the Module Outline. Be sure to allow ample time to prepare your own training or lesson plan and gather all required materials and equipment.

MODULE OVERVIEW

This module discusses the proper handling, loading, unloading, and securing procedures of mobile cranes and their components. Information is also presented on driver requirements and requirements for securing mobile cranes for transport.

PREREQUISITES

Prior to training with this module, it is recommended that the trainee shall have successfully completed *Core Curriculum; Mobile Crane Operations Level One; Mobile Crane Operations Level Two;* and *Mobile Crane Operations Level Three,* Modules 21301-05 through 21307-05. This is an elective module.

OBJECTIVES

Upon completion of this module, the trainee will be able to do the following:

1. Describe the proper handling of crane components.
2. Describe the proper procedures for securing the crane and its components for transport.
3. Describe the proper procedures for loading and unloading mobile crane components.
4. Properly load a crane and its components onto a flatbed trailer.
5. Properly secure a load on a flatbed trailer.
6. Recognize applicable local, state, and federal requirements.

PERFORMANCE TASKS

Under the supervision of the instructor, the trainee should be able to do the following:

1. Properly load a crane and its components onto a flatbed trailer.
2. Properly secure a load on a flatbed trailer.

MATERIALS AND EQUIPMENT LIST

Overhead projector and screen
Transparencies
Blank acetate sheets
Transparency pens
Whiteboard/chalkboard
Markers/chalk
Pencils and scratch paper
Manufacturer's literature on loading, unloading, and securing cranes
Mobile crane
Flatbed trailer and ramp
Blocking
Tie downs, binders, or other strapping
Tools and equipment to load, secure, and unload a crane
ASME B30.5
49 CFR Subpart I
23 CFR Sections 127 and *658*
Federal Motor Carrier Safety Regulations Sections 390.21 and *393*
Copies of site safety manual or procedures
Module Examinations*
Performance Profile Sheets*

*Located in the Test Booklet.

SAFETY CONSIDERATIONS

Ensure that the trainees are equipped with appropriate personal protective equipment and know how to use it properly. This module requires trainees to operate a crane. Review site safety procedures and site evacuation procedures. Ensure that all trainees are familiar with hand signals and other site communication procedures. Brief trainees on pinching and crushing hazards attendant with crane operations. This module may require trainees to visit construction sites. Brief trainees on site safety procedures.

TEACHING TIME FOR THIS MODULE

An outline for use in developing your lesson plan is presented below. Note that each Roman numeral in the outline equates to one session of instruction. Each session has a suggested time period of 2½ hours. This includes 10 minutes at the beginning of each session for administrative tasks and one 10-minute break during the session. Approximately 10 hours are suggested to cover *Load Charts*. You will need to adjust the time required for hands-on activity and testing based on your class size and resources. Because laboratories often correspond to Performance Tasks, the proficiency of the trainees may be noted during these exercises for Performance Testing purposes.

Topic	Planned Time
Sessions I and II. Introduction, Loading and Unloading Cranes	
A. Introduction	____________
B. Loading	____________
C. Unloading	____________
D. Securing the Load	____________
E. Laboratory – Trainees practice properly loading a crane and its components onto a flatbed trailer. This laboratory corresponds to Performance Task 1.	____________
F. Laboratory – Trainees practice securing the load on a flatbed trailer. This laboratory corresponds to Performance Task 2.	____________
Session III. Transporting Cranes	
A. Transporting Cranes	____________
B. Transport Regulations	____________
Session IV. Review, Module Examination, and Performance Testing	
A. Review	____________
B. Module Examination	____________
1. Trainees must score 70 percent or higher to receive recognition from NCCER.	
2. Record the testing results on Craft Training Report Form 200, and submit the results to the Training Program Sponsor.	
C. Performance Testing	____________
1. Trainees must perform each task to the satisfaction of the instructor to receive recognition from NCCER. If applicable, proficiency noted during laboratory exercises can be used to satisfy the Performance Testing requirements.	
2. Record the testing results on Craft Training Report Form 200, and submit the results to the Training Program Sponsor.	

Module 21308-05

Transporting Requirements

Instructor's Notes:

COURSE MAP

This course map shows all of the modules in the third level of the *Mobile Crane Operations* curriculum. The suggested training order begins at the bottom and proceeds up. Skill levels increase as you advance on the course map. The local Training Program Sponsor may adjust the training order.

Assign reading of Module 21308-05.

MOBILE CRANE OPERATIONS LEVEL THREE

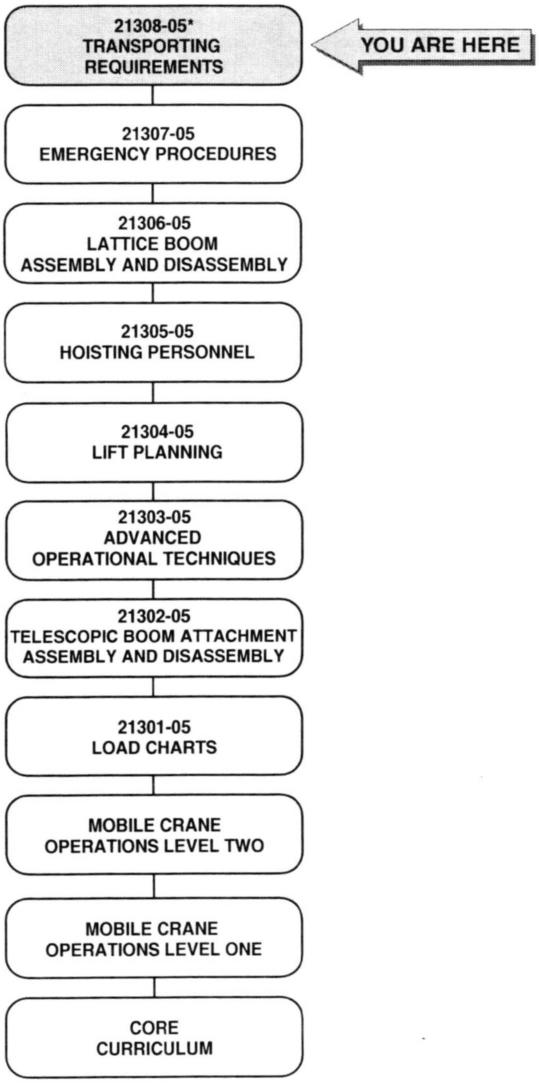

*Please note that 21308-05 is an elective module. It is not necessary to complete this module in order to advance in your training.

308CMAP.EPS

Copyright © 2005 National Center for Construction Education and Research, Gainesville, FL 32614-1104. All rights reserved. No part of this work may be reproduced in any form or by any means, including photocopying, without written permission of the publisher.

TRANSPORTING REQUIREMENTS 8.iii

Instructor's Notes:

MODULE 21308-05 CONTENTS

Figures

TRANSPORTING REQUIREMENTS 8.v

Instructor's Notes:

MODULE 21308-05

Transporting Requirements

Objectives

When you have completed this module, you will be able to do the following:

1. Describe the proper handling of crane components.
2. Describe the proper procedures for securing the crane and its components for transport.
3. Describe the proper procedures for loading and unloading mobile crane components.
4. Properly load a crane and its components onto a flatbed trailer.
5. Properly secure a load on a flatbed trailer.
6. Recognize applicable local, state, and federal requirements.

Prerequisites

Before you begin this module, it is recommended that you successfully complete *Core Curriculum; Mobile Crane Operations Level One; Mobile Crane Operations Level Two;* and *Mobile Crane Operations Level Three,* Modules 21301-05 through 21307-05.

Required Trainee Materials

1. Pencil and paper
2. Appropriate personal protective equipment

1.0.0 ◆ INTRODUCTION

The proper loading, unloading, and transport of crane components is just as important as any aspect of lifting. Knowledge of the appropriate handling guidelines, safety guidelines, and Department of Transportation (DOT) regulations for interstate and intrastate travel is essential for safely transporting crane components to the job site in an undamaged condition. If components are damaged, they are of no use at the job site. This results in lost productivity, decreased safety, and increased costs. Also note that company policies and procedures for transporting may differ.

2.0.0 ◆ LOADING AND UNLOADING

The procedures involved in getting a mobile crane ready to move from one site to another, and the precautions that must be observed while transporting it, can vary depending on the number of crane attachments and the size and weight of the crane. There are, however, general recommendations to follow. The most important consideration in loading and unloading a mobile crane is to avoid damaging it.

2.1.0 Loading

When loading crawler cranes onto a flatbed, the loading ramp must be long enough to make the ramp angle low (see *Figure 1*). During loading from the rear of a trailer, it may be necessary to use cribbing under the rear of the trailer frame to prevent overloading the trailer suspension and tires. If the crane is carrying boom sections, the boom should kept as low as possible. This will prevent kickback or overturning.

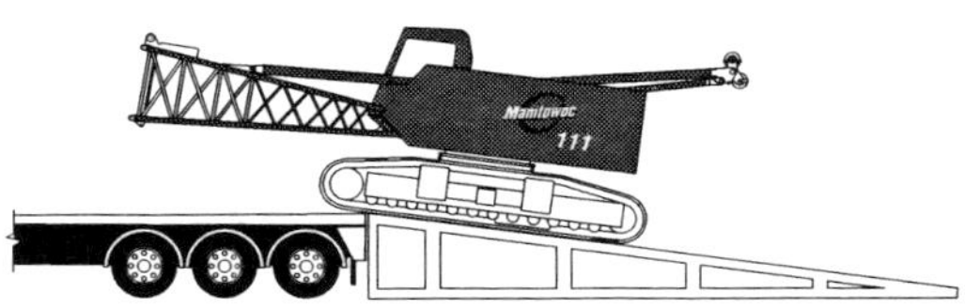

308F01.EPS

Figure 1 ◆ Ramp configuration.

Ensure that you have everything required to teach the course. Check the Materials and Equipment list at the front of this module.

See the general Teaching Tip and the Teaching Tip for Section 2.0.0 at the end of this module.

Explain that terms shown in bold are defined in the Glossary at the back of this module.

Show Transparency 1, Objectives, and Transparency 2, Performance Tasks. Review the goals of the module, and explain what will be expected of trainees.

Discuss the importance of understanding how to load, secure, and unload cranes. Explain that cranes are easily damaged during loading and transport.

Show Transparency 3 (Figure 1). Explain that the ramp must be long enough to keep the angle low.

Explain how to securely shackle the crane to the vehicle.

Show Transparencies 4 and 5 (Figures 2 and 3). Discuss different configurations for loading cranes.

Explain how to avoid damaging boom sections when loading and securing them.

Show Transparencies 6 and 7 (Figures 4 and 5). Explain how slings are used to lift boom sections. Discuss the importance of blocking to support boom sections.

Explain the dangers of working on top of unsecured cranes.

Provide manufacturer's literature on loading, unloading, and securing cranes for the trainees to examine.

The crane must be securely shackled to the vehicle. The crane brakes must not be relied on to keep the crane from moving during transport. You must also cross-chain the upperworks of the crane against rotation. Both the swing brake and the swing lock of the crane should be set.

Some smaller lattice boom cranes are designed so that the boom sections travel with the crane as shown in *Figure 2*. The whole crane can fit on a single trailer. When transporting the whole crane on one truck, you must check maximum weight restrictions.

Most cranes must be disassembled and transported on separate vehicles because the overall weight is more than maximum allowed for highway transportation by a single vehicle. The boom sections are detached and transported separately from the main body of the crane. For the largest cranes, the crawler treads are also removed and shipped separately. Boom sections can be stacked for transportation as shown in *Figure 3*. The optimal configuration for transportation is usually provided by the crane manufacturer.

The weight of the boom should not rest on a metal cradle (support). This could cause crushing of the tubular sections and result in boom failures. Use suspension ropes and the boom hoist cable to secure and support the base section of the boom.

If the load is delivered for use at a later date, **dunnage** should be placed under the landing gear so that the trailer does not sink into the ground. The driver should be made aware of this condition. Other suggested guidelines to remember when loading boom sections are as follows:

- When lifting boom sections onto flatbeds (*Figure 4*), do not attach the slings to the lattice members. The lattice will bend. Slings must be attached to the fittings at the ends of the main chords.
- Personnel should mount and dismount the trailer using the grab irons while maintaining three-point contact.

WARNING!

Working on top of an unsecured load should be avoided. The load could slip and cause property damage and personnel injury.

- The boom sections must be supported by blocking (*Figure 5*). If the sections are stacked, there must be blocking between them, with a 4" × 4" minimum size.

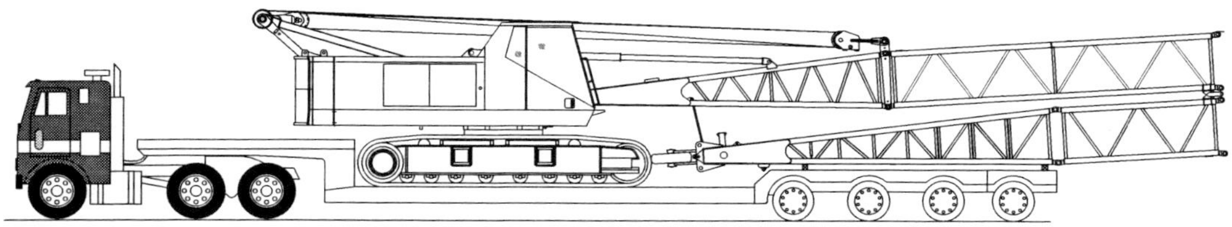

308F02.EPS

Figure 2 ◆ Lattice boom crane configured for transport.

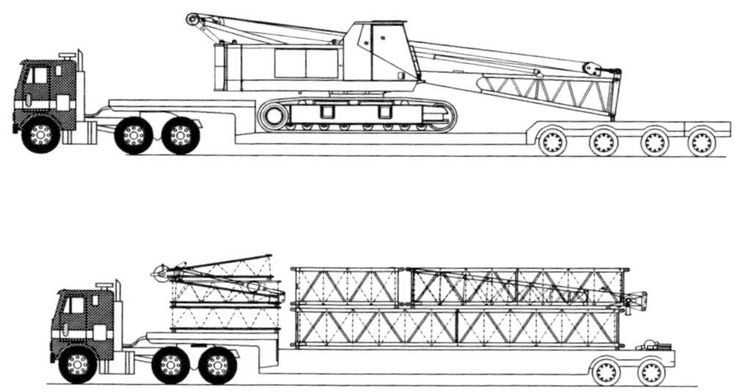

308F03.EPS

Figure 3 ◆ Transport configuration for stacked boom sections.

Instructor's Notes:

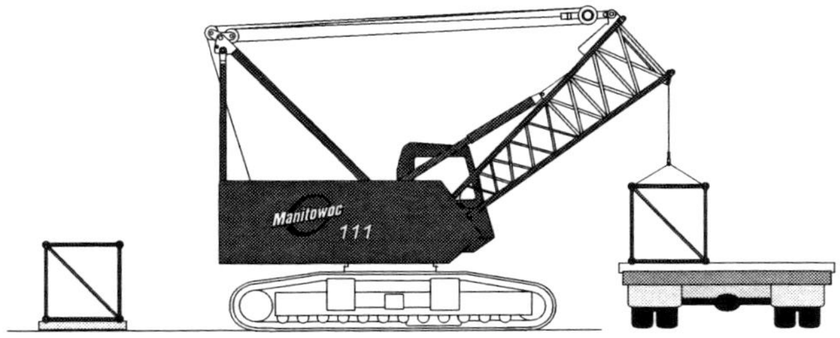

Figure 4 ◆ Crane loading boom sections onto a flatbed.

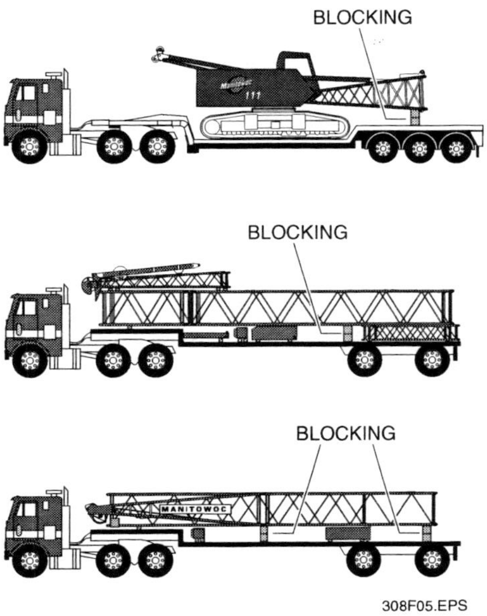

Figure 5 ◆ Crane and boom sections transported separately.

- The boom sections should be secured to the trailer using nylon straps over the boom chords at both ends only.
- Chains should not be used to secure boom sections.
- Depending on the manufacturer, counterweights and crawler side frames are allowed under the boom. Do not tie the boom to these other loads, and do not stack loads on top of boom sections.
- Nylon straps must be kept very tight. Check strap tightness often.
- Any boom extensions or jibs stacked together on a trailer should be of equal length.

2.2.0 Unloading

If lever- or ratchet-type **binders** are used to secure the straps, the unloading personnel must remember that they multiply stored energy in the load. Lever types multiply this stored energy by a factor of 25, whereas ratchet types multiply it by a factor of 50. For example, if the lever is applying a force of 100 pounds before transit, that same force may become 2,500 pounds when the load has reached its destination. Never use a cheater pipe or handle extension to release the binder; a steel bar should be used. Have the crane operator take up any slack on the sling before you release the binder.

> **WARNING!**
> Always use extreme caution when unloading. It is possible to tip the crane over when unloading, which will severely damage the crane and can injure personnel.
> Releasing binders can cause kickback, which can injure personnel.

3.0.0 ◆ TRANSPORTING CRANES

According to *ASME B30.5 5-1.5.3(d)*, commercial truck vehicle-mounted cranes shall meet the requirements of the U.S. Department of Transportation Standards. This includes both rough-terrain and all-terrain cranes when used on public roads.

Although rough-terrain (RT) cranes are, in most cases, roadworthy, it is not feasible to drive these cranes any great distance. Rough-terrain cranes are primarily designed to travel from one location to another on a job site. In many instances, there are state or local restrictions applied to their use on public thoroughfares. Always check with local authorities before using

Discuss general guidelines for loading boom sections onto a flatbed. Explain how to release strapping and unload a crane safely.

Discuss the possible hazards during unloading and how to prevent them.

Demonstrate how to load and secure a crane on a flatbed trailer.

Have trainees practice loading and securing a crane on a flatbed trailer. This laboratory corresponds to Performance Tasks 1 and 2.

Have trainees review Sections 3.0.0–4.1.0.

Ensure that you have everything required for teaching this session.

Discuss transporting rough-terrain cranes.

TRANSPORTING REQUIREMENTS 8.3

Provide a copy of *ASME B30.5* for the trainees to examine.

Discuss transporting all-terrain cranes and explain general guidelines for transporting truck-mounted cranes.

Describe required checks during a pre-transport inspection.

Explain how to prepare a boom truck for transport.

See the Teaching Tip for Section 4.0.0 at the end of this module.

Explain that transporting equipment is subject to federal, state, and local regulations.

this type of crane in transit from one job site to another. For longer distances, they should be loaded and secured on a flatbed trailer (*Figure 6*). Always follow the manufacturer's recommendations for loading and transporting.

All-terrain (AT) cranes are a cross between rough-terrain cranes and truck cranes. They are designed to travel at highway speeds from job site to job site. The unit is normally configured with the boom carried to the rear for weight distribution. Always consult the operator's manual of an all-terrain or truck crane prior to driving it from one job site to another. These types of cranes may have options for on-road and off-road travel.

When transporting a truck crane, keep the boom in line over the front with the forward direction of the crane. Secure the upper structure by applying the house lock to prevent any unwanted swinging of the rotating superstructure. Always remove the load block or headache ball whenever possible, or restrain the block from swinging during transit (*Figure 7*).

308F06.EPS

Figure 6 ◆ Rough-terrain crane loaded for transport.

308F07.EPS

Figure 7 ◆ All-terrain crane secured for transport.

Before driving a truck crane, the operator should inspect the vehicle. The following items requirements must be met:

- Oil level and condition in the carrier engine are satisfactory.
- The electrical connection to the superstructure must be set up (if applicable).
- The superstructure must be locked with the pin inserted.
- Check the oil in the automatic gearbox.
- Marker light is plugged in at the boom head and connected electronically.
- The main boom is set down on the boom transport.
- Telescopic parts are retracted and locked in place.
- The battery master switch has been switched on.
- Outrigger cylinders and carriers are retracted, outrigger beams are secured with pins, and outrigger pads are removed.
- Applicable counterweight is removed.
- Tire pressure and all necessary fluids are at recommended levels.

Any additional requirements or precautions must be observed as recommended by the manufacturer. These include operational controls, driving in adverse weather, or driving on uphill or downhill slopes.

Before transporting a boom truck, you must prepare it for travel. The operator/driver must make sure that:

- The outriggers are stowed properly.
- The tires are properly inflated.
- The load line, hook, boom, and any loose items on the truck bed are secured; traveling is not permitted with the load on the hook.
- The power take off (PTO) is disengaged.
- The parking brake is released.

No personnel are allowed to ride on the equipment when it is being transported. The driver should make sure there is enough overhead clearance before entering any underpass. A signal person may be required when moving or backing the unit in areas where visibility is limited.

4.0.0 ◆ TRANSPORT REGULATIONS

Equipment transportation is regulated under federal, state, and local laws and ordinances. Every vehicle must be operated in accordance with the laws, ordinances, and regulations of the jurisdiction in which it is being operated. In addition, the Department of Transportation (DOT) has regulations regarding equipment transportation,

Instructor's Notes:

including those under the Federal Motor Carrier Safety Administration (FMCSA) and the Federal Highway Administration (FHA). If a DOT regulation imposes a higher standard of care than a local law, ordinance, or regulation, the stricter DOT standard must be followed.

The DOT regulations are contained in *Section 49* of the *Code of Federal Regulations (CFR)*. The general rules for loading and securing cargo are contained in *Subpart I - Protection Against Shifting or Falling Cargo, Sections 393.100 –393.136.*

The specific rules for how cranes and boom sections must be secured are contained in *Section 393.130* and is called *What are the Rules for Securing Heavy Vehicles, Equipment, and Machinery?* This section applies to heavy vehicles or equipment weighing 10,000 pounds or more. The load must be secured in the following manner:

- Accessory equipment, such as hydraulic shovels, must be completely lowered and secured to the vehicle.
- Articulated vehicles (vehicles that can pivot in the middle) shall be restrained in a manner that prevents articulation while in transit.
- Heavy equipment or machinery with crawler tracks or wheels must be restrained against movement in the lateral, forward, rearward, and vertical directions using a minimum of four tiedowns.
- Each of the tiedowns must be affixed as close as practicable to the front and rear of the vehicle or to mounting points on the vehicle that have been specifically designed for that purpose.

There are also size and weight restrictions that must be adhered to when transporting boom sections or driving a truck crane or all-terrain crane. These regulations are found in *23 CFR Sections 127* and *658*. Maximum limits include the following:

- Weight of 20,000 pounds on a single axle (*Figure 8*)
- Weight of 34,000 pounds on a tandem axle
- Legal weight of 80,000 pounds for road travel for a truck and trailer combination on an external bridge
- Length of the straight truck-trailer combination of 65 feet
- Height set at 13 feet 6 inches
- Width of the truck/load on a road with traffic lanes 12 feet wide or more is 8 feet 6 inches

If any of these limits are exceeded, then the Overweight/Oversize Division of the FHA must be contacted for permits. A pre-transport route plan is also essential in the transporting process. The above numbers are guidelines for road travel, but the job assignment may require transporting the crane on a back road or other terrain that may not have the weight capacities as listed above. Personnel involved in the transport must take into consideration any variance in weight limitations.

Internal bridge weight limits (see *Figure 9*) apply to how much your truck and trailer combination can legally weigh when you are traveling on the Interstate Highway System. To determine your limit, you must consider two factors: the truck bridge weight and the trailer bridge weight.

Consideration should also be given to lighting and marking devices needed on the trailer, identification markings, and any emergency equipment, flags, and signals. These are covered under *Sections 390.21* and *393* of the Federal Motor Carrier Safety Regulations. Contact the FMCSA to determine what stipulations apply to your particular situation.

Crane manufacturers provide carrier specifications with the operator's manual and load/capacity charts. These specifications provide information about the amount of weight on the axles in various carrier configurations, such as those shown earlier in this module. It is important that the operator/driver review this information.

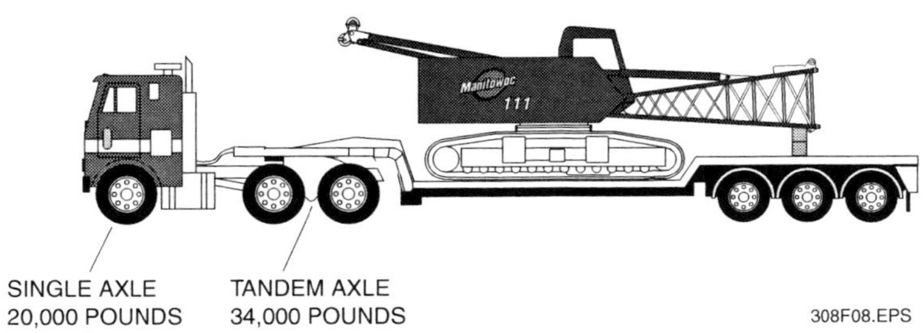

Figure 8 ◈ Axle weights.

Provide a copy of *49 CFR Subpart I* and *23 CFR Sections 127* and *658* for the trainees to examine.

Explain the different federal regulations that control equipment transportation.

Have trainees review the federal rules for securing equipment for transport, including size and weight restrictions.

Show Transparencies 8 and 9 (Figures 8 and 9). Explain axle weight limitations and bridge weight restrictions.

Discuss lighting, markings, flags, and other devices that may be required.

Provide copies of *Federal Motor Carrier Safety Regulations Sections 390.21* and *393*.

TRANSPORTING REQUIREMENTS 8.5

Explain that drivers must meet specific requirements and may need a commercial driver's license to transport equipment.

Summarize the major concepts presented in the module.

Have trainees complete the Review Questions, and go over the answers prior to administering the Module Examination.

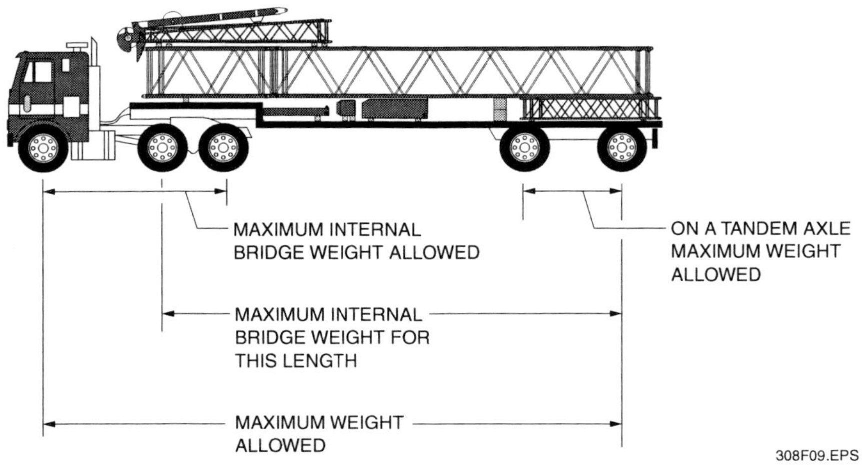

Figure 9 ◆ Bridge weights.

4.1.0 Driver Requirements

According to *FHA 383.23*, no person shall operate a commercial vehicle unless that person has taken and passed written and driving tests which meet the standards for the vehicle the person operates or plans to operate. These standards include *Subpart F – Vehicle Groups and Endorsements (383.91, 383.93,* and *383.95), Subpart G – Required Knowledge and Skills (383.110-383.121* and *Appendix),* and *Subpart H – Tests (383.131, 383.133, 383.135,* and *383.141).* Depending on the situation, the person may also be required to meet the requirements of *Subpart J – Commercial Driver's License Document (383.151, 383.153,* and *383.155).*

Drivers must pass a DOT physical exam every two years. Persons with a history of substance abuse or certain chronic illnesses or disability may not be able to obtain a commercial license.

Summary

There are numerous safety and handling concerns that apply to unloading and loading boom sections and crane parts. It is important that everyone involved in the crane or boom transport operation is familiar with DOT and FHA guidelines.

Review Questions

1. The most important consideration in loading and unloading a mobile crane is _____.
   a. having the maximum weight for each load
   b. speed in setting it up
   c. to avoid damaging it
   d. having the fewest loads

2. Loading a crane with the boom as low as possible will _____.
   a. prevent kickback
   b. overload the trailer
   c. cause the load to become unstable
   d. speed up the loading process

3. Most cranes must be disassembled and transported in pieces because _____.
   a. they cannot be secured properly
   b. they weigh too much for a single load
   c. federal regulations do not permit the transport of assembled cranes
   d. it is less expensive

4. The optimal configuration for loading a crane for transportation is supplied by _____.
   a. the crane manufacturer
   b. OSHA
   c. the Federal Motor Carrier Safety Administration
   d. the Federal Highway Administration

Instructor's Notes:

5. Lever-type binders multiply stored energy by a factor of _____.
   a. 10
   b. 15
   c. 20
   d. 25

6. If boom sections are stacked on a trailer, there must be _____ between them.
   a. suspension ropes
   b. binders
   c. blocking
   d. chains

7. Cranes designed to travel at highway speed from job site to job site are called _____ cranes.
   a. DOT approved
   b. ASME approved
   c. rough-terrain
   d. all-terrain

8. Even if tightly secured, it is forbidden to transport with _______ attached to a crane.
   a. extended outriggers
   b. a load
   c. a load block
   d. headache ball

9. Regulations for the transportation of cranes are covered by the _____.
   a. National Safety Administration
   b. Department of Transportation
   c. Federal Transit Authority
   d. Occupational Health and Safety Administration

10. The regulations under *Section 393.130* apply to vehicles and equipment that weigh more than ____ pounds.
    a. 2,000
    b. 5,000
    c. 10,000
    d. 20,000

11. Heavy equipment or machinery must be restrained with a minimum of _____ tiedowns.
    a. two
    b. four
    c. six
    d. eight

12. The maximum load that is allowed on a single axle is _____ pounds.
    a. 10,000
    b. 20,000
    c. 34,000
    d. 44,000

13. When you are traveling on the Interstate Highway System, _____ apply to your truck and trailer combination.
    a. internal bridge weight limits
    b. steering radius dimensions
    c. internal leverage weights
    d. internal rail weights

14. A straight truck-trailer combination that exceeds 65 feet may not be driven on public highways without a special permit.
    a. True
    b. False

15. In order to drive a commercial vehicle you must pass _____.
    a. a written test only
    b. a DOT physical exam every two years
    c. the CDL Class A Advanced Training test
    d. the Professional Driver Certification exam

Administer the Module Examination. Record the results on Craft Training Report Form 200, and submit the results to the Training Program Sponsor.

Administer the Performance Test, and fill out Performance Profile Sheets for each trainee. If desired, trainee proficiency noted during laboratory sessions may be used to complete the Performance Test. Record the results on Craft Training Report Form 200, and submit the results to the Training Program Sponsor.

TRANSPORTING REQUIREMENTS 8.7

Instructor's Notes:

GLOSSARY

Trade Terms Introduced in This Module

Binder: A mechanical device for securing a load for shipment.

Dunnage: Loose material laid beneath or wedged among objects to prevent damage. Also called blocking.

TRANSPORTING REQUIREMENTS 8.9

Instructor's Notes:

Figure Credits

Link-Belt Construction Equipment Company	308F02, 308F03, 308F07
Manitowoc Crane Group	308F04, 308F05
Ceco Ltd.	308F06

TRANSPORTING REQUIREMENTS 8.11

MODULE 21308-05 — TEACHING TIPS

The following are suggested activities or instructional methods to help you teach the material in this module.

General

When you call on someone to answer a question, the rest of the class relaxes or even tunes out because they expect that the question and answer will take place only between you and the trainee you called on. Instead, use this technique to involve more trainees in answering questions and to keep them on their toes.

1. Ask trainees to define a term or explain a concept.
2. After one trainee has answered, ask a trainee seated nearby if the answer is right. Then ask whether a trainee in the back of the room agrees.
3. Ask trainees to explain why they think an answer is right or wrong.
4. Use the session to clear up incorrect ideas and encourage trainees to learn from their mistakes.

Section 2.0.0 *Loading and Unloading Cranes*

Each type of crane is loaded and secured differently. Trainees will need appropriate personal protective equipment, pencils, and paper. Arrange for an opportunity to observe several types of cranes being loaded or unloaded; alternatively, have a manufacturer's representative give a presentation on loading different types of cranes. Allow 30 to 60 minutes for this exercise.

1. Before the tour discuss the different types of cranes and their configurations with trainees.
2. Identify different methods for loading and securing cranes and attachments. Have trainees take notes and write down questions during the tour.
3. After the tour answer any questions the trainees may have.

Section 4.0.0 *Transport Regulations*

Transport requirements vary from state to state. Trainees will need pencils and paper. Arrange for a representative from the Department of Motor Vehicles or a commercial driving school to give a presentation on transport limitations and driver qualifications. Allow 30 to 45 minutes for this exercise.

1. Before the presentation discuss transport and driver requirements with trainees.
2. Have a representative from the Department of Motor Vehicles give a presentation covering driver requirements and vehicle restrictions.
3. After the presentation have the presenter answer any questions the trainees may have.

MODULE 21308-05 — ANSWERS TO REVIEW QUESTIONS

Answer	Section
1. c	2.0.0
2. a	2.1.0
3. b	2.1.0
4. a	2.1.0
5. d	2.2.0
6. c	2.2.0
7. d	3.0.0
8. b	3.0.0
9. b	3.0.0
10. c	4.0.0
11. b	4.0.0
12. b	4.0.0
13. a	4.0.0
14. a	4.0.0
15. b	4.0.0

CONTREN® LEARNING SERIES — USER UPDATE

The NCCER makes every effort to keep these textbooks up-to-date and free of technical errors. We appreciate your help in this process. If you have an idea for improving this textbook, or if you find an error, a typographical mistake, or an inaccuracy in NCCER's Contren® textbooks, please write us, using this form or a photocopy. Be sure to include the exact module number, page number, a detailed description, and the correction, if applicable. Your input will be brought to the attention of the Technical Review Committee. Thank you for your assistance.

Instructors – If you found that additional materials were necessary in order to teach this module effectively, please let us know so that we may include them in the Equipment/Materials list in the Annotated Instructor's Guide.

Write: Product Development and Revision
National Center for Construction Education and Research
P.O. Box 141104, Gainesville, FL 32614-1104

Fax: 352-334-0932

E-mail: curriculum@nccer.org

Craft ______ Module Name ______

Copyright Date ______ Module Number ______ Page Number(s) ______

Description ______

______

______

______

(Optional) Correction ______

______

______

(Optional) Your Name and Address ______

______

______

Index

Index

Q

R

S

T